Science: Philosophy, History and Education

Scope of the Series

This book series serves as a venue for the exchange of the complementary perspectives of science educators and HPS scholars. History and philosophy of science (HPS) contributes a lot to science education and there is currently an increased interest for exploring this relationship further. Science educators have started delving into the details of HPS scholarship, often in collaboration with HPS scholars. In addition, and perhaps most importantly, HPS scholars have come to realize that they have a lot to contribute to science education, predominantly in two domains: a) understanding concepts and b) understanding the nature of science. In order to teach about central science concepts such as "force", "adaptation", "electron" etc, the contribution of HPS scholars is fundamental in answering questions such as: a) When was the concept created or coined? What was its initial meaning and how different is it today? Accordingly, in order to teach about the nature of science the contribution of HPS scholar is crucial in clarifying the characteristics of scientific knowledge and in presenting exemplar cases from the history of science that provide an authentic image of how science has been done. The series aims to publish authoritative and comprehensive books and to establish that HPS-informed science education should be the norm and not some special case. This series complements the journal Science & Education http://www.springer.com/journal/11191 Book Proposals should be sent to the Publishing Editor at bernadette.ohmer@springer.com.

More information about this series at http://www.springer.com/series/13387

Maria Elice de Brzezinski Prestes
Cibelle Celestino Silva
Editors

Teaching Science with Context

Historical, Philosophical, and Sociological Approaches

Springer

Editors
Maria Elice de Brzezinski Prestes
Institute of Biosciences
University of São Paulo
São Paulo, SP, Brazil

Cibelle Celestino Silva
Institute of Physics of São Carlos
University of São Paulo
São Carlos, SP, Brazil

ISSN 2520-8594 ISSN 2520-8608 (electronic)
Science: Philosophy, History and Education
ISBN 978-3-030-08905-4 ISBN 978-3-319-74036-2 (eBook)
https://doi.org/10.1007/978-3-319-74036-2

Printed on acid-free paper

This Springer imprint is published by the registered company Springer International Publishing AG part of Springer Nature.
The registered company address is: Gewerbestrasse 11, 6330 Cham, Switzerland

Foreword

I would never be able to read a book titled "Aprendendo ciência e sobre sua natureza: abordagens históricas e filosóficas" [Learning Science and About Its Nature: Historical and Philosophical Approaches], edited by Cibelle Celestino Silva and Maria Elice Brzezinski Prestes, much as my knowledge of French allows me to understand a bit of Portuguese and Spanish when I see it. Nevertheless, my understanding would be limited as to what the book is about, as it would be impossible for me to read it in detail and appreciate its contents.

But it so happened that I received a review of that book by André Ferrer Pinto Martins, arranged by our book review editor Charbel El Hani for *Science & Education*. While reading the review I realized that the reviewed book must be good and that it is a pity that it is inaccessible to an international audience, as most chapters were written in either Spanish or Portuguese. Therefore, I decided to find a solution to this problem. I immediately contacted the editors and suggested to them to edit a revised, updated, and expanded version of that book, which we would have published in the series I am editing. The outcome is the book that you now hold in your hands.

Whether we like it or not science, and all serious science-related scholarship, is done in English. This does not mean that the scholarship produced in other languages is not serious. However, any scholar who has important results to report or a good point to share will write an article or book in English because this is the best way to reach a broader audience. This is why I wrote my first book ever, *Understanding Evolution,* in English and a few years later had the bizarre experience to read its translation into Greek, my native language. I found myself reading my ideas in Greek, even though I had not written myself that particular text I was reading. But the goal had been accomplished. It is in the same spirit, but in the opposite way, that I strived to make the present book accessible to a broader audience.

The language barrier is an important issue. English is a foreign language for me and I had to learn it because, despite its historical and cultural importance, Greek is not really used outside my country – and it is certainly much less used than Portuguese or Spanish. But writing and publishing in English is the best way for scholarly work to reach a broader audience. Therefore, we need to encourage, and support, scholars from all over the world to achieve this. It is a loss to the authors themselves and to the respective scholarly communities if important works that might have broader significance do not reach an international audience. Springer, our publisher, made a special arrangement and investment to ensure that the language quality of the final book would be as high as possible, and we appreciate this.

Most appreciable is of course the work of the editors of the present book, Maria Elice Brzezinski Prestes and Cibelle Celestino Silva, who accepted the onerous task to re-work on the first book and to bring together scholars from several countries to contribute to this international edition. I do hope that the present book will motivate authors from other countries who are not native English speakers to also publish their work in English and bring it to the attention of scholars all over the world.

University of Geneva
Switzerland

Kostas Kampourakis
Series Editor

Introduction

Eight years have passed since the first Latin American joint conferences devoted to the contextual teaching of science, through historical, philosophical, and sociological approaches, in Maresias Beach, in the state of São Paulo, Brazil. The intention was that the presence of such an interdisciplinary group of researchers, coming from all over the world, would encourage the development of the area on our continent. This book is evidence that it was a proficuous idea.

Over 5 days, from 16 to 21 August 2010, a total of 184 works representing research carried out in 13 different countries were presented at the two simultaneous events, the *1st Latin American Conference of the International History, Philosophy and Science Education Group* (1st LA-IHPST) and the *8th International Conference on the History of Science in Science Education* (8th ICHSST). The organization of subsequent IHPST conferences on the continent (Mendoza, Argentina, 2012; Santiago, Chile, 2014; and Rio de Janeiro, Brazil, 2015) showed that there was indeed a vibrant and active community in the area, willing to promote international dialogue.

As important as the conferences were the publications derived from the two simultaneous events in Maresias. The first product was the publication of a special issue of *Science & Education* (Volume 21, number 5), in 2012, including seven works developed in Brazil, Mexico, Canada, and Turkey. A second publication arrived in 2013, with 38 selected works from both Maresias conferences in the book *Aprendendo ciência e sobre a sua natureza: abordagens históricas e filosóficas*, organized by Cibelle Celestino Silva and Maria Elice de Brzezinski Prestes (São Carlos: Tipographia, 2013). The chapters, developed by researchers from different countries, were published in Portuguese, Spanish, and English.

In 2016, the editor of the book series "Science: Philosophy, History and Education," Kostas Kampourakis, invited the Brazilian editors to make a new publication for a broader audience, in English. Promptly embracing the proposal, we intended to offer a comprehensive and updated illustration of Latin American contributions to and perspectives of the field. The adhesion of different collaborators was also immediate and lively. As a result, the present publication comprises 26 works discussing a broad range of topics under different methodological, epistemological, and didactic approaches, reflecting the richness and robustness of the dynamic growth of the area of research expanding to Ibero-America.

The book brings together theoretical and empirical studies on diverse research subjects, such as teacher education, learning, didactic materials, and official documents carried out in the different disciplines, such as biology, psychology, geology, chemistry, and physics. Although there is at least one work engaged with the agenda of the sociology of science, most of the works emphasize the history or philosophy of science – including those on the nature of science (NOS). The book is organized into five parts: the first one includes nine chapters on a NOS research program, followed by 13 chapters focused on the history of science, two chapters on the philosophy of science, and one final chapter on the sociology of science.

Working as an overall introduction to the book, the first chapter corresponds to a reasoned exposition of the principles of the area. In "Critical and Transformative Teachers: A Rationale for History and Philosophy of Science in Teacher Education," Breno Moura and Cibelle Celestino Silva discuss the concept of the critical and transformative education of teachers and the role of history and philosophy of science to enhance it. Based on the ideas of three educators close to critical pedagogy – Paulo Freire, Henry Giroux, and João Zanetic – the authors claim that a critical and transformative teacher is someone who establishes a critical dialogue with the world surrounding him or her, problematizes, and transforms it with his or her actions and beliefs. History and philosophy of science are essential to promote critical and transformative education because it illuminates scientific concepts and metascientific aspects involved in scientific enterprise and reveals broader issues and aims of education in general, approximating teachers and students to the development of science and enhance its human, complex, and temporary nature. With this discussion, the authors intend to present a rationale for the introduction of historical and philosophical content in teacher education, favoring the dialogue among educators, historians and researchers.

The sequence of nine chapters on NOS starts with a methodological contribution to the construction of questionnaires, reflecting the product that Nathália Helena Azevedo and Daniela Lopes Scarpa developed for the assessment of conceptions of the NOS. The chapter "Contextualized Questionnaire for Investigating Conceptions of the Nature of Science: Procedure and Principles for Elaboration" signalizes the importance of research on NOS, illustrating how it is maturing in Brazil. The work reports on the procedure employed during the elaboration of a contextualized instrument using ecology as a model for investigating NOS conceptions among biological science undergraduates. The authors propose orientation and principles that contribute toward research procedures, justifying the epistemological and methodological decisions. They also present strategies used for evaluating instrument efficacy, with a question grid constructed with the procedure and the principles adopted. This strategy furnished an explicit and critical manner of creating new questionnaires that deepened the analysis of the students' concepts on NOS.

A general contribution to the development of NOS research comes in the chapter "Consensus and Dissent Around the Concept of Nature of Science in the Ibero-American Community of Didactics of Science". Rafael Yecid Amador Rodríguez and Agustín Aduriz-Bravo follow the purpose of presenting some of the agreements and disagreements around the concept of NOS that arise among specialists in didactics of science (i.e., science education as an academic discipline) in Ibero-America. They focus on researchers from Spain, Portugal, and countries in Latin America and the Caribbean whose main official language is Spanish or Portuguese. The authors take into account academic production in the form of research articles published in specialized journals between 2000 and 2009. They selected articles that purposefully state a substantive relationship with the research line on NOS, and analyze how authors conceptualize NOS. As a result, some explicit definitions of NOS came from the corpus of articles. The main finding is that NOS constitutes a recognizable and developed field of reflection and innovation for researchers in didactics of science in Ibero-America, especially in Spain and Argentina. The extended agreement can be identified on the "list" of NOS ideas that should be taught, whereas there is disagreement on which theoretical perspectives from the philosophy of science have most educational value.

The next two chapters are devoted to teachers' conceptions of NOS. In "Scientific Skills in Secondary Education: A Study of Curriculum Expectations and Teachers' Thinking," María Teresa Guerra-Ramos and José García-Horta note that the Mexican secondary science curriculum incorporated scientific process skills among other traditional content. This implies that teachers already face the challenge of teaching scientific skills as part of scientific practice. This study combines document analysis identifying curriculum expectations, and an empirical study exploring the representations of 22 secondary science teachers with regard to scientific skills. A hypothetical pedagogical scenario was used to include descriptions of actions, and teachers were asked to argue whether or not they would consider them an example of scientific skills in a science lesson, providing arguments for doing so. Their responses were incorporated and explored in detail through individual semi-structured interviews. The documentary analysis provided evidence for a lasting presence of problems regarding the development of skills by students. The qualitative analysis of teachers' responses revealed a tendency to consider the descriptions provided as examples of the use of some of the scientific skills. To argue whether or not a scientific skill was involved, teachers referred to the scientific or nonscientific domain and the investigative or non-investigative purpose. The use of disciplinary knowledge to illustrate their responses was infrequent. Data suggest that teachers' representations point towards an image of scientific skills as mechanical actions, independent of context. The authors argue that teachers' representations inevitably interact with curriculum innovations and that a characterization of them such as that provided in this study can suggest some implications for teacher education.

Focusing on the context of biology teachers, the chapter "Theory, Evidence, and Examples of Teaching the Nature of Science and Biology using the History of Science: a Chilean Experience" was the result of the research carried out by Paola Núñez, Hernán Cofré, David Santibáñez, José Pavez, and Claudia Vergara. According to the authors, a significant number of science educators have recognized the importance of the history of science (HOS) in understanding NOS and scientific content. However, there is little empirical evidence for this effect in the South

American educational context. This article shows empirical data about the contribution of HOS to the enhancement of in-service biology teachers' understanding of NOS and the effect of HOS on enhancing the understanding of evolution and NOS in high school students. The authors used the VNOS-D+ questionnaire to assess teachers' and students' views of NOS at the beginning and the end of interventions. The inclusion of writing artifacts such as lesson tickets-out, content tests, and lesson plans for teachers enriched the analysis. The students' understanding of evolutionary theory was assessed using the Assessment of Contextual Reasoning about Natural Selection questionnaire. Some of the most important results of the project are the significant improvements observed in teachers' understanding of NOS, although they assigned different levels of importance to HOS in these improvements, and a significant effect of HOS with students' understanding of NOS. There was no significant difference between students' understanding of evolution in treatment and control classes. The authors make suggestions regarding science teacher education and future research to improve the effect of HOS on students' and teachers' understanding of NOS and scientific content.

María B. García, Silvia Vilanova, and Sofía Sol Martín in "Epistemological Conceptions of University Teachers and Students of Science" present an empirical study comparing NOS conceptions of undergraduate students and teachers of the Faculty of Exact and Natural Sciences of the National University of Mar del Plata, Argentina. Using an *ex post facto* design, the study was carried out in two stages: one aimed to describe the conceptions each group has, and another targeting their comparison. The results obtained from both groups in the descriptive stage show a tendency toward relativist conceptions about the nature of knowledge and more intellectual ones about the acquisition of knowledge. The comparative study revealed that students' conceptions on "acquisition of knowledge" differ from those of their university teachers.

Focused on NOS conceptions of students, the two following chapters developed empirical research in undergraduate courses on physics and psychology.

The first is "Is the electron real? Who discovered the expanding Universe? Debating Nonconsensus topics of Nature of Science in Science Classrooms." After presenting a summary of recent criticisms of the so-called consensus view of the nature of science, the authors André Noronha, Alexandre Bagdonas, and Ivã Gurgel analyze two educational activities based on studies of history and philosophy of science applied in physics classes. The objective was to promote critical reflections of science, with a focus on controversial aspects of NOS. Initially, the authors present an educational game designed to teach cosmology in high school, which allowed the teacher to discuss with the students social and cultural influences on the construction of cosmological models of the universe during the first half of the twentieth century. Second, they present the results of a survey on physics undergraduate students' conceptions of the scientific realism debate during lectures about the evolution of physics concepts. These two examples illustrate the possibility that controversial aspects of NOS might be discussed with an increasing level of complexity, from the first years of basic education until preservice teacher training courses. Finally, the authors defend that proposals regarding controversial topics of NOS may bring a more critical and reflective understanding of NOS in science education.

The second is "Undergraduate Psychology Students' Conceptions of Scientific Knowledge and Psychology-Specific Epistemological Beliefs" by Zuraya Monroy-Nasr, Rigoberto León-Sánchez, Kirareset Barrera García, and Germán de León Alvarez-Díaz. The work starts by considering that epistemological beliefs are supposed to change and influence the way in which students understand and explain reality. These changes seem to go from a straightforward and direct "objective comprehension" of reality to a "subjective comprehension." The research was carried out with a nonprobabilistic, intentional sample of 156 undergraduate psychology students of different semesters to assess and compare their conceptions of scientific knowledge and psychology-specific epistemological beliefs. The study used two questionnaires: The Psychology-Specific Epistemological Beliefs (Psych-SEBS), designed by McMahan et al., and a modified questionnaire inspired by Raviolo et al. The authors adapted these questionnaires by focusing mainly on the students' understanding of the nature of scientific knowledge, the concepts of scientific models and explanations, and the role of social and historical context in scientific production. The results do not support the expected difference between personal epistemologies of the participants, demonstrating that changes in epistemological beliefs are not linear. Moreover, beginner students tend to place greater value on the functional aspects of scientific theories (explanation and prediction) whereas advanced students more frequently acknowledge the influence of context in the generation of scientific knowledge; and the data suggest that both groups of students might coordinate the objective and subjective aspects of knowledge.

The last two chapters in the group on NOS focuses on didactic materials in the area of biology.

The first, on NOS in biology materials, is the contribution of Sílvia Regina Groto and André Ferrer P. Martins in "The 'Science' as Portrayed in Documents of the Biological Evolution Versus Intelligent Design Debate." They focused on biological evolution by identifying and discussing aspects of the ideas about science expressed by groups/individuals involved in the debate between biological evolution and intelligent design in Brazil. Using the methodology of discursive textual analysis, they analyzed specific documents produced by organizations and individuals belonging to both groups, totaling 11 documents produced by the evolution group, and 21 documents produced by the intelligent design group. The ideas of the existence of scientific proofs and truths are present in both groups, the latter being more prevalent in the intelligent design group than in the evolution group. Both also advocate the existence of a scientific method, although the biological evolution group presents a broader view of what method means, whereas the intelligent design group brings a narrower view. Both groups interpret the idea of scientific temporariness in various ways. Notably, while the biological evolution group understands it as an extension of knowledge, the design group perceives it as changing of knowledge. The epistemology of the philosopher Karl Popper seems to support the arguments used by both groups, although the latter produces shifts from Popper's original ideas. The authors conclude by emphasizing the relevance of the debate in science education, and the need for discussion of questions about NOS both in basic education biology classes and in undergraduate courses in biological sciences.

The second, by Alcira Rivarosa and Carola Astudillo, is entitled "Multiple Narratives as Cognitive and Political Bridges to Understanding the Nature of Scientific Knowledge." The authors present the arguments, theoretical frameworks, and characteristics of specific educational materials designed to promote reflection on the nature of scientific knowledge in the training of researchers and teachers of biological sciences. They selected and designed these resources with the intention of activating a critical reflection on the *cuisine* of scientific research. The authors are mainly concerned with the processes of reflection, reasoning, and conceptual arguments that real scientists conduct during their practice. Rivarosa and Astudillo reflect on contexts, highlighting their rich connections to other fields, such as politics, economics, religion, art, and technology. Finally, with specific activities, they focus on understanding the processes of the construction of explanatory models in biological sciences. They found that the students improve their skills in overcoming cognitive and epistemic obstacles regarding the work of research as an institutional business, in dialogue with society, history, and culture. Along the same lines, they progress in the denaturalization of absolutist conceptions of scientific knowledge. Finally, the work shows that students recognize the collaborative nature of scientific research and methodological pluralism, including processes of creative imagination and ethical considerations.

The third group of chapters mainly emphasizes the HOS approach, reflecting its development in the different areas. There are two chapters devoted to episodes of the history of biology, four to the history of chemistry, and seven to physics. Some of them bring empirical analysis of applications of historical cases in classrooms.

The chapter by Filipe Faria Berçot, Eduardo Cortez, and Maria Elice Brzezinski Prestes, "Abraham Trembley (1710–1784) and the Creature That defies Classification: Nature of Science and Inquiry through a Historical Narrative," presents a a case study of the history of biology. The narrative tells the story of studies on freshwater "polyps" (*Hydra*) conducted by the Genevan naturalist Abraham Trembley, in the first half of the eighteenth century. The historical narrative is part of the instructional resources of a teaching learning sequence (TLS) using an inquiry approach produced for biology teacher training courses and secondary school students. The objectives of the TLS are to facilitate the learning of conceptual, procedural, and attitudinal content related to the current knowledge on reproductive biology and to foster the development of informed conceptions about the nature of science through an explicit and reflective approach. The authors used design-based research and the achievement of design principles to develop the narrative and the TLS. Considering the state of the art in the area, the narrative proposed here is an educational innovation that can be extended to similar contexts, promoting more engagement and motivation to learn science through an authentic and historical context and inquiry learning.

The chapter by Anna Cassia Sarmento, Cláudia Sepúlveda, and Charbel N. El-Hani deals with the "Historical Reconstruction of Membrane Theoretical Models: An Educative Curriculum Material." The authors present an educative material for the teaching of cell biology in the 10th grade, attended by students ranging from 15 to 16 years of age. The material, composed of a teaching sequence and associated extracts from primary sources, addresses the historical development of membrane theoretical models, with the goal of promoting learning about membrane structure and functions, and a more informed view of science. This material is part of a broader study, conducted according to the theoretical and methodological underpinnings of educational design research, in which the teaching sequence and source text were planned, constructed, and tested through an investigation of their outcomes in the science classroom.

The chapter "The History of Chemistry in Latin America" by José A. Chamizo opens the subsection on the history of chemistry. Chamizo advocates that teaching chemistry means something more than teaching an undifferentiated mass of names and dates. The chapter reconstructs the history of chemistry in five revolutionary moments. These moments are considered regarding the Kuhnian notion of "exemplar," rather than "paradigm" and enable the incorporation of instruments and concepts in the Latin American context under the umbrella of scientific pluralism.

Luciana Zaterka and Ronei Clécio Mocellin, in "Natural History, Chemistry, and Teaching in Modern Scientific Culture" consider that a historically delimited study can offer the teachers the opportunity to problematize with students the historicity of scientific knowledge and its social, economic, pedagogical, and philosophical implications. In this chapter, the authors intend to draw attention to the centrality of chemical knowledge in modern scientific culture and how chemistry teaching is a manifestation of that social interest. Francis Bacon's new program for natural history in the seventeenth century, more specifically his view of the history of arts, was adopted as the core idea of the work. The first goal is to describe the reasons that made chemistry the fundamental knowledge of Bacon's new program. The purpose is to point out the shared objectives between the natural history of arts and chemistry teaching, as laid out in the manuals and courses inspired by the lessons about the Baconian chemical philosophy of the physician–chemist Herman Boerhaave and his French "disciples." The authors highlight the case of the chemist Guyton de Morveau and his group at the Dijon Academy, not only for their pedagogical innovations, but, above all, for their central role in French chemistry at the end of the eighteenth century and their active participation in the reformulation of chemical nomenclature. However, they point out that this linguistic–conceptual revolution also marked a pedagogical rupture, in which the history of chemistry was no longer necessary for the learning of this science.

Aligned with teacher education is the chapter entitled "The History of Science in Teacher Training Programs: A Series of Contributions and Debates for the Teaching of Electrochemistry" by Johanna Camacho González, Mercè Izquierdo, and Núria Solsona. The authors investigate a course for a pre-service teacher training program in Chile. The theoretical support of the course is Stephen Toulmin's pragmatic and natural view of science, and the contributions of HOS to science teaching. The main results show that in a training process based on HOS, it is possible to innovate and contribute positively to the teaching of the electrochemical theory, generating metacognitive skills.

Dulce María López-Valentín presents "An Empirical View of the Teaching of the Chemical Element Concept," arguing that if teachers know about the history of the chemical element concept, they understand students' difficulties and can promote a better teaching of this concept. The objective of her research was to survey if the empiricist view of science prevails in the teaching of chemical elements at pre-university level. Forty-eight in-service Mexican teachers participated in this study, and a sample of 30 chemistry textbooks was reviewed and analyzed using two convergent experimental designs. The results show that teachers lack knowledge about the history and philosophy of science and a critical view of the development of the chemical element concept.

The part dealing with the history of physics starts with "An Educational Blend of Pseudohistory and History of Science and Its Application in the Study of the Discovery of Electromagnetism," by Roberto de Andrade Martins. His work describes a strategy for teaching HOS that contrasts different types of historical resources, including pieces of pseudohistory, with primary and secondary sources. This approach provides lessons and stimulates attitudes that are very difficult to overcome using other tactics. Notably, one of the aims of this method is to teach the students how to distinguish a well-grounded historical narrative from an unreliable one and to develop a critical attitude concerning the flawed stories that proliferate on the internet and in popular science books. Besides describing the general procedure used at a Brazilian university, the paper presents its exemplification in the study of Ørsted's discovery of electromagnetism.

The chapter "Isaac Newton and 'Hidden Forces' in Universal Gravitation: Delimiting an Approach for Teacher Training", by Thaís Cyrino de Mello Forato, presents a didactical proposal addressing Newton's peculiar neo-Platonist conception. To provide elements for critical approaches to NOS, the didactical plan includes elements to discuss physics concepts, and epistemic and non-epistemic aspects of science in physics teacher training courses. The contextualized focus on Newton's involvement with different fields of knowledge, such as alchemy, mystical knowledge, and religion, is aimed at discussing stereotypes and naive views on NOS, still perpetuated in science teaching. The discussion of the influences of the biblical prophecies on Newton's thoughts and alchemical ideas in the conception of action at a distance allows critical and explicit reflections on how personal values of a thinker or group of thinkers also have an impact on science. In addition, the chapter intends to bring the topic "science and religion" to teacher training, offering subsidies to deal with intolerances, prejudices, controversial and fundamentalist's positions. Religious conceptions or metaphysical ideas are part of the culture, and they can be understood as social and historical knowledge, regardless of the personal belief of each teacher. Reflections about such content can promote the respect for diversity, as a way of approaching human rights in classrooms.

Marco Braga, Andreia Guerra, and José Claudio Reis consider that schools and science teaching have a culture composed of practices, content, and rituals. In "The Enlightenment *Paideia*: The French Origins of Modern Science Teaching," they state that, in general, this culture represents an obstacle to implementing pedagogical practices in an historical–philosophical approach. Therefore, it is relevant to question the science teaching culture while introducing HOS into science teaching. The problem is that the guiding principles of science teaching culture are hidden and do not appear in the actions or material elements of a classroom and school. Everything seems to be natural for teachers and students. In this paper, the authors discuss some of the issues concerning the origins of science teaching culture, focusing on the *Polytechnique* model of science teaching. They show how several peripheral nations influenced by French culture replicated this model. Dogmatic and instrumental characteristics marked the *Polytechnique* model, which was in harmony with the objectives of education for a new kind of engineer in the context of the French Revolution, a professional who should be less of a craftsman and more of a technological master, mixing practical and theoretical knowledge.

Ana Paula Bispo da Silva, José Antonio Ferreira Pinto, and Éwerton Jéferson Barbosa Ferreira contributed the chapter "Design and Implementation of a Lesson Plan for High School Students: A Case Study with Oersted's Experiment." According to the authors, one of the most common classroom activities related to electromagnetism is to use the "right-hand rule" to describe the direction of the electromagnetic field lines around the wire or the direction of the electric current in the wire. However, the historical episode involved in this simple action was complex. It can work as a good example of how scientific concepts and scientists' beliefs are related. During the nineteenth century, experiments with electricity and magnetism were very common. Although many philosophers did not know how to explain their results, those familiar with *Naturphilosophie*, such as Hans Christian Oersted, hypothesized that electricity and magnetism were related. This philosophical background helped Oersted to interpret his experiments and the action of the electric current over a compass. The authors used the historical episode in an inquiry lesson plan including hands-on activities, texts about the historical episode, and reports produced by the students, with the aim of satisfying pedagogical requirements and current historiographical recommendations.

In "Investigating the Didactic Use of Primary Sources on the History of Vacuum and Atmospheric Pressure," by Juliana M. Hidalgo F. Drummond, Wesley Costa de Oliveira, and Deyzianne Santos Fonseca present empirical research referring to the application of a historical–philosophical didactic sequence in a short course taught to High School students in a Brazilian public school. The authors' goal was to contribute to the understanding of the concept of atmospheric pressure through the interpretation of excerpts from primary sources in a dialogical approach. Beginning with explanations for an every-day event – the sipping of liquid with a straw – students were engaged in an investigative process. With the participation of a mediating teacher, students were encouraged to point out the similarities (and differences) between their initial explanations and those supported by past thinkers. Students also got in touch with the process of construction of the concept of atmospheric pressure, noticing the inexistence or the fragility of the "atmospheric pressure" factor (as a theoretical concept) in their initial explanations of the straw-sipping phenomenon and the possibility of reconsidering them.

Sonia Maria Dion's contribution completes the fifth empirical study on physics. Her chapter, entitled "The Status of the Lines of Force in Michael Faraday's

Thought: History and Philosophy of Science in the Classroom," lies in the interface among science, history, and philosophy. Using primary sources, she presents a study on the change in the ontological status of the lines of force in the thought of Michael Faraday (1791–1867), who moves from an instrumentalist to a realist view on this subject. This historical case is then further examined to show the context surrounding his ideas and his reasons justifying them. From this point on, the chapter suggests a reading approach to introducing a philosophical question into the teaching of physics. To this end, some connections are established between Faraday's ideas and students' views about lines of force, without turning into the so criticized parallelism between ontogenesis and phylogenesis. Results from the literature, interpreted in the light of Bachelard's ideas, indicate the existence of a widespread realist perspective among students. Sonia Dion suggests that this realism identified in the historical and pedagogical contexts can work as the communication link between reader and text considering the given differences between their contexts.

The subsection on the history of physics ends with a more general methodological and epistemological contribution. Sandra Regina Teodoro Gatti and Roberto Nardi, in "Analysis of Pedagogical Practices Carried Out in Continuing Education Activities for Physics Teachers: Limits and Possibilities," consider that researchers into science teaching seem to point to a consensus on the importance of the historical and philosophical approach to teaching. Nevertheless, the pedagogical practice developed in the classroom hardly ever embodies such recommendations. In this research, the authors planned a continuing education course and followed the development of five in-service physics teachers during the school year of 2008. The researchers discussed with teachers, focusing on their experiences and seeking to assist them in the construction of alternatives of what it means to observe and understand the students' work. The authors seek to revisit the training actions by discussing the propositions and perceptions of two of the participants and the respective impacts of the experience, 3 years later, when one of them became a supervisor to a pre-service teacher.

In a philosophical approach, the chapter "Philosophy of Science in Science Teacher Education: Meeting Some of the Challenges" by Ana C. Couló states that a reasonable familiarity with philosophical content, skills, and attitudes can contribute to a science teacher becoming an educator rather than an instructor or a mere supplier of content. However, lately, philosophy and other foundational studies courses have lost their status in many teacher-training curricula – philosophy courses are frequently accused of being irrelevant, unnecessarily obscure or merely an old-fashioned ornamental addition to teacher culture. These criticisms can be refuted by the discussion of the many relevant issues that philosophy can bring to the development of an educator. However, this chapter does not focus on the philosophical content to be taught (*what* is taught), but rather on the way in which that content is brought to the classroom (*how* it is taught). Couló discusses different approaches to teaching philosophy and suggests some of the ways in which a philosophy course could be aimed at becoming a more significant experience for science teacher education.

M. Eugenia Seoane, Irene Arriassecq, and Ileana M. Greca, in "Epistemological Debate Underlying Computer Simulations Used in Science Teaching: The Designers' Perspective," develop a more specific theme imbricating philosophical and pedagogical approaches. According to the authors, many research areas widely use computer simulations, and their role in the production of scientific knowledge is nowadays the subject of debate in philosophy of science. This work presents the results of a phenomenographic case study involving three researchers who design and use computer simulations in physics. The study analyzes these designers' view on simulations and the role of simulations in physics teaching. The results show that they agree on the fact that computer simulations have changed the way we do science and that they share many characteristics with the classical models: they derive from theories, they help to predict and explain phenomena, and their results need to be empirically validated. They consider simulations used in science teaching – which differ from those used in research in their objectives and in their design – to be beneficial as they allow students to visualize and work on a phenomenon from the viewpoint of the mathematical model, the physical, and the virtual one in an interrelated way. In general, the designers' views on simulations and their use in science and education were more complex and meaningful than those conveyed by novice researchers in science teaching or found in research articles on secondary education that look at the same subject.

The chapter addressing sociology of science is by Ana M. Morais, Sílvia Castro, Sílvia Ferreira, and Isabel P. Neves. In the work "The Nature of Science in Secondary School Geology: Studying Recontextualizing Processes," the authors analyze official documents and content syllabi of geology teaching in Portuguese secondary school (age 15–16). In epistemological terms, the study uses John M. Ziman's conceptualization of science construction and in sociological terms uses Basil Bernstein's model of pedagogical discourse. Methodologically, the study combines quantitative and qualitative methods of analysis. The results show that recontextualizing processes did occur within the curricula and between the curricula and textbooks. It influences teachers' perceptions of the messages contained in these texts and may have consequences for their pedagogical practices.

We are thankful to the authors who collaborated on this work. We believe that the book gives not only an inspiring illustration of the research carried out in Ibero-America, but also a robust contribution to the research of history, philosophy, and sociology of science in science teaching around the world. We also thank the referees, whose detailed analysis has enriched not only each chapter, but has contributed to a more organic and coordinated presentation of all the contributions gathered here.

Contents

Part I
Rationale

Chapter 1
Critical and Transformative Teachers: A Rationale for History and Philosophy of Science in Teacher Education

Breno Arsioli Moura and Cibelle Celestino Silva

1.1 Introduction

Throughout the last few decades, the necessity of enhancing students' and teachers' views on science was introduced into educational guidelines in several countries.[1] The claim is supported by the idea that teachers must be able to promote discussions about science with their students, to criticize and analyze aspects of the development of scientific knowledge, to relate science to social, cultural, political and religious contexts, among other topics. From this perspective, teachers are not merely transmitters of knowledge, but producers of contextualized school knowledge, and the students are not receivers, but active agents of the teaching and learning process. Accordingly, it is relevant to reflect on what should guide teacher education in preparation.

Being a teacher is not an easy task. Teachers interact with a complex, changeable and unpredictable environment in their everyday work and have the responsibility of guiding their students toward a broad comprehension of the world. Among other formative needs, teachers must place themselves in new roles in the classroom and face problems that were unknown to them while students (Bejarano and Carvalho

[1] In Brazil, for instance, several official documents refer to a critical view of science from students and teachers: *Parâmetros Curriculares Nacionais* (National Curriculum Parameters) (Brasil 1997, 2002), *Diretrizes Curriculares Nacionais para a Formação de Professores* (National Curriculum Guidelines for Teacher Education) (Brasil 2015), and the *Plano Nacional de Educação* (National Education Plan) (Brasil 2001, 2014).

B. A. Moura (✉)
Center of Natural Sciences and Humanities, Federal University of ABC, Santo André, SP, Brazil
e-mail: breno.moura@ufabc.edu.br

C. C. Silva
Institute of Physics of São Carlos, University of São Paulo, São Carlos, SP, Brazil
e-mail: cibelle@ifsc.usp.br

M. E. B. Prestes, C. C. Silva (eds.), *Teaching Science with Context*, Science: Philosophy, History and Education,
https://doi.org/10.1007/978-3-319-74036-2_1

2003). Furthermore, many common-sense conceptions about teaching – for instance the idea that a good knowledge of content and a sort of natural talent for "playing the teacher role" are enough to teach – should be abandoned. The recommendation for them to be more critical adds more preconditions, which make teachers' education challenging and severe work. In the case of science teachers, they face other challenges to combine theoretical frameworks with disciplinary contents and methodological approaches. For instance, physics teachers may know the conceptual basis of the theory of relativity, but struggle to develop ways of teaching it. If teaching scientific content in a decontextualized way is already challenging, the discussion of their historical, philosophical, and epistemological aspects may be even more difficult (Höttecke and Silva 2011, p. 302).

The above aspects, aims, and perspectives of science teacher education form what we call critical and transformative teacher education. From this viewpoint, critical and transformative science teachers are professionals who foster critical dialogue with the world, have the ability to opine with coherent and well-based arguments, comprehend their role and responsibilities as educators, and understand science as a product of cultural, social, and political contexts. Critical and transformative science teachers are constantly teaching and learning with their students and comprehend that their role is to criticize, change, and transform education. They understand the schools as a locus of the development of ideas and of personal, professional, and political emancipation.

In the present paper, we advocate that the incorporation of history and philosophy of science (HPS) into teacher education can contribute to promoting critical and transformative teacher education. On the one hand, it illuminates scientific concepts and meta-scientific aspects involved in scientific enterprise, being a powerful resource for discussing the nature of scientific knowledge (Allchin 2013; Matthews 2015; Dagher and Erduran 2016). On the other hand, it is also connected to broader issues and aims of education in general, as indicated before, as it introduces teachers and students to the development of science and enhances its human, complex, and temporary nature (Zanetic 1989; Martins 2006). To establish this connection and to delineate the conception of critical and transformative education, we visited frameworks close to critical pedagogy: the dialogical education proposed by Paulo Freire (1921–1997), the intellectual transformative teacher discussed by Henry Giroux (1943–) and the ideas of science as part of culture, presented by João Zanetic (1941–).

The Brazilian educator and philosopher Paulo Freire understands education as a pathway to freedom and conceives the educator as the one who problematizes knowledge, teaches and learns with the students. Teachers should interact with the reality that surrounds them, perceiving themselves as part of it and capable of transforming it. The American and Canadian educator Henry Giroux, one of the main developers and defenders of critical pedagogy in North America, works with three basic concepts that act directly against the technocratic and traditional pedagogies: the intellectual transformative teachers, the school as a place of opposition to archaic ideas, and radical pedagogy as a form of cultural politics. Finally, the Brazilian scholar João Zanetic claims the necessity of discussing science as part of human culture. In the next sections, supported by the ideas of these three intellectuals, we discuss the concept of critical and transformative teachers and the rationales for the use of HPS in teacher education considering critical pedagogy as a framework.

1.2 The Dialogical Education of Freire

Paulo Freire defends education as an instrument of the liberation of the oppressed people. According to him, there is no dichotomy between teaching and learning, as both teacher and student learn and teach. Therefore, teachers cannot put themselves in a position of those who know more, those who transmit knowledge to the students, the "empty bank accounts." On the other hand, teachers should adopt the position of learners, who learn and teach at the same time *with* the students in a process mediated by the world. In his book *Pedagogy of the Oppressed*, first published in 1970, the author criticizes "banking education [...] in which students are the depositories, and the teacher is the depositor" of knowledge (Freire 2005a, p. 72). For Freire, teaching is not an act of transferring knowledge from the teacher and learning is not an act of receiving knowledge by the student. The author asserts that "banking" education hampers creative and active thinking because it considers students to be ignorant, and should adapt to the orders of the teachers, representatives of the dominant system. Teachers, in turn, tend to reinforce these roles, as any critical attitude from the students can undermine their dominant position. He summarizes:

> The teacher presents himself to his students as their necessary opposite; by considering their ignorance absolute, he justifies his own existence. The students, alienated like the slave in the Hegelian dialectic, accept their ignorance as justifying the teacher's existence – but, unlike the slave, they never discover that they educate the teacher. (Freire 2005a, p. 72)

Freire also argues the value of teachers' words and attitudes in the classroom. For him, teachers' discourse reveals their praxis, understood to be the combination of teachers' attitudes – the action – and their evaluation of them – the reflection. Teachers' words and attitudes are consequences of their praxis. If the teaching practice excludes action, it becomes verbalism. If it excludes reflection, it becomes activism. Therefore, "human beings are not built in silence, but in word, in work, in action-reflection" (Freire 2005a, p. 87–88).

In the essay *Extension or Communication?*, first published in Chile in 1969 during Freire's exile due to the dictatorship in Brazil, he claims that whereas communication is characterized as liberating, extension represents enslaving and domesticating, thereby hindering human consciousness and impeding enduring changes. Freire argues that education's purpose is humanizing and transforming the world and the simple transmission of knowledge is a cultural invasion in which "the invader dictates; the invaded patiently accept what is dictated" (Freire 2005b, p. 103). The author criticizes the excess of discredit upon the students, claiming that it is based on preconceptions and exaggerated pessimistic views: "If people are assumed to be absolutely ignorant, there must be people who think of them this way" (Freire 2005b, p. 107). Freire proposes a form of education that favors the dialogue between teacher and student. When communicating with each other, teacher and student construct a critical view of the world together, that frees them from oppressive ideas and behaviors.

Freire claims that problematizing scientific knowledge is essential for the development of a more critical view of the world. The problematization of scientific knowledge can only be achieved when we comprehend it as a human activity.

> [...] if scientific knowledge and the formulation of disciplined thought cannot be separated from a problematic approach, then the apprehension of this scientific knowledge and of this disciplined philosophical thought cannot be separated from a problematic approach to the very learning which the educatee must absorb. (Freire 2005b, p. 112)

Although the transmission of knowledge is a passive process, problematization brings students to a leading position, under teachers' guidance. By considering that no knowledge is absolute, but subject to criticism and reflection, the students assume a questioning attitude, arguing, thinking, and structuring their body of arguments and ideas. As Freire proposes, problematization offers an opportunity to enhance a critical view of the world, because "no one can present something to someone else as a problem and at the same time remain a mere spectator of the process" (Freire 2005b, p. 135).

In the book *Pedagogia da Autonomia* (Pedagogy of Autonomy), of 1996, Freire presents a corpus of knowledge that is necessary to create a dialogical praxis. The author advocates the importance of freeing society from fatalistic ideas about education. He believes that not fighting against an archaic school system is to renounce the most fundamental role of a teacher, that is, to promote criticism, ethics, and the fulfillment of the rights and duties of each citizen. Another essential element toward a critical education is the transformation of discourse into real practice. It is not enough that the teacher says he or she is ethical and honest; their words must be coherent with their actions: according to Freire, "it is not possible for the teacher to think that he thinks right, but at the same time ask if the student 'knows with whom he is talking'" (Freire 1996, p. 35).

Also, teachers must always aim for students to overcome a superficial curiosity, transforming it into epistemological curiosity, another fundamental concept in Freire's ideas. He believes that a true passage from common sense knowledge to scientific understanding cannot be achieved spontaneously, but only with a deeper reflection on the studied subject. Scientific understanding is related to the "critical ability [of the individual] to take distance from the object, of observing it, delimiting it, dividing it, 'enclosing' the object or making methodical approximations; the capacity of comparing, asking" (Freire 1996, p. 85). According to Freire, teaching also implies an awareness of the incompleteness, as it is always possible to learn more. Also, another necessary attribute for better teaching is to be critical of oneself; therefore, he affirms the importance of teachers reflecting on their own praxis. This will enable teachers to learn more about themselves, their attitudes, and ideologies, what went right and wrong in their classes, to understand their teaching praxis and practice.

Freire's ideas contribute to forming teachers who not only comprehend themselves as educators, but are also capable of transforming themselves and their surroundings. The emancipator (or freer) educator composes one of the elements that we believe are essential for a progressive conception of teacher education. Freire's pedagogy has been developed by Henry Giroux and others in a movement known as "critical pedagogy" (Giroux 2011).

1.3 The Intellectual Transformative of Giroux

Supported on three pillars – the teacher as a transformative intellectual; the school as a public locus strictly associated with questions of power and democracy, and radical pedagogy as a form of cultural policy – Henry Giroux considers educators not only guides toward a critical perspective, but also toward transformative practices. Giroux understands teachers agents who help students to analyze, modify, and construct knowledge from a critical standpoint.

The transformative intellectual teacher plays the role of making the pedagogical more political and the political more pedagogical, considering that "the schooling, the critical reflection, and the action become fundamental parts of a social project to help students to develop a profound and unshakable faith on the fight to overcome the injustices and to change themselves" (Giroux 1992, p. 32). In Giroux's understanding of the school as a locus for the development of critical political notions, teachers contribute to an intellectual enhancement of the students. Additionally, making the pedagogical more political means "to use forms of pedagogy that treat students as critical agents, problematize knowledge, dialogue and make knowledge meaningful, in such a way to make it critical in order to be emancipatory" (Giroux 1992, p. 33). Thus, "instead of considering the knowledge as objective, as something to be merely transmitted to the students, teachers can demonstrate how it is constructed by a selective process of emphasis and exclusions" (Giroux 1992, p. 42).

Giroux assumes that schools should be intrinsically associated with issues related to power and democracy and that the teachers, as transformative intellectuals, can avoid this separation. Thus, teaching situations are not limited to the classroom space, but echoes in the whole school and its community, in students' relations with their families and friends, and in public policies for education. As transformative intellectuals, teachers allow the students to have an active voice, helping "to establish the pedagogical conditions in which they express themselves" (Giroux 1992, pp. 47–48), and contributing to the construction of part of the cultural capital that fosters students' emancipation.

Giroux criticizes teacher education based on technical rationality (Schön 1983) and what he calls "proletarianization of teacher work" (Giroux 1992, p. 10), the tendency to reduce teachers to the status of specialized technicians within the school bureaucracy, whose function becomes simply managing and implementing standardized procedures. According to him, this model of education does not effectively prepare teachers for the reality of school, as it banishes the intellectual work of problematizing knowledge. Like Freire, Giroux proposes that an uncritical transmission of knowledge might not promote the intellectual emancipation of students.

1.4 Science as Part of Culture

The Brazilian physicist and educator João Zanetic has been claiming for more than 20 years the necessity for scientists and teachers to understand physics as part of human culture, in the same sense as music, painting, literature, and other forms of expression. In his doctoral thesis, entitled *Física também é cultura!* (Physics is also culture!), Zanetic argues that contemporary citizens are taught "that science is an esoteric subject that has nothing related to people's real life and that does not belong to their cultural background" (Zanetic 1989, p. 146). Therefore, they do not see themselves as involved in it or that science reflects societies' values, questions, and styles. Contrary to art, which is generally associated with people's expressions about the world, physics seems to be restricted to a few outsiders. However, Zanetic claims that science is also embedded in cultural, political, economic, and social contexts. Thus, science is part of the cultural background of modern societies; it reflects people's values, beliefs, and objectives. Zanetic argues about this idea for physics, but his words are also adequate to describe a desired view of science:

> Certainly, a physics that involves emotions, the comings and goings of the great generative ideas present in significant problems, the use of rational discourse, the role of discourse and conceptualization from the magicians, the fantastic ideas of scientific thinkers that built the great theories that had dominated or still dominate the everyday of physicists. Ultimately, this physics is incomparably richer than the physics essentially formal, non-historic, filled with exercises, distant, either from popular culture, either from scientific culture, part of contemporary intelligent life. (Zanetic 1989, pp. 61–62)

According to Zanetic, physics teaching "cannot exclude [...] the theoretical conceptualization, the experimentation, the history of physics, the philosophy of science and its links with society and other areas of culture" (Zanetic 2005, p. 21). This argument evokes a problematizing view of education, as seen in Freire and Giroux, as it suggests a broad approach to teaching scientific content. According to the author, "physics should be present in cultural education of contemporary citizen, independently of individual differences of interest and the most varied academic and professional motivations" (Zanetic 2006, p. 41). Therefore, the understanding of science as part of culture is not only relevant for scientists, but for everyone. Zanetic asserts that the approximation of physics and culture encourages an intelligent dialogue between students and the world, once it incites the curiosity and arouses the feeling that knowledge can lead to an emancipation of human beings. In this perspective, he argues about the value of HPS to this approximation:

> [...] we can say that the cultural education of any person will become enriched if science teaching takes into consideration elements of history and philosophy of science, of social studies of science and their relationship to other areas of knowledge, in particular literature. (Zanetic 2009, p. 288)

Zanetic claims that the study of historical and philosophical aspects of science shows how science is close to the social and cultural dimensions of society. Physics and culture are inseparable, because the development of scientific knowledge – as of music and poetry – belongs to human activity. The idea of science as part of culture sustains a different view of teacher education because it implies the comprehension of temporality, historicity, and the mutual influence of scientific knowledge and society. Teachers, as historical–social beings, become conscious of their role in the process of developing scientific knowledge, not considering it to be something separate from their daily life, but part of it.

1.5 The Critical and Transformative Education of Teachers and the Role of HPS

In the previous sections, we presented the ideas that form the basis for what we call the critical and transformative teacher being someone who establishes a critical dialogue with the world surrounding him or her, problematizes, and transforms it with his or her actions and beliefs. The critical dialogue implies reciprocity. Teachers do not receive information passively and withhold it, but dialogues and communicates with it. By communicating, teachers create their judgement of matters of education and science. By assuming the necessity of allowing their students to have an active voice, critical and transformative teachers offer the indispensable elements of a wider and reflexive comprehension of science and its relation to social, cultural, and political contexts.

Critical and transformative teacher education emphasizes the active role of teachers in the education of their students. It involves the comprehension that any process of teaching and learning implies an intervention in human relations. Hence, "it is not possible to exercise teaching as if nothing happens to us" (Freire 1996, p. 96). As he or she understands himself or herself to be an essential part of the education process, the critical and transformative teacher accepts the responsibility to foment critical attitudes of his students. It is not related to indoctrination, but to the expansion of intellectual curiosity, of a new and motivating knowledge.

From this perspective, the critical and transformative teachers respect the autonomy of their students as unique and complex individuals, who have feelings, personalities, and different desires. They assume it to be "an ethical imperative and not a favor that we can or cannot concede to one another" (Freire 1996, p. 59). By considering students as essential parts of the educational process, the critical and transformative teacher enhances both theirs and their students' roles as citizens, in a broader sense. As they are individuals that dialogue, reflect, agree, and disagree, they are more capable of making better decisions, to understand the importance of social interactions with people who think differently and to promote social justice and democracy. A critical and transformative education of teachers also allows the transformation of the school into a locus of debate and inclusion, fostering its pedagogical and political roles (Giroux 1992).

Critical and transformative teachers conceive scientific knowledge as part of human culture and are aware of its epistemological, ontological, and axiological particularities. They understand science as a collective, diverse, changeable, and historically constructed knowledge. Therefore, critical and transformative teachers promote a more dynamic educational process, in which the traditional ways of understanding, teaching, and learning science are problematized. They provide moments for students to experience being part of science, as historical elements, and capable of changing it. They emphasize the role of science among the cultural productions of humanity and its importance to the development of intellectuality and autonomy of individuals (Zanetic 1989).

What, then, are the roles of HPS in the critical and transformative education of teachers?

The importance of HPS in teacher education is almost consensual nowadays, at least in science education community. The introduction of meta-scientific content into science classes may lead to a better comprehension of the scientific enterprise, its actors, its relation to social and cultural contexts, and its influence on our daily lives. Thus, considering the critical and transformative education of teachers, HPS is necessary because every human being is a historical being. Disassociating humanity from history is impossible. Humans act in the world, as the world acts upon us. Freire (2005b, p. 94) states: "thus it is impossible to dichotomize human beings and the world, since the one cannot exist without the other". Hence, it follows that studying the historical and philosophical basis of scientific knowledge is studying the cultural history of humankind. In this context, there is no place for a "banking" conception of teaching, in which human beings are removed from their historical and social roles and considered as mere receivers of preconceived knowledge.

The study of science should also be about the process, not only the product. The latter is ended, limited, and closed to itself. The process, otherwise, is dynamic and revealing, as it shows human attempts to understand the world. By studying science as a process, teachers create their history and apprehend their transformative actions on the world, reality, and humanity. Additionally, from HPS, teachers see cases that illustrate science as part of human culture, promoting a better understanding of its relationships with politics, religion, arts, and other forms of human production (Zanetic 1989).

The problematization of the scientific knowledge is the key to make teaching and learning a reciprocal development of intellectuality, not a unilateral communication. As mentioned before, Freire (2005b, p. 135) asserts that putting the world as a problem to the students is to make them and the teacher see it in a critical way. The critical and transformative education of teachers implies teaching with HPS because an educational process that promotes a reflection of the ideas, attitudes, and praxis of science teachers cannot be adequate without discussing the historical and philosophical development of what is taught:

> History, as a period of human events, is made by human beings at the same time as they "make" themselves in history. If the work of education, like any other human undertaking, cannot operate other than "within" the world of human beings (which is a historical-cultural world), the relations between human beings and the world must constitute the starting-point for our reflections of that undertaking. (Freire 2005b, p. 131)

The consideration of HPS as a crucial element of critical and transformative education guides us to some reflections. First, it is important to consider that including HPS in teacher education is not simple. It is necessary to admit that the mere creation of historical and philosophical courses is not enough. We claim that teacher-training programs analyze the objectives and content of these courses, in addition to the experience of who is teaching. These issues can be overcome, but involve a rethinking of teacher education as a whole. It implies the revision of the goals of these programs, the curriculums, and, more importantly, the education of teacher trainers. For our community, which is in the interface between HPS and education, it is obvious that science is historically constructed. But we also know that teacher education involves other agents that come from different areas and perspectives. We should look to them, to their understanding of HPS, to their doubts, beliefs, and positions. Fostering critical and transformative teacher education should be a collaborative work and we must look forward to encouraging other individuals to join it.

If the critical and transformative education involves the comprehension that scientific knowledge is historical, HPS must be included throughout the whole education of teachers. This means that it cannot be restricted to specific HPS courses. We do not claim that we should have a "historical" teacher education, but that the notion that science is a historical enterprise must be present in different phases of teacher education. Besides presenting the historical and philosophical perspective of science, teacher educators should reserve moments when pre-service teachers think about the incorporation of HPS into their practices. Moments like these are essential to the development of critical and transformative thought, as it allows pre-service teachers to put into action their attitudes, know-how, and ideas.

1.6 Final Remarks

Critical and transformative teacher education summarizes the current aspects, aims, and perspectives that claim a liberating pedagogy and teachers' awareness of their historical and social roles as agents of transformation. HPS can foster critical and transformative teacher education because it portrays an open and critical view of science, by disclosing its processes and relating it to cultural, social, political, and other contexts of society. However, if critical and transformative teacher education is important and if HPS can contribute to promoting it, how can we establish a fruitful connection between them? In other words, how can we make HPS a powerful resource for fostering critical and transformative teacher education?

The answers to the aforementioned questions depend on many factors. First, it is necessary to elaborate on proposals for classrooms that are founded on the principles of the critical and transformative education of teachers. This has been a constant effort by the authors of this chapter and some of the results are available in Moura and Silva (2013). Second, it is important to promote new forms of teacher education. We are certain that a critical and transformative teacher implies a reshaping of traditional forms of teacher education. A critical and transformative education of teachers cannot emerge from traditional forms of teacher education, in which, for instance, there is the idea that a good teacher is the one who knows all the content that he or she passively transmits to the student. This is a more complex issue, as it involves a transformation in the whole educational system and the values that each community imputes to the teachers. Third, it implies the awareness that changes in education are difficult, gradual, and slow, though far from unachievable.

Therefore, there is much to be explored in the relationship between HPS and critical and transformative teacher education. Introducing and improving future interventions could demonstrate how powerful the historical and philosophical content can be in enhancing teachers' views on their importance to the education of their pupils.

References

Allchin, D. (2013). *Teaching the nature of science: Perspectives and resources*. Saint Paul: SHiPS Education Press.

Bejarano, N. R. R., & de Carvalho, A. M. P. (2003). Tornando-se professor de ciências: crenças e conflitos. *Ciência e Educação, 9*(1), 1–15.

Brasil. (2015). Resolução n° 2, de 1° de Julho de 2015. Define as diretrizes curriculares nacionais para a formação inicial em nível superior (cursos de licenciatura, cursos de formação pedagógica para graduados e cursos de segunda licenciatura) e para a formação continuada. Brasília: DOU.

Brasil. Ministério da Educação. (2001). *Plano Nacional de Educação – PNE*. Brasília: Inep.

Brasil. Ministério da Educação, Secretaria de Educação Média e Tecnológica. (1997). *Parâmetros Curriculares Nacionais: ensino médio*. Brasília: Ministério da Educação.

Brasil. Ministério da Educação. Secretaria da Educação Média e Tecnológica. (2002). *PCN+ Ensino médio: orientações educacionais complementares aos Parâmetros Curriculares Nacionais. Ciências da natureza, matemática e suas tecnologias*. Brasília: MEC/SEMTEC.

Brasil. Ministério da Educação. Secretaria de Articulação com os Sistemas de Ensino. (2014). *Planejando a próxima década: conhecendo as 20 metas do Plano Nacional de Educação*. Brasília: MEC/SASE.

Dagher, Z., & Erduran, S. (2016). Reconceptualizing the nature of science for science education: Why does it matter? *Science & Education, 25*(1), 147–164.

Freire, P. (1996). *Pedagogia da autonomia*. São Paulo: Paz e Terra.

Freire, P. (2005a). *Pedagogy of the oppressed*. New York: The Continuum International Publishing Group Inc..

Freire, P. (2005b). *Education for critical consciousness*. New York/London: Continuum.
Giroux, H. (1992). *Escola crítica e política cultura*. São Paulo: Cortez/Autores Associados.
Giroux, H. (2011). *On critical pedagogy*. New York: The Continuum International Publishing Group.
Höttecke, D., & Silva, C. C. (2011). Why implementing history and philosophy of science in school science education is a challenge: An analysis of obstacles. *Science & Education, 20*(3–4), 293–316.
Martins, R. A. (2006). Introdução: a história das ciências e seus usos na educação. In C. C. Silva (Ed.), *Estudos de história e filosofia das ciências: subsídios para aplicação no ensino* (pp. xvii–xxxx). São Paulo: Editora Livraria da Física.
Matthews, M. (2015). *Science teaching: The contribution of history and philosophy of science* (2nd ed.). New York: Routledge.
Moura, B. A., & Silva, C. C. (2013). Multi-contextual approach of history of science (MCAHS): Introducing historical content in science teacher training. In *Proceedings of the 12th biennial conference of the international history, philosophy and science teaching group*.
Schön, D. A. (1983). *The reflective practitioner: How professionals think in action*. New York: Basic Books.
Zanetic, J. (1989). *Física também é cultura!*. PhD Dissertation. Faculdade de Educação, Universidade de São Paulo, São Paulo, Brazil.
Zanetic, J. (2005). Física e cultura. *Ciência e Cultura, 57*(3), 21–24.
Zanetic, J. (2006). Física e arte: uma ponte entre duas culturas. *Pró-posições, 17*(1), 39–57.
Zanetic, J. (2009). Física ainda é cultura! In A. F. P. Martins (Ed.), *Física ainda é cultura?* (pp. 176–189). São Paulo: Editora Livraria da Física.

Part II
Nature of Science

Chapter 2
Contextualized Questionnaire for Investigating Conceptions of the Nature of Science: Procedure and Principles for Elaboration

Nathália Helena Azevedo and Daniela Lopes Scarpa

2.1 Introduction

Over the last few decades, there has been an increase in research in the field of Science Education dedicated to the investigation of Nature of Science (NOS) conceptions (Azevedo and Scarpa 2017), with special attention paid to the inclusion of scientific metacognition in curriculums at all educational levels (Lederman 1992; Azevedo and Scarpa 2017). Part of this preoccupation resides in the need to develop individuals who can critically discuss socio-scientific issues. In large part because of this, the increasing collective efforts of researchers in the area of science education interested in aspects of the history and philosophy of science are well known.[1]

Apart from the importance of studying students' conceptions of the natural sciences, the lack of consensus on the definition of *what is science* resulted in questioning how to evaluate these conceptions. In this regard, Alters (1997) investigated agreements between the most common criteria of NOS used in research on education and the criteria of philosophers of science. In this author's study *Whose Nature of Science?*, it was shown that, in general, there is more agreement than disagreement among the criteria investigated, even though there is no consensus on aspects relevant to a definition of science. Concordant with this viewpoint, some of the significant authors in the area of Science Education[2] have defended the use of the most frequently accepted characteristic aspects of science, and have given value to the criticalness that is to be expected in science education. These authors emphasized

[1] Such as: Robinson (1965), Duschl (1985), Lederman (1992), Matthews (1992), Monk and Osborne (1997), McComas et al. (1998), Seroglou and Koumaras (2001), Abd-El-Khalick (2012), and Allchin (2013).

[2] Such as: Osborne et al. (2003), Lederman (2007), and Paraskevopoulou and Koliopoulos (2011).

N. H. Azevedo (✉) · D. L. Scarpa
Institute of Biosciences, University of São Paulo, São Paulo, SP, Brazil
e-mail: helena.nathalia@usp.br; dlscarpa@usp.br

M. E. B. Prestes, C. C. Silva (eds.), *Teaching Science with Context*,
Science: Philosophy, History and Education,
https://doi.org/10.1007/978-3-319-74036-2_2

that lists of NOS aspects are useful for reflecting on the view of science that is generally taught (Matthews 2012).

Although some authors[3] consider a consensual pedagogical view of science, which includes seven NOS aspects (e.g., Lederman et al. 2002), there is a diversity of other aspects – at least 25, according to the systematic review by Azevedo and Scarpa (2017) – related to the various sciences, and which are not contemplated in the view that is taken to be consensual. Even if there is certain agreement on NOS aspects that should be examined in basic education (Lederman 2007), and as a consequence, of NOS conceptions that should be investigated, the listing of such aspects comes under hard criticism and the joint debate about *what*, *how*, and *why* NOS aspects need to be taught has been increasing in the literature recently (Azevedo and Scarpa 2017).

This debate is relevant, especially when considering that the literal use of lists can compromise critical learning. Allchin (2011) pointed out that there are central questions on scientific practice that are often omitted in a noncontextual approach to NOS. Among the most pointed criticism, according to Allchin, is that if the lists of NOS aspects are not contextualized, and are merely aimed at making personal and social decisions that involve scientific topics, these lists would bring to the fore an idealized vision of science. From this viewpoint, we understand that *to contextualize* means to expose the students to scientific themes and contexts likely to be identified as real, to enable them to make well-informed decisions, taking into consideration those characteristics that are relevant throughout the production of scientific literacy.

Following this line of thought, and even though with no objection to a list of NOS aspects, Irzik and Nola believe that the items presented should be better understood. In their view, lists can "portray a too narrow image of science", as the rules and objectives of science are rarely exploited (Irzik and Nola 2011, p. 592). Furthermore, definition of the scientific method involved is generally extinguished by the statement that there are no unique methods in science, with no explanation whatsoever of the variety of methodological practices that could clarify the limits of this statement. Without specifying prevalent methodologies of science, it becomes difficult to evaluate, for example, how science might or might not be "self-corrective." This is a point exploited by some authors.[4]

The relevance of discussion about lists of NOS conceptions becomes even more manifest when we consider that many of the concepts usually investigated (present both in research instruments and in classroom assessment) bring to the fore questions that are open to debate, especially by demonstrating discrepancies in some sciences, as in the case of biological sciences. There is ample debate about the "scientism" of biology and its sub-areas, such as Ecology, because of nonconformity to

[3] Such as: Lederman (1992, 2007), McComas et al. (1998), Stanley and Brickhouse (2001), Osborne et al. (2003), McComas (2008), and Abd-El-Khalick (2012).

[4] Such as: Laudan (1981), Allchin (2015), and Romero (2016).

scientific concepts bolstered in such areas as physics and chemistry (Mayr 2004; Rosenberg 2008). Worthy of note, for example, is the discussion on the existence or not of laws in Ecology, as demonstrated in studies by McIntosh (1987), Murray (1992, 2001), Sober (1997), and Lawton (1999), and brought about by the difficulty of forecasting in this area. Another point often questioned is the capacity for generalization in ecological studies, which is confronted by the high number of case-studies and by the difficulty of replication.

Consideration of these characteristics in Ecology opens the dialogue on the position of Irzik and Nola (2011), about lists of NOS aspects generally dealing with characteristics related only to the production of scientific knowledge. With such a focus, practices rooted in scientific investigation, viz., data collection, analysis, and classification, which constitute everyday methods in Ecology, are set aside. These practices also need to become part of scientific education, by being prerequisites both in the comprehension of NOS and in how to do science.

Many of the instruments used for investigating NOS conceptions introduce conceptual questions that are perhaps incomprehensible, because they are very generic or vague for students (Azevedo and Scarpa 2017). Some questions require definitions of laws, theories, and experiments to check whether the meaning of the postulate is clear. In some instruments, students are even requested to justify answers by means of examples. Even so, we believe that questions placed in the context may have a heavy impact on the replies obtained. Thus, during investigation of the efficiency of NOS conceptions, attention must be paid to the capacity of the student to evaluate the presence of NOS aspects in contexts that involve scientific practice. Instruments comprising contextual situations of everyday science may be strategic for identifying the real student conceptions of NOS and evaluating their critical abilities.

Although themes are of fundamental importance, in general, efforts dedicated to the study of NOS conceptions have been concentrated on investigating concepts in basic education (students and teachers), with little attention being paid to university courses associated with the area of Biological Sciences (Azevedo and Scarpa 2017), whence the gap in scientific literacy among pre-service biology teachers and biology undergraduate students. This leads us to reflect on the type of scientific education that this population has received. Owing to the importance of the theme and the absence of contextualized instruments for investigating NOS conceptions (Azevedo and Scarpa 2017), and in spite of the amount of research that has been developed over the last few years (Neumann et al. 2011; Azevedo and Scarpa 2017), our intention herein was to:

1. Place evidence for methodological practice during the generation, authentication, and application of a contextualized research instrument, where Ecology has been used as the basic model.
2. Facilitate reflection on the research instruments usually employed in research into NOS conceptions.

2.2 Producing the Instrument

The VENCCE (Students' Views of the Nature of Science by way of Contextualization in Ecology, originated from the acronym in Portuguese) questionnaire was elaborated with the aim of investigating NOS conceptions from undergraduate bioscience students. During elaboration, decisions were taken in relation to:

1. The NOS aspects to be dealt with
2. The inclusion of open or closed questions
3. The nature of their items
4. The scale and form of evaluating these items
5. The size and language of statements
6. The presence of authentic situations
7. Authenticity and reliability of the results.

Based on analysis of the post-application results, the strategies employed for generating the questionnaire and evaluating efficiency can be extrapolated when generating others. Even though the theme does not terminate here, our procedure was organized in the form of a question grid (Fig. 2.1), and we hope that being critically answered and expanded by other researchers can contribute toward the execution of further studies that require the objective and clear construction of an instrument. In the sequence, we explain some of our decisions.

2.3 Theoretical Dimensions of the Instrument

On developing a questionnaire for biological science undergraduates, the aim was to outline aspects that could really influence "doing science" among biologists, and that needed to be understood by biology teachers (replying to question 1 of Fig. 2.1). The reason for this decision is that, in Brazil, undergraduate courses in Biological Sciences are intended for future biology researchers *and* biology teachers. In general, the curricula of the two courses are quite similar, but pre-service biology teachers also attend disciplines in education. We believe that respondents' interpretations reflect the real level of NOS conceptions, more than agreement or disagreement with noncontextualized statements, such as the question "what is a law?," which is frequent in questionnaires on NOS conceptions. Thus, we opted for a thematic questionnaire that could explore authentic situations during undergraduate education and performance of these professionals.

Epistemological choices were based on a survey carried out during systematic revision (Azevedo and Scarpa 2017) that included studies by Aikenhead and Ryan (1992), Alters (1997), McComas and Olson (1998), Gil-Pérez et al. (2001), Osborne et al. (2003), Lederman (2007), Abd-El-Khalick (2012), and DiGiuseppe (2014), and a suggestion as presented by Allchin (2011). As we agree with the latter author's viewpoint, the choice was to build a widely contextualized instrument for investigating NOS conceptions (question 2 of Fig. 2.1). For the write-up of VENCCE situations and statements (Table 2.1), it was considered that students need to under-

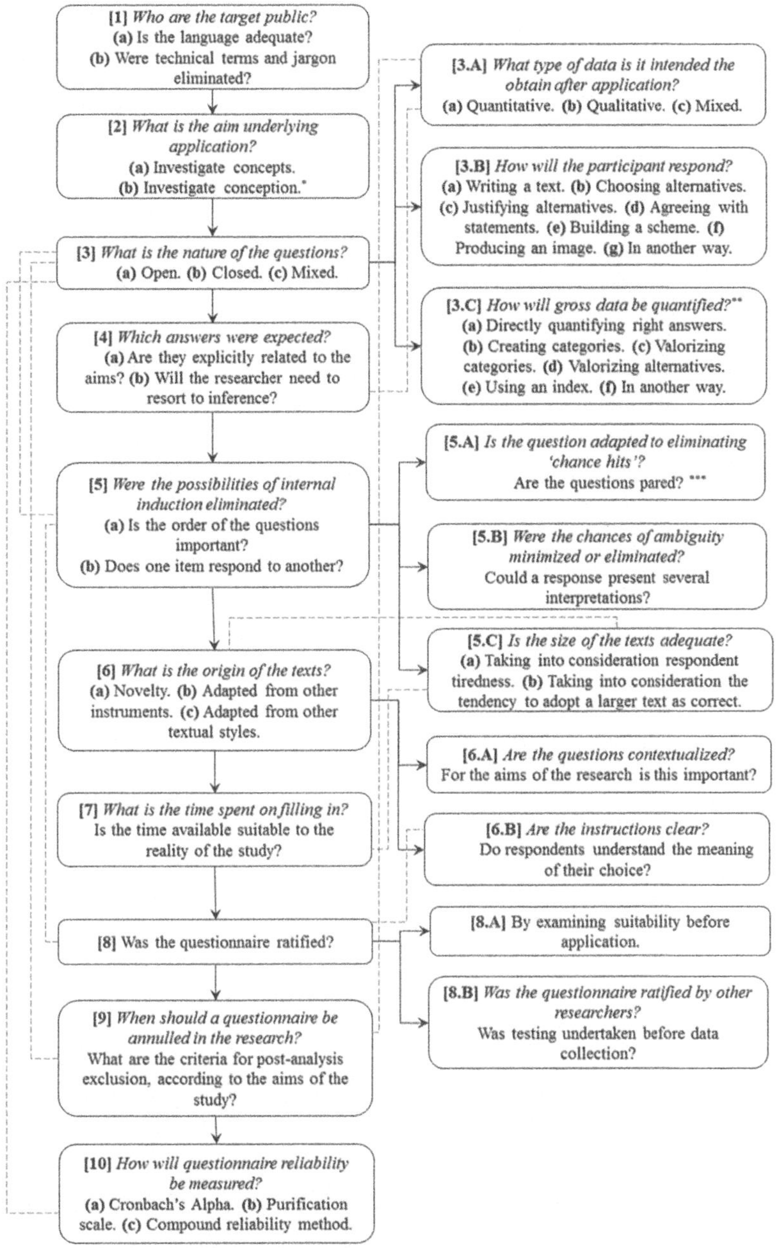

Fig. 2.1 Diagram of questions containing considerations for the elaboration of a research questionnaire, based on procedures for developing Students' Views of the Nature of Science by way of Contextualization in Ecology (VENCCE). Solid lines indicate steps followed during the development and implementation of the instrument. Dashed lines show correlated actions and decisions that were continuously weighted during the study. (*) The presence of greater subjectivity, which requires theoretical care. A possible solution for overcoming the problem would be to link points of agreement between authors. (**) *A priori* or *a posteriori*. (***) Especially for questionnaires containing closed questions

Table 2.1 An example of a situation and the statements contained in the Students' Views of the Nature of Science by way of Contextualization in Ecology (VENCCE) questionnaire, according to the form of application

SITUATION 5. *To check whether the size of a certain clam would interfere in its chance of being preyed upon, a group of scientists took samples. They collected shells from randomly marked-out parcels on a beach, calculated the area of each, and recorded whether they showed any marked characteristic of predation.*
Take into consideration the episode as described, along with your conceptions of science. Read the following statements and *mark the degree of your agreement with each.*

		Low			Medium			High			Don't know	Cla.
A	Describe a scientific practice with systemized collection	1	2	3	4	5	6	7	8	9		WI
B	As no experiment was carried out, it cannot be considered scientific practice	1	2	3	4	5	6	7	8	9		LI
C	The conclusion can only be generalized if data collection is repeated in several locations	1	2	3	4	5	6	7	8	9		PI
D	The conclusion can only be generalized if the collection of data is undertaken with a very large sample	1	2	3	4	5	6	7	8	9		PI
E	For scientists to reach a conclusion, they should take into consideration the possibility of predation being by chance	1	2	3	4	5	6	7	8	9		WI
F	The results of the study will only be accepted by the scientific community if they are presented in graphic form	1	2	3	4	5	6	7	8	9		LI
G	The results of the study will be accepted by the scientific community, even when due care was not taken in the collection of data	1	2	3	4	5	6	7	8	9		LI
H	The results of the study will be accepted by the scientific community if data receive some statistical treatment	1	2	3	4	5	6	7	8	9		WI
I	The choice of statistical treatment could interfere in the final results	1	2	3	4	5	6	7	8	9		WI
J	Describes a scientific practice, as scientists have formulated a testable hypothesis that has been confronted with the observations	1	2	3	4	5	6	7	8	9		WI

The last column shows the classification (*Cla.*) of statements as Well-informed (*WI*), Partially informed (*PI*) or Less-informed (*LI*), which was not available for respondents stand scientific practice, but not in an abstract manner or merely philosophically. We endorse the need for a functional understanding of NOS (Allchin 2013), thereby facilitating student analysis of day-to-day scientific statements, not only as citizens, but also as teachers and scientists (questions 6 and 6A of Fig. 2.1).

The starting point for elaborating situations and statements was based on Allchin's (2011) proposal of relevant NOS aspects. With alterations considered convenient for Biological Science undergraduates, attention was then paid to elaborating the largest number of statements on those NOS aspects that appeared with greater frequency in surveys (Azevedo and Scarpa 2017). Attention was also paid to aspects incongruous with the peculiarities of Ecology as a science. This aspect is exploited, for example, in statement B of situation 5 of the questionnaire (Table 2.1). For a well-informed viewpoint of science, aspects such as financing, scientist motivation, revision by pairs, cognitive vices and frauds, in addition to the ratification of new methods, were also included in questionnaires. Always where possible, the attempt was made to supply information toward a well-informed analysis of the situations and focused on the student's ability to analyze (Allchin 2011). Thus, the student was not expected to know details about the epistemology of the science for his viewpoint to be considered adequate (question 1 of Fig. 2.1).

2.4 Instrument Format

Open-ended questions have the advantage of fewer induced responses, leaving respondents freer to express themselves (Gil 1999; Bell and Lederman 2003). They also allow a greater depth of analysis, which can be inaccessible when closed questions are asked (Bardin 2009). However, with open-ended questions, there is the disadvantage that the resulting data present a greater difficulty of treatment and comparison with other studies, as the texts produced by the participants are subject to a greater degree of arbitrariness or subjectivity on the part of the researcher, leading to difficulties in replicating the instrument in other contexts (Bardin 2009).

A questionnaire composed of closed questions permits the inclusion of a larger number of items, thereby facilitating pairing and the inclusion of more aspects for analysis. Furthermore, there is both less effort involved for participants during filling in and facilitated processing of data for the researcher (question 3 of Fig. 2.1). Such aspects were taken into consideration during the conscientious choice of closed questions, as was the perspective of applying a questionnaire to a more ample sample, thereby giving greater statistic potential to analysis and further possibilities of testing hypotheses, because of a conceivable increase in generalization (question 3A of Fig. 2.1). The decision regarding the format of the questionnaire items is also associated with the response format, and how the analysis should be carried out (questions 3 and 4 of Fig. 2.1). Likewise, these decisions involve the establishment of data exclusion criteria (question 9 of Fig. 2.1).

Together with the choice of the nature of statements (statements were elaborated, not questions), it was necessary to adopt a method of valorizing and attributing a score. After revising the literature, the index used in the work of Aikenhead (1973), Aikenhead and Ryan (1992) and employed in the questionnaire of Manassero and Vázquez (2001) was chosen, as this appeared to be adequate for the intended statistical tests following application. The index emerged from an adaptation of Likert's scale of agreement (Likert 1932) and takes into consideration the classification of a statement to give it values according to a punctuation scale. This index (which we called $VENCCE_{index}$) reflects the approximation of respondents with what is considered a well-informed (or valuable) viewpoint of the concepts presented. The closer to +1, the more well-informed the NOS conceptions; the closer to −1, the less well-informed they were (question 3C of Fig. 2.1).

To facilitate the use of $VENCCE_{index}$, it was necessary to create statements that could be classified as *well-informed, partially informed* or *less-informed*, because for calculation, statements would receive distinct scores accordingly. The aim was to elaborate these statements based on the conformity of aspects (or conceptions) associated with some theme of NOS, such as the diversity of methods, the capacity for generalization, the role of laws in the organization of knowledge, or mutual correlations. Hence, it was established that a statement considered to be *well-informed* would present one or more correct concepts and a correlation, when it exists, between these, would also be considered correct. On the other hand, a *less-informed* statement would present incorrect concepts and no mutual correlations. Finally, in the case of statements classified as *partially informed*, the procedure would be twofold:

1. A correct and an incorrect concept were presented, or
2. Two correct concepts without mutual correlations

Although statements were elaborated for each of these categories, the classification of statements into *well-informed, partially informed* or; *Less-informed* was adjusted to the answers obtained in the two stages of validation. This is explained later.

Five problem situations were generated and for each a series of statements (question 6 of Fig. 2.1). Each statement is followed by a scale varying from 1 to 9, in which respondents should note their degree of agreement (question 3B of Fig. 2.1). This model facilitates the evaluation of the agreement with each of the statements, which is impossible, for example, when using a one-response model (a question with many alternatives). This is so, because by showing numbers by agreement scale (1–9) instead of category (total, partial, low, none, for example), respondents would have access to a scale that facilitates making comparisons and better evaluating their own placing. Thus, with this model of response, access to that which it is intended to measure is better, as instead of a categorical response (yes or no), there is quantifiable information for each statement. With this model, it is also possible to identify contradictable replies from the students, in addition to "chance hits." With this in mind, paired statements were elaborated that could be excluded from analysis in these cases (question 5A of Fig. 2.1).

When producing statements, attention was given to reducing textual ambiguity. Part of this problem was overcome by amplifying the contextualization of the questionnaire, especially by this being its main characteristic in comparison with other existing questionnaires for sampling NOS conceptions (questions 5 and 6A of Fig. 2.1). Attention was also paid to the language used in statements, with special care being taken to ensure that it was appropriate for biology undergraduates. Thus, the use of characteristic terms from the philosophy of science was avoided, and possible jargon employed among researchers in the area of Science Education (Lederman and O'Malley 1990; Allchin 2011), and even in Ecology (question 1 of Fig. 2.1). This care was reflected during the critical phase of validation, in which various personal and professional viewpoints regarding the instrument were assessed.

As for the general format of the VENCCE, attention was paid to the possibility of conducting answers by the formulation of short statements of a similar size, so that the respondent was not led toward marking those thought to be "apparently more correct" because of any difference in size (question 5 of Fig. 2.1). Statements were randomized and two versions of the questionnaires used, as a way of reducing some of the possible problems during application, such as tiredness, that could prejudice the authenticity of answers in the final stage, and cheating among students (questions 5A and 5C of Fig. 2.1). Finally, and with the aim of obtaining only answers that represent a real degree of agreement on the part of the respondent, the choice was made to include the option "don't know." This was a way of reducing ambiguity and eliminating one more of the possible sources of doubt regarding choice, thereby guaranteeing no obligation to mark a number from the scale, even when unable to assume a position with regard to a statement (questions 5B and 6B of Fig. 2.1).

2.5 Ratification of the Instrument

Following revision by the researchers, the VENCCE passed through two stages of ratification. The first was undertaken among the researchers of the authors' research group. This group was composed of individuals at different types of qualification, comprising science initiation, undergraduates, and PhD students, in addition to PhDs. The second stage was among professors of the institute and scientists from the different areas of biology and education (question 8 of Fig. 2.1).

In the first stage, the attempt was to:

1. Point out problems regarding instructions on filling in the form (question 6B of Fig. 2.1)
2. Identify general faults in writing up that could lead to a bad interpretation of situations and statements
3. Estimate the average time needed to fill in the questionnaire
4. Check the efficacy of classification (*well-informed*, *partially informed* or *less-informed*) of the statements

5. Evaluate the adequacy of the size of the instrument (question 7 of Fig. 2.1). For this, the 11 members of the group received the questionnaire and replied to it individually, without intervention. Comments were noted down in the questionnaires, which were then handed in. Pertinent criticism and suggestions were incorporated into a new version that was used during the second stage of ratification.

In the second stage, systematic individual conversations were made with eight professors from the Biosciences Institute of the University of São Paulo. All are also researchers in the areas related to the themes dealt with in the VENCCE (Ecology, Science Education, and the Philosophy of Sciences). At this stage, statements and situations were discussed one by one, and conceptual, theoretical, and formal adjustments were afterward put into effect.

2.6 Application and Evaluation of Reliability

The VENCCE was applied in printed form to a sample of 691 biological science undergraduates from 14 Brazilian universities. Of these, 78.6% were from the south-east, 7% the north-east, 7% the north, and 7% the south. Of the 14 universities, 78% were public and 22% private.

The reliability of the instrument was estimated after application, according to Cronbach's alpha coefficient (α) (Cronbach 1951, 2004), to check the capacity of the VENCCE for measuring or inferring proposed propositions (question 10 of Fig. 2.1). Consideration was given to the variance obtained for each item of the questionnaire (in our case, each statement), in addition to individual replies. Values fell within the interval (0–1), reliability being considered higher the closer α is to 1. Use was made of the values of the index associated with the VENCCE following application, thereby receiving a value of $\alpha = 0.912$. Thus, the reliability of the instrument could be considered adequate according to established parameters.[5] Differentiation between participant NOS conceptions was possible; thus, the aim was considered to have been reached.

In the literature on the methodology for the ratification of questionnaires, some authors recommend alterations in the research instruments when the values for α are very low (less than 0.7) or very high (more than 0.95, when there is an excess of redundancy) (Parassuraman et al. 1985; Oviedo and Campo-Arias 2005). In these cases, successive application may be necessary, with the insertion or elimination of items until an adequate value for α has been reached. Such a procedure is often called the purification stage. In the case of the VENCCE, purification was not called for, because with the first application it was possible to reach an adequate value of reliability, which could be associated with the critical stage of validation through which it had passed.

[5] Such as: Gliem and Gliem (2003), Kline (2005), and Maroco and Garcia-Marques (2006).

Considering the 691 questionnaires resulting from the application, an average of 2.936 ± 4.961 statements with "don't know" were obtained per student. By being below the 10% of statements in the VENCCE, this value could be considered adequate, thereby indicating that, even with the option of marking "don't know," students most often chose to mark their degree of agreement. Moreover, the lack of this option could have led students to mark a degree of agreement that was incompatible with their NOS conceptions, thereby biasing future conclusions on these conceptions in the sample.

2.7 Final Considerations

The development of new instruments reflects the need for adequate evaluation by statistical quantification and content analysis, which could be applied on a large scale, and would reflect a better understanding of the real NOS conceptions from the students (Abd-El-Khalick 2014). However, as the criteria adopted in generating an instrument are not always explained in the literature, an attempt was made to clearly show the practices adopted throughout the elaboration of the VENCCE. Two of these practices that could make a significant contribution toward other research projects were:

1. The ratification stages, in which an expressive number of researchers was included, to thus reduce the possible sources of bias
2. The adoption of indexes (such as $VENCCE_{index}$), both for measuring the replies in the instrument and for checking reliability.

The use of reliability indexes and the adoption of a purifying stage for questionnaires in the area of science education, is not, as yet, common. It is hoped that on showing our experience, it will be possible for other researchers to consider the use of reliability methods throughout their research, with due adaptation. We point out, for example, that the purifying stage is not a substitute for qualitative analysis of the instrument. Although a reduction of the number of items and the elimination of those that could bias instrument reliability values are important for a more critical use of the results obtained from the questionnaires, the possibility of losing information through the purely mechanical practice of this methodology should be considered. In the VENCCE, repetition of certain themes throughout the questionnaire was a conscientious choice, with the aim of obtaining more reliable data on students' NOS conceptions included in the sample. The repetition of some themes was intended to evaluate the consistency of replies, concerning to the themes that were considered more relevant, in view of the aims of the research. Thus, it was considered that the use of reliability indices and purifying methodology needs to be simultaneous with researcher evaluation of the relevance of certain items for the questionnaire. We also sought to highlight the dependence of the decisions taken on the development of a research instrument, as indicated by the dashed lines in Fig. 2.1.

The present work is a further contribution to the already ample discussion of the literature on the format of questionnaires for diagnosing NOS conceptions (Abd-El-Khalick 2014), and the results thus obtained (Lederman and Lederman 2014; Azevedo and Scarpa 2017). When presenting the Views on Science–Technology–Society questionnaire, Aikenhead and Ryan (1992) pointed out that this instrument is different from the rest by presenting questions that were derived empirically. For the authors, this is a strategy for reducing possible ambiguities generated by differences in the language of researcher and researched, especially when the scale used to assess replies is based on the Likert scale. We have already reported the care we have taken to reduce ambiguities during the ratification phases, although we recognize that better reliability of NOS conceptions could be obtained through a comparison of the results obtained from other sources, such as open questions applied to the same sample.

Also, as regards instrument format and the reliability of results, in a revision of instruments for evaluating NOS conceptions, Abd-El-Khalick (2014) pointed out that a restriction of the number of instruments for investigation could increase comparison robustness. Furthermore, the author enhanced the importance of context when evaluating conceptions. We agree and therefore wish to point out that differences between instruments need to be taken into account in studies (Azevedo and Scarpa 2017). Thus, explaining the steps and the decisions adopted throughout the elaboration of the research instruments is relevant to facilitating critical conceptions on NOS survey analysis and dialogue between studies.

References

Abd-El-Khalick, F. (2012). Examining the sources for our understandings about science: Enduring conflations and critical issues in research on nature of science in science education. *International Journal of Science Education, 34*(3), 353–374.

Abd-El-Khalick, F. (2014). The evolving landscape related to assessment of nature of science. In S. K. Abell & N. G. Lederman (Eds.), *Handbook of research on science education* (Vol. II, pp. 621–650). New Jersey: Lawrence Erlbaum Associates.

Aikenhead, G. S. (1973). The measurement of high school students' knowledge about science and scientists. *Science Education, 57*(4), 539–549.

Aikenhead, G. S., & Ryan, A. G. (1992). The development of a new instrument: "Views on Science–Technology–Society" (VOSTS). *Science Education, 76*(5), 477–491.

Allchin, D. (2011). Evaluating knowledge of the nature of (whole) science. *Science Education, 95*, 518–542.

Allchin, D. (2013). *Teaching the nature of science: Perspectives and resources*. Saint Paul: SHiPS Education Press.

Allchin, D. (2015). Correcting the 'Self-correcting' mythos of science. *Filosofia e História da Biologia, 10*, 19–35.

Alters, B. J. (1997). Whose nature of science? *Journal of Research in Science Teaching, 34*(1), 39–55.

Azevedo, N. H., & Scarpa, D. L. (2017). A systematic review of studies about conceptions on the nature of science in science education. *Revista Brasileira de Pesquisa em Educação em Ciências, 17*(2), 621–659.

Bardin, L. (2009). *Análise de conteúdo*. Lisboa: Edições 70.
Bell, R. L., & Lederman, N. G. (2003). Understanding of the nature of science and decision making on science and technology based issues. *Science Education, 87*, 352–377.
Cronbach, L. J. (1951). Coefficient alpha and the internal structure of tests. *Psychometrika, 16*(3), 297–334.
Cronbach, L. J. (2004). My current thoughts on coefficient alpha and successor procedures. *Educational and Psychological Measurement, 64*(3), 391–418.
DiGiuseppe, M. (2014). Representing nature of science in a science textbook: Exploring author–editor–publisher interactions. *International Journal of Science Education, 36*(7), 1061–1082.
Duschl, R. A. (1985). Science education and philosophy of science: Twenty-five years of mutually exclusive development. *School Science and Mathematics, 85*(7), 541–555.
Gil, A. C. (1999). *Métodos e técnicas de pesquisa social*. São Paulo: Editora Atlas.
Gil-Pérez, D., Montoro, I. F., Alís, J. C., Cachapuz, A., & Praia, J. (2001). Para uma imagem não deformada do trabalho científico. *Ciência and Educação, 7*(2), 125–153.
Gliem, J. A., & Gliem, R. R. (2003). Calculating, interpreting and reporting Cronbach's alpha reliability coefficient for Likert-type scales. In *Midwest research to practice conference in adult, continuing, and community education* (pp. 82–88). http://www.alumni-osu.org/midwest/proceeding.html. Accessed 20 Jul 2016.
Irzik, G., & Nola, R. (2011). A family resemblance approach to the nature of science for science education. *Science & Education, 20*, 591–607.
Kline, R. B. (2005). *Principles and practice of structural equation modeling*. New York: The Guilford Press.
Laudan, L. (1981). *Science and hypothesis: Historical essays on scientific methodology, The University of Western Ontario Series in Philosophy of Science* (Vol. 19). Dordrecht: Springer.
Lawton, J. H. (1999). Are there general laws in ecology? *Oikos, 84*, 177–192.
Lederman, N. G. (1992). Students' and teachers' conceptions of the nature of science: A review of the research. *Journal of Research in Science Teaching, 29*(4), 331–359.
Lederman, N. G. (2007). Nature of science: Past, present and future. In S. K. Abell & N. G. Lederman (Eds.), *Handbook of research on science education* (pp. 831–880). New Jersey: Lawrence Erlbaum Associates.
Lederman, N. G., & Lederman, J. S. (2014). Research on teaching and learning of nature of science. In S. K. Abell & N. G. Lederman (Eds.), *Handbook of research on science education* (pp. 600–620). New Jersey: Lawrence Erlbaum Associates.
Lederman, N. G., & O'Malley, M. (1990). Student's perceptions of tentativeness in science: Development, use and sources of change. *Science Education, 74*(2), 225–239.
Lederman, N. G., Abd-El-Khalick, F., Bell, R. L., & Schwartz, R. S. (2002). Views of nature of science questionnaire: Toward valid and meaningful assessment of learners' conceptions of nature of science. *Journal of Research in Science Teaching, 39*(6), 497–521.
Likert, R. (1932). A technique for the measurement of attitudes. *Archives of Psychology, 140*, 1–55.
Manassero, M. A., & Vázquez, A. A. (2001). Instrumentos y métodos para la evaluación de las actitudes relacionadas con la ciencia, la tecnología y la sociedad. *Enseñanza de las Ciencias, 1*(20), 15–27.
Maroco, J., & Garcia-Marques, T. (2006). Qual a fiabilidade do alfa de Cronbach? Questões antigas e soluções modernas. *Laboratório de Psicologia, 4*(1), 65–90.
Matthews, M. R. (1992). History, philosophy and science teaching: The present rapprochement. *Science & Education, 1*(1), 11–48.
Matthews, M. R. (2012). Changing the focus: From nature of science (NOS) to features of science (FOS). In M. S. Khine (Ed.), *Advances in nature of science research: Concepts and methodologies* (pp. 3–26). Dordrecht: Springer.
Mayr, E. (2004). *What makes biology unique? Considerations on the autonomy of a scientific discipline*. Cambridge: Cambridge University Press.

McComas, W. F. (2008). Seeking historical examples to illustrate key aspects of the nature of science. *Science & Education, 17*(2–3), 249–263.
McComas, W. F., & Olson, J. K. (1998). The nature of science in international science education standards documents. In W. F. McComas (Ed.), *The nature of science in science education: Rationales and strategies* (pp. 41–52). Dordrecht: Kluwer.
McComas, W. F., Almazora, H., & Clough, M. P. (1998). The nature of science in science education: An introduction. *Science & Education, 7*(6), 511–532.
McIntosh, R. P. (1987). Pluralism in ecology. *Annual Review of Ecology and Systematics, 18*, 321–341.
Monk, M., & Osborne, J. (1997). Placing the history and philosophy of science on the curriculum: A model for the development of pedagogy. *Science Education, 81*(4), 405–424.
Murray, B. G. (1992). Research methods in physics and biology. *Oikos, 64*, 594–596.
Murray, B. G. (2001). Are ecological and evolutionary theories scientific? *Biological Reviews, 76*, 255–289.
Neumann, I., Neumann, K., & Nehm, R. (2011). Evaluating instrument quality in science education: Rasch-based analyses of a nature of science test. *International Journal of Science Education, 33*(10), 1373–1405.
Osborne, J., Collins, S., Ratcliffe, M., Millar, R., & Duschl, R. (2003). What "ideas-about-science" should be taught in school science? A Delphi study of the expert community. *Journal of Research in Science Teaching, 40*, 692–720.
Oviedo, H. C., & Campo-Arias, A. (2005). Aproximación al uso del coeficiente alfa de Cronbach. *Revista Colombiana de Psiquiatría, 34*(4), 572–580.
Paraskevopoulou, E., & Koliopoulos, D. (2011). Teaching the nature of science through the Millikan-Ehrenhaft dispute. *Science & Education, 20*(10), 943–960.
Parassuraman, A., Zeithaml, V. A., & Berry, L. L. (1985). A conceptual model of service quality and its implications for future research. *Journal of Marketing, 49*(4), 41–50.
Robinson, J. T. (1965). Science teaching and the nature of science. *Journal of Research in Science Teaching, 3*(1), 37–50.
Romero, F. (2016). Can the behavioral sciences self-correct? A social epistemic study. *Studies in History and Philosophy of Science, 60*(A), 55–69.
Rosenberg, A. (2008). Biology. In M. Curd & S. Psillos (Eds.), *The Routledge companion to philosophy of science* (pp. 511–519). London: Routledge.
Seroglou, F., & Koumaras, P. (2001). The contribution of the history of physics in physics education: A review. *Science & Education, 10*(1–2), 153–172.
Sober, E. (1997). Two outbreaks of lawlessness in recent philosophy of biology. *Philosophy of Science, 64*, 458–467.
Stanley, W. B., & Brickhouse, N. W. (2001). Teaching science: The multicultural question revisited. *Science Education, 85*(1), 35–49.

Chapter 3
Consensus and Dissent Around the Concept of Nature of Science in the Ibero-American Community of Didactics of Science

Rafael Yecid Amador-Rodríguez and Agustín Adúriz-Bravo

3.1 Introduction

When taking into account the views and policies for science education proclaimed nowadays, is it enough to introduce the well-known notions of *scientific literacy* and *science–technology–society* (STS)? Or, formulated in a slightly different way, is equipping citizens with a general preparation in science and technology enough to reach the very ambitious goals that modern societies set for their children, adolescents, and young adults in the twenty-first century? In this chapter, we adhere to the idea that, when aiming at a *quality science education for all*, it is necessary to explicitly introduce into science classes some *instrumental* content from the philosophy, history, and sociology of science. Such content is what in our field of didactics of science is currently called the "nature of science" (NOS) (McComas 1998; Flick and Lederman 2004; Acevedo Díaz et al. 2005).

American scholar William McComas (1998) depicts NOS as a "fertile hybrid arena" that combines elements from various meta-sciences (that is, scientific disciplines that establish second-order, "meta"-discourse *on* science). He proposes that contributions to NOS would mainly come from the philosophy, history, and

R. Y. Amador-Rodríguez (✉)
GEHyD, Group of Philosophy, History and Didactics of the Natural Sciences, CeFIEC, Research Institute Center for Training and Research in Science Teaching, Faculty of Exact and Natural Sciences, University of Buenos Aires, Buenos Aires, Argentina

Institute of Educational Studies-IESE, University of Norte, Barranquilla, Colombia
e-mail: ryamador@uninorte.edu.co

A. Adúriz-Bravo
GEHyD, Group of Philosophy, History and Didactics of the Natural Sciences, CeFIEC, Research Institute Center for Training and Research in Science Teaching, Faculty of Exact and Natural Sciences, University of Buenos Aires, Buenos Aires, Argentina
e-mail: aadurizbravo@cefiec.fcen.uba.ar

M. E. B. Prestes, C. C. Silva (eds.), *Teaching Science with Context*, Science: Philosophy, History and Education,
https://doi.org/10.1007/978-3-319-74036-2_3

sociology of science. Thus, NOS, from the point of view of didactical research and innovation, can be understood as a line emerging from new curriculum needs: in the formation of citizens, science teaching should combine two important aspects – knowing science and knowing *about* science.

From a curricular viewpoint, we broadly conceptualize NOS as a selection of meta-scientific ideas with value for science education (Adúriz-Bravo 2005a, b; 2009). When constructing NOS, we give priority to the *didactical transposition* (i.e., functional adaptation for teaching purposes) of models from the philosophy of science, highlighting the aim of allowing citizens to generate their own answers to questions such as: What is science? How does it change over time? What are its relationships with the context framing it?

In this chapter, our general purpose is to examine NOS representations (linked to curriculum and research) expressed in academic publications in the field of didactics of science in Ibero-America (that is, Spain, Portugal, and the Spanish- and Portuguese-speaking countries in Latin America and the Caribbean), with the aim of ascertaining the extent of consensus among "didacticians of science" (i.e., science education researchers) in our region around this research line. We want to review locally which ideas guide the inclusion of the nature of science in didactics of science, and how researchers in our field assume the relationship between these two scientific corpuses. We also want to inquire after the links that are established in Ibero-American literature between NOS and other lines of didactical research.

3.2 Conceptual Framework

In our opinion, generating changes in science teaching practices, in the construction of science curricula, in science teachers' professional development, and in other fields of interest to current didacts of science requires explicitly assuming a *recognizable* position from the philosophy of science that expresses our views on scientific activity and its practitioners, on the derived products of such activity, and on its social implications. Quality research on didactics of science requires *epistemological positioning*.

Philosophical positions to which specialists in didactics of science refer in their work could be located in three big "stages" of the philosophy of science of the twentieth century (Adúriz-Bravo 2004):

1. Logical positivism and "received view" (1920–1970)
2. Critical rationalism and "new philosophy of science" (1930–1985)
3. Post-Kuhnian and contemporary views (1970 to today).

In this article, we use this broad periodization of the philosophy of science to map the conceptions of the nature of science that we science education researchers hold.

With regard to our understanding of the history of science as a meta-science, for the purposes of this study, we agree with Camacho and Quintanilla (2008), who depict scientific activity as a rich process of individual and social construction that takes place within a set of complex cultural relations. These authors claim that neither discovery nor justification of scientific knowledge should be seen "a-historically" in science education. They argue that the history of science could be understood as a "cartography" of science, providing us with a map in which the modeled objects (scientific discoveries, inventions, paradigms, theories, controversies, etc.) *exist* (i.e., take meaning) only in relation to other objects that are part of the system represented. Therefore, raw historical data are not interpretable in isolation, taken out of the scientific communities and their modes of production. They must be related to one another and with the interpretive theoretical models provided by the history and philosophy of science. The "network" of science is a function of the conceptual moves and value-laden aims of each historical setting (place and time). The history of science, according to this "diachronic" conception, takes into account the social, political, economic, and cultural contexts in which scientific knowledge is developed.

3.3 Methods

Figure 3.1 presents an overview of the methodological path that we pursued to study the Ibero-American academic production around NOS.

The central goal of our study was to characterize some of the representations of the concept of NOS among researchers in our regional community of didactics of science. We limited the study in time and space, restricting it to the academic production that was issued between 2000 and 2009 in Ibero-America. For the purposes of data collection, we constructed a corpus consisting of research articles included

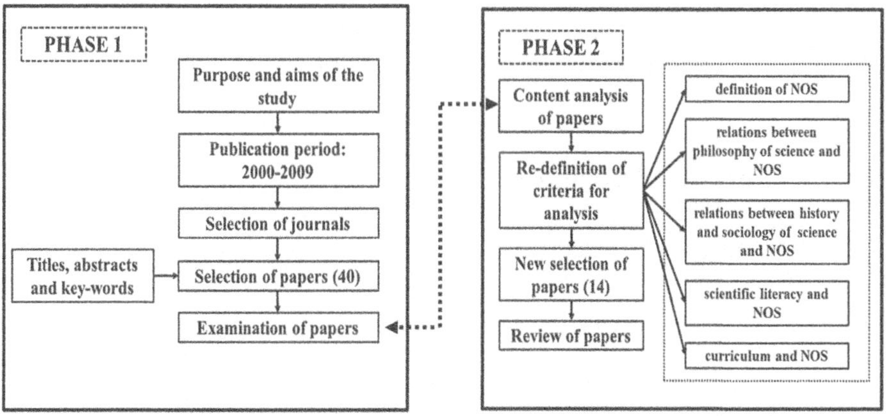

Fig. 3.1 Main methodological elements of our study

Table 3.1 List of the articles further selected for the second phase of the study using more refined criteria

Articles in the second phase of the study (full references in the appendix)
Arriassecq and Greca (2002), Fernández et al. (2002), Díaz and Jesús (2002), Colombo de Cudmani (2003), Guridi and Arriassecq (2004), Acevedo Díaz et al. (2005), Adúriz-Bravo (2005a, b), Viau et al. (2006), Acevedo Díaz et al. (2007a, b), Guisasola and Morentin (2007), Praia et al. (2007), Acevedo Díaz (2008) and Fernández-González (2008)

in journals *completely devoted to science education* that are published in the region under study. We thus excluded journals on more general educational studies or those from other disciplines.

The journals we investigated were *EC-Enseñanza de las Ciencias* (Spain), *REEC-Revista Electrónica de Enseñanza de las Ciencias* (Spain), *Eureka-Revista Eureka sobre Enseñanza y Divulgación de las Ciencias* (Spain), *REF-Revista de Enseñanza de la Física* (Argentina), *CE-Ciência & Educação* (Brazil), *IEC-Investigações em Ensino de Ciências* (Brazil) and *TED-Tecné, Episteme y Didaxis* (Colombia); all of these rank among the most prestigious and best-diffused in Ibero-America.

Once the selection of journals was made, we proceeded to collect all the articles that contained the descriptors "nature of science", "epistemology", "philosophy of science", or "history of science" in their titles, abstracts or key-words (in Spanish and Portuguese). With this criterion, we obtained 40 items (first phase of the study).

After rapidly revising the content of these documents from the first selection, we re-defined the analytic criteria. In this sense, the mere mention of the concepts above was not sufficient for our study. Upon this second, more specific reading, we constructed a new set of criteria. The articles should explicitly address one or more of these topics: definition of NOS; relations between NOS and philosophy of science; relations between NOS and history and sociology of science; NOS and scientific literacy; NOS and curriculum. With these new criteria, we proceeded to a second selection of articles, in which remained 14 out of the original 40. Our content analysis was conducted on this new corpus (second phase of the study).

Table 3.1 lists the articles that belonged to the second phase of the study. We performed our content analysis on these selected 14 texts.

3.4 Results and Discussion

3.4.1 General Analysis

This first subsection presents the general features of the sample: the journals that included NOS articles and the countries of the authors of those articles. Figure 3.2 shows, in two graphs, the journals and numbers of the articles on NOS for each phase of the study.

Figure 3.2 shows a preference for Spanish journals (*Enseñanza de las Ciencias*, *Eureka* and *REEC*). *Eureka* (with five articles in the second phase) is the publication that contributed most to our sample, but the number of papers published in the period is too small to establish a tendency.

Figure 3.3 shows (in two graphs, one for each phase of the study) the countries of affiliation of the authors of the articles that were selected as empirical material for our investigation. The figure also shows the number of authors for each country. We selected main/first/correspondence authors and counted each recurring author only once.

We can see that, in both the first and second phases, researchers with the most interest in the field of NOS are from Spain (13 and 5 main authors respectively). Among these, the group led by José Antonio Acevedo Díaz is the most productive on the topic of NOS. The second community of researchers in didactics of science most frequently represented in the journals selected for our study is from Argentina: nine main authors in the first phase and five in the second. Therefore, in the line of NOS, and for the parameters selected, most of the production comes from these two countries. Each of them produces the same or a greater amount of literature than the rest of the Ibero-American countries taken together, and therefore constitutes a regional reference in the topic.

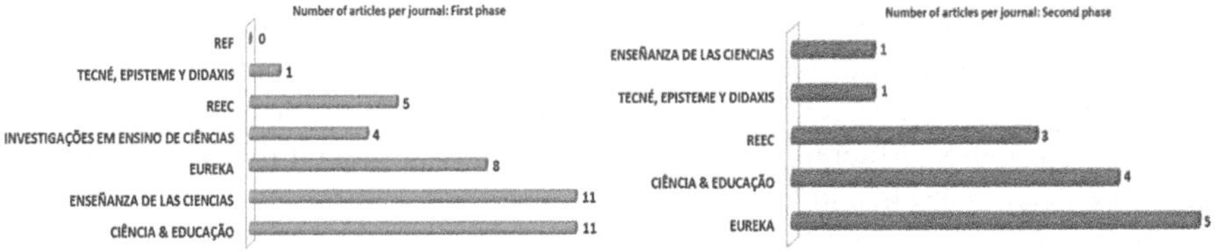

Fig. 3.2 Contribution of each of the journals analyzed to the corpuses of the first and second phases of the study

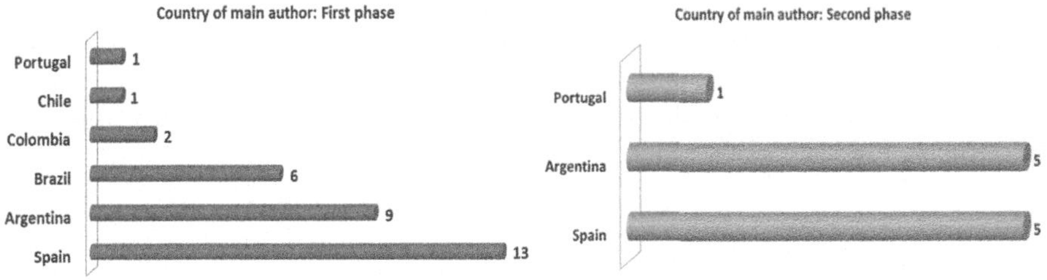

Fig. 3.3 Countries and numbers of main authors of the articles included in the corpuses of the first and second phases of the study. Each main author with more than one paper was only counted once

3.4.2 Analysis of Titles, Abstracts, and Key-Words

This second subsection is devoted to an examination of the titles, abstracts, and key-words of the papers. Figure 3.4 shows the concepts related to NOS that can be found in the *titles* of the articles in each phase of the study.

In the titles of the 40 articles analyzed in the first phase, the most frequent concepts are epistemology and philosophy of science, coupled with constructs such as teaching, scientific education, and science. In the 11 articles that contain none of our "course" descriptors (the bar "other concepts" in the left graph of Fig. 3.4), more general expressions related to views on science and to meta-scientific reflection are used (e.g., attitudes towards science, aims of science education, constructivism, models).

In the second phase of the study, the construct of NOS is preeminent (five titles), followed by epistemology + philosophy of science (four titles). This is coherent with our conceptual framework: most researchers in the field of didactics of science in our region adhere to the thesis that the most important component of NOS should be philosophical.

Table 3.2 shows the results of our analysis of the *abstracts* of the 14 articles of the second phase of the study. We established the frequency of occurrence of some compound enunciations. Table 3.2 ranks those enunciations in decreasing frequency. The most frequent enunciation is around the *inclusion of NOS in science curricula*. Thus, the problem of including NOS as a new curriculum component seems to be the main concern of the research community. The issue of finding consensuses around NOS in didactics of science only appears in two articles; which contrasts with the international literature, where this issue has generated lively discussion in the last decade (Duschl and Grandy 2013).

Table 3.3 presents, in order of decreasing frequency, the *key-words* included in the 14 articles of the second phase of the study. As expected, the most recurring collocation is "nature of science." The rest of the key-words show us how NOS is related to other "big" issues in didactics of science. A major concern seems to be the contribution of NOS to scientific literacy, especially in the Spanish literature.

The chart on the right of Fig. 3.4 together with Tables 3.2 and 3.3 show how the concept of NOS, which we used as the main criterion for selection for the second phase, is included in titles, abstracts, and key-words and linked to other important constructs of didactics of science.

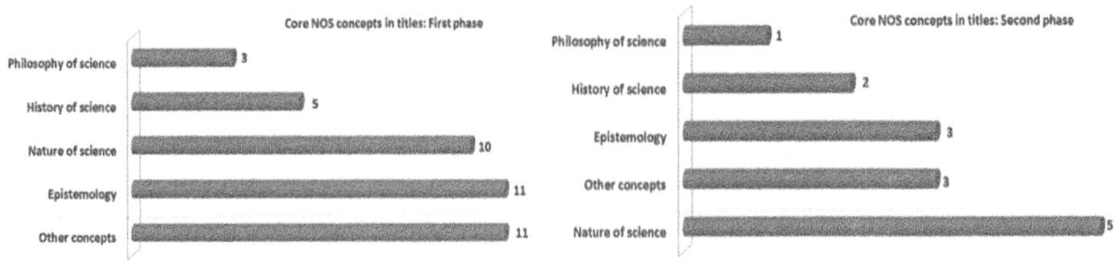

Fig. 3.4 Main concepts related to the nature of science (NOS) that appear in the titles of the articles included in the first and second phases of our study

Table 3.2 Most relevant thematic aspects around nature of science (NOS) that are enunciated in the abstracts of the 14 articles of the second phase of the study. They are organized in order of decreasing frequency

Relevant thematic aspects around NOS as enunciated in the abstracts
1. Inclusion of NOS in the science curricula
2. Relationship between NOS and didactics of science
3. Relationship between NOS and scientific literacy
4. Incorporation of the philosophy of science into science education
5. Incorporation of the history of science into science education
6. Consensuses around NOS in didactical research

Table 3.3 Most frequent key-words in the 14 articles of the second phase of the study

Most frequent key-words in the corpus of the second phase
1. Nature of science
2. Scientific literacy
3.a. Philosophy of science/3.b. Science–technology–society (equal frequencies)
4.a. Science curriculum/4.b. Science teacher education (equal frequencies)
5.a. Conceptions of science/5.b. History of science (equal frequencies)

3.4.3 Explicit Definitions of the Concept of NOS

This subsection is devoted to presenting the explicit "definitions" of NOS that we found in the corpus; our intention was to ascertain the conceptual representations on this topic that circulate among didactics of science researchers in Ibero-America.

José Antonio Acevedo Díaz and his colleagues (2007a) state that, in the community of didactics of science, the concept of NOS is dialectical and complex; this would make precisely defining and reaching a consensus difficult to do. A theoretical proposal from this Spanish research group consists in assuming NOS as meta-knowledge about science arising from the interdisciplinary reflections made from the philosophy, history, and sociology of science (Acevedo Díaz et al. 2005, 2007b; Acevedo Díaz 2008).

A more refined definition is provided by Guisasola and Morentin (2007): NOS combines aspects of different areas of knowledge such as history, sociology, and philosophy of science with the aim of exploring what science is, how scientists work as a social group, and how society manages the problems derived from the scientific activity.

Agustín Adúriz-Bravo (2005a) characterizes NOS as a set of meta-scientific content with value for science education. In his article, he argues in favor of this "broad" definition, which may be appropriate for three reasons. First, because it locates NOS in the field of meta-sciences, which are disciplines of scientific character, compatible with the science to be taught and "teachable" within the same curriculum space. Second, because the expression "NOS" does not separate the various sources of the ideas to teach; such ideas come mainly from the philosophy, history, and sociology of science, three disciplines between which strict demarcation is disputed even among their own specialists. Third, because when we speak of the *educational* essence of NOS, its construction requires genuine didactical transpositions, and not straightforward incorporation of "foreign" content.

3.4.4 Teaching NOS: Experts' Views

In this subsection, we systematize the NOS that Ibero-American specialists in didactics of science suggest should be introduced into compulsory science education. In the first place, we collected the general, theoretical considerations of experts in the corpus on how to construct a solid NOS to be taught. After that, we retrieved the content of NOS that, according to those experts, should be taught.

Table 3.4 contains the theoretical discussion around the issue of how didactics of science can construct valid NOS for science education. Four out of the 14 articles in the corpus of the second phase address the classic curriculum questions: *why*, *what*, and *how* to teach.

Regarding which *specific* philosophical models are preferred in Ibero-America when constructing the NOS, most papers that we studied chose the so-called "new philosophy of science" of the 1950s–1970s, citing authors such as T. Kuhn or I. Lakatos, followed by critical rationalism, mainly in K. Popper's version (Amador-Rodríguez and Adúriz-Bravo 2016). More recent philosophical accounts of science (e.g., contextualism, new experimentalism, the semantic view of scientific theories) appear much less frequently.

Table 3.5 contains the most relevant NOS content that is suggested in the corpus, organized into *topics*. It is constructed following the principles suggested by the research group led by Acevedo Díaz: as the main organizers of the table, we used a re-phrasing of the *aspects* proposed by this group, and we included the other authors' claims in this general structure.

Aspects and topics in Table 3.5 are defined in the most general and abstract way possible, strongly independent of the different philosophical schools of the twentieth century, which, when discussing such topics, constructed specific philosophical *ideas*. Among the most frequently cited of such ideas in Ibero-American NOS are: theory-ladenness of observation, tentativeness, non-linearity of scientific progress,

Table 3.4 Theoretical discussion around the construction of NOS that can be found in the corpus of the second phase

Theoretical issues	Authors' suggestions
What are the sources of NOS?	NOS should include ideas from the philosophy, sociology, and psychology of science (Acevedo Díaz 2008)
	NOS should mainly be a philosophical reflection on science, "set" in the history of science, and "warned" by the sociology of science against scientism (Adúriz-Bravo 2005a)
Constraints for the construction of NOS	NOS should be aimed at providing an image of science that is moderately realist and rationalist, balancing the presentation of the intellectual feats and limitations of science as a human enterprise (Adúriz-Bravo 2005a)
	NOS content should be "tuned into" scientific and didactical content (Adúriz-Bravo 2005a)
Didactical considerations	Researchers, innovators, and teachers should reflect on the following questions: What NOS should we be teaching? Why should NOS be taught? How can it be taught to different target audiences? (Acevedo Díaz et al. 2005, 2007a; Adúriz-Bravo 2005a; Praia et al. 2007)

Table 3.5 Suggestions of the authors from the corpus of the second phase of the study regarding what NOS topics should be taught in science education. The left column adapts an organization of "aspects" proposed by Acevedo Díaz and colleagues

Aspects of NOS	Topics of NOS
Methodology	Methods that validate scientific knowledge (Adúriz-Bravo 2005a; Acevedo Díaz et al. 2007b; Acevedo Díaz 2008)
	Internal and external perspectives on the functioning of science as an activity (Acevedo Díaz et al. 2007b; Acevedo Díaz 2008)
Sociology and axiology	Values involved in scientists' activity (Adúriz-Bravo 2005a; Acevedo Díaz et al. 2007b; Acevedo Díaz 2008; Fernández-González 2008)
	Nature and characteristics of the scientific community (Arriassecq and Greca 2002; Acevedo Díaz et al. 2007b; Acevedo Díaz 2008; Fernández-González 2008)
Interactions	Relationship between science and technology (Acevedo Díaz et al. 2007b; Acevedo Díaz 2008)
	Relationships among science, society, and culture (Adúriz-Bravo 2005a; Acevedo Díaz et al. 2007b; Acevedo Díaz 2008)
Theories	"What is this thing called science?" (Arriassecq and Greca 2002; Adúriz-Bravo 2005a; Viau et al. 2006; Acevedo Díaz et al. 2007b; Praia et al. 2007; Acevedo Díaz 2008)
	How is scientific knowledge constructed? (Arriassecq and Greca 2002; Adúriz-Bravo 2005a; Viau et al. 2006; Acevedo Díaz et al. 2007b; Praia et al. 2007; Acevedo Díaz 2008; Fernández-González 2008)

falsifiability, paradigms, social constraints of the scientific activity, temperate realism, and the methodological plurality of science. Most of these are from the new philosophy of science that came about in the three decades after World War 2.

According to our results, the diversity of theoretical issues that arise when discussing the *construction* of NOS is high. Along these lines, the regional community of didactics of science is far from reaching a consensus. One shared aspect, though, is the selection of the philosophy of science as a principal component of NOS.

As for the content that should be taught, Ibero-American authors seem to have a greater consensus. They share a "standard" representation of NOS that, in its basics, is not too far from the international proposals (McComas 1998). In such a representation, the nature and function of scientific theories and the methods used to construct or validate them are central.

3.4.5 A Point of Disagreement

In this subsection, we discuss one overarching idea in the articles of the second phase of the study that may be indicating dissent or disagreement within the Ibero-American community of experts working on NOS. The analysis that we propose allows us to identify *positions* within NOS research in our region, each with its own theoretical and methodological nuances. In the same spirit, Acevedo Díaz and colleagues (2007a) point at the complexity and dynamicity of the representations of NOS that are available in the Ibero-American didactics of science community. According to these authors, there is a broad spectrum of NOS conceptions among specialists, which differ conceptually and methodologically, and compete with one other.

Within this spectrum, we concentrate here on two distinct positions around the place of the philosophy of science in NOS. This identifiable disagreement refers to the possibility of including elaborate NOS content transposed from the meta-sciences in the first stages of science education (i.e., primary or lower secondary). In this respect, Acevedo Díaz and colleagues (2005) deem that the inclusion of philosophical reflection with students does not constitute a "reasonable" teaching objective, owing to the complexity and abstraction of such reflection. They suggest that the main aim of NOS should be empowering students with a sound understanding of contemporary science and technology, which could be better achieved through the inclusion of *content from the area of STS*.

Adúriz-Bravo (2005a), on the other hand, advocates the use of *models from the philosophy of science*, adequately transposed, to attain three main "finalities" for NOS in the classroom:

1. An *intrinsic* finality, where the nature of science serves as a critical reflection on science
2. A *cultural* finality, where NOS can establish synergies with other curriculum areas (philosophy, history, social science, mathematics, language, etc.) to highlight the value of science as a human creation
3. An *instrumental* finality, where NOS is used as a tool to improve the teaching of science content.

This dissent on the main "internal constitution" of NOS reflects on the philosophical schools and topics chosen in these two papers: historicist accounts and models from the sociology of science in the first case, and post-Kuhnian accounts and authors within the "semanticist family" in the second case.

3.5 A Possible Organization of Lines of Research Within NOS

With the descriptors that we used for the second phase of the study, and the results of the content analysis of the papers, we propose a preliminary organization of the field of NOS in Ibero-America into four lines of research.

3.5.1 Constructing NOS: Contributions of the Philosophy of Science

Researchers into NOS in Ibero-America agree with the thesis that contemporary science education should give priority to the aim of developing an "image of science" that is more valid from the epistemological point of view, i.e., more consistent with current scientific practices (Fernández et al. 2002; Acevedo Díaz et al. 2007b). A question that then arises is *which* epistemological view we should assume for the construction of an educationally valid NOS. Acevedo Díaz and colleagues (2007b) state that nowadays, in the field of NOS, epistemologies of all types can be found, among them: realism, positivism, pragmatism, relativism, and almost any possible combination thereof, such as critical rationalism, evolutionism, constructive realism, constructive empiricism, social constructivism, weak relativism, etc. Authors in our community have produced a variety of arguments around which of those is more pertinent for NOS (Vázquez Alonso et al. 2001); the preferred choice has been, as we stated above, toward the new philosophy of science.

Another discussion presented by Acevedo Díaz (2008) refers to the constitution of NOS: should it be restricted to the philosophy of science, or should it be opened up to contributions of the internal and external sociology of science? Guisasola and Morentin (2007) and Adúriz-Bravo (2005a) give pre-eminence to a "pure" philosophy of science, favoring the treatment of epistemological questions in the classroom: the way in which science constructs knowledge, its methods, the assumptions, and underlying beliefs.

In Table 3.6, we compare the positions in two papers (Acevedo Díaz et al. 2007b; Guisasola and Morentin 2007) in which philosophical aspects should be the core of NOS. Acevedo Díaz bases his choices on the work of philosophers (Eflin et al. 1999), whereas Guisasola and Morentin refer to a famous international study from didactics of science aimed at reaching consensus around NOS content (Osborne et al. 2002).

The matrix of Table 3.6 shows a high degree of agreement between two distinct lists of NOS ideas, even though these were constructed from very different theoretical foundations. Another feature of the lists is that the ideas come mainly from the philosophy of science, despite the aforementioned debate on the most appropriate sources of NOS.

3.5.2 A Place for the History and Sociology of Science in NOS

The paper by Irene Arriassecq and Ileana María Greca (2002) is the one from the corpus that gives most salience to the history of science in the constitution of NOS. These authors state that the history of science can serve several important purposes in science education, e.g.:

1. Ascertaining the difficulties and obstacles that had to be overcome to formulate new theories
2. Determining the cultural, epistemic and technological contexts in which such formulations took place
3. Depicting science as a human activity performed by men and women that contribute collectively
4. Warning us against anachronistic readings of historical episodes, reminding us that scientists from the past resorted to the logical, epistemological, and methodological tools available in their contexts.

3.5.3 The Role of NOS in Scientific Literacy

Experts in didactics of science argue that, to achieve citizen participation, it is indispensable to include some content of NOS in the science curriculum. In most of the papers in the corpus, this is the most important reason pointed out to support the curricular inclusion of NOS: a better understanding of the nature of science enables students – as future citizens – to make more informed decisions on public technoscientific issues, and this entails more significant and more responsible participation in these matters (Acevedo Díaz et al. 2005). For instance, Guridi and Arriassecq (2004) suggest that knowing some ideas from the philosophy of science around the topics of evidence, method or explanation might help students with informed decision-making, which is a central feature of democratic societies.

In spite of this extensive agreement, Acevedo Díaz (2008) advocates for more empirical research in didactics of science to determine to what extent a proper understanding of NOS can contribute to the construction of more appropriate beliefs

Table 3.6 Core ideas of NOS according to two articles in the second phase. Acevedo and colleagues base their selection on arguments by philosophers of science; Guisasola and Morentin, base theirs on a Delphi study on positions in the field of didactics

		Acevedo Díaz et al. (2007b)	Guisasola and Morentin (2007)
Relations between NOS and philosophical questions	Epistemic features	The main aim of science is to acquire knowledge on the physical world	The aim of science is to provide explanations for natural phenomena
			Science is an activity that involves creativity and imagination; some scientific ideas are great intellectual achievements
		Science is tentative and dynamic	Scientists' work requires a continuous and cyclic process of making questions and giving answers that lead to new questions. Therefore, scientific knowledge is tentative
		Knowledge generation relies on theoretical commitments and depends on contextual factors (social and historical)	Science is immersed in a socio-cultural context, and is therefore influenced by shared values, subjectivity, and established knowledge
	Methods	There is no single scientific method	Science uses a characteristic methodology, different from those of other forms of knowledge
			Science relies on empirical evidence to validate its ideas, but scientific knowledge does not arise from data. Science requires processes of data interpretation and of theory construction
			Science uses a great variety of methods
			Scientists develop hypotheses and predictions regarding phenomena that are empirically tested
	Realism	The truth of scientific theories is determined by some aspects of the real world that have independent existence (ontological realism)	Current scientific knowledge is the best that we have, but it can be modified in the future owing to new evidence or new interpretations
		There is fundamental order in the real world, which science is intended to describe in the simplest and most comprehensive ways	

and attitudes toward science among students, and eventually to the achievement of a more robust scientific literacy.

All 14 papers of the second phase agree with the mandate that citizens in the twenty-first century should not only know science, but also how it is created and validated, how it changes through history, and how it relates to its social and cultural milieu. In some cases, authors suggest that this aim might be achieved through the inclusion of the perspective STS.

3.5.4 Introducing NOS in the Science Curriculum

Acevedo Díaz and colleagues (2005) indicate that traditional science curricula are heavily focused on conceptual content governed by the internal logic of science; reflection *on* science has been consistently left aside. These authors intend to highlight the educational value of questions such as:

What is the internal and external functioning of science as an activity?
How is scientific knowledge built, developed, and evaluated?
What are the aims, methods, and values guiding the scientific enterprise?
What are the main features of the scientific community?
What are the relationships between science and the techno-scientific system in contemporary societies?
What are the main contributions of science to human culture and to the progress of society?

Treatment of the aforementioned questions would constitute NOS as an emerging curriculum component.

In another article, Acevedo and colleagues (2007a) state that, as a curricular component of scientific literacy, NOS constitutes a major innovation in science education, and is therefore not easy to implement. In its implementation, two major obstacles would appear. The first one is on-going dissent within the community around "basic agreements" on NOS. The second one is the need to adapt NOS content to a diversity of students (with different ages, backgrounds, requirements). Didactical transposition of NOS, i.e., designing appropriate activities to convey meanings, conceptions, theories, models, etc., is still a great challenge for researchers into the didactics of science.

Fernández-González (2008) contributes with two axes that could help to operationalize NOS for curriculum purposes. NOS teaching should:

1. Help students to develop a "scientific culture" for citizenship by examining issues coming from everyday life and from acute social debates
2. Enable students to better grasp the world they live in by becoming aware of technological, environmental and ethical concerns in science.

According to the experts in the corpus, contemporary NOS is consistently expanding the pool of schools and authors in meta-scientific reflection so as to ensure increased educational value and better adjustment to the targeted audiences. Didactics of science is seeking answers to the question of *what* NOS we should teach in each specific context of science education (Adúriz-Bravo 2005a).

3.6 Conclusive Remarks

The data discussed in this article permit the inference that there is relative agreement among authors in the sample around two main issues:

1. That NOS should mainly comprise philosophical ideas
2. The list of main NOS "topics" that should be taught in compulsory science education.

Disagreement, on the other hand, emerges around two important questions:

1. What schools, models, ideas, authors, texts from the philosophy of science should be considered when constructing NOS for the classrooms?
2. Which meta-sciences or meta-scientific studies should accompany the philosophy of science in the constitution of NOS?

All authors in the corpus agree that NOS is not restricted to the philosophy of science, but also includes ideas from the history and sociology of science and from other meta-scientific studies. In spite of this agreement, experts diverge on which views from these academic fields are most suitable for NOS reflection in the classrooms.

We believe that the results of our study have shown that the interface between the philosophy and history of science and the didactics of science is a developed area of research in Ibero-America, but the specific perspective called "nature of science" is more restricted to Spain and to Argentina.

Our study also identified *lines* within NOS research and innovation in our regional community. These lines reveal that there are extended consensuses and, at the same time, noteworthy theoretical disagreements among experts. More theoretical and empirical work is needed to advance our understanding of the state of the art of this international area of research in our region.

References

Acevedo Díaz, J. A. (2008). El estado actual de la naturaleza de la ciencia en la didáctica de las ciencias. *Revista Eureka sobre Enseñanza y Divulgación de las Ciencias, 5*(2), 134–169.

Acevedo Díaz, J. A., Alonso, Á. V., Martín, M., Oliva, J. M., Romero, P. A., Paixão, M. F., & Mas, M. A. M. (2005). Naturaleza de la ciencia y educación científica para la participación ciudadana: Una revisión crítica. *Revista Eureka sobre Enseñanza y Divulgación de las Ciencias, 2*(2), 121–140.

Acevedo Díaz, J. A., Alonso, Á. V., Mas, M. A. M., & Romero, P. A. (2007a). Consensos sobre la naturaleza de la ciencia: Fundamentos de una investigación empírica. *Revista Eureka sobre Enseñanza y Divulgación de las Ciencias, 4*(1), 42–66.

Acevedo Díaz, J. A., Alonso, Á. V., Mas, M. A. M., & Romero, P. A. (2007b). Consensos sobre la naturaleza de la ciencia: Aspectos epistemológicos. *Revista Eureka sobre Enseñanza y Divulgación de las Ciencias, 4*(2), 202–225.

Adúriz-Bravo, A. (2004). Apuntes sobre la formación epistemológica de los profesores de ciencias naturales. *Pedagogía y Saberes, 21*, 9–19.

Adúriz-Bravo, A. (2005a). ¿Qué naturaleza de la ciencia hemos de saber los profesores de ciencias?: Una cuestión actual de la investigación didáctica. *Tecné, Episteme y Didaxis* Número extra: 23–33.

Adúriz-Bravo, A. (2005b). *Una introducción a la naturaleza de la ciencia: La epistemología en la enseñanza de las ciencias naturales*. Buenos Aires: Fondo de Cultura Económica.

Adúriz-Bravo, A. (2009). La naturaleza de la ciencia "ambientada" en la historia de la ciencia. *Enseñanza de las Ciencias, número extra*, VIII Congreso Internacional sobre Investigación en Didáctica de las Ciencias (pp. 1178–1181).

Amador-Rodríguez, R. Y., & Adúriz-Bravo, A. (2016). Concepciones de naturaleza de la ciencia en textos de enseñanza de la química: Indagando por los tópicos epistemológicos de correspondencia y racionalidad. In *XII Jornadas Nacionales. VII Congreso Internacional de Enseñanza de la Biología. III Congreso Internacional de Enseñanza de las Ciencias (CIEC)*. Argentina: Ciudad Autónoma de Buenos Aires.

Arriassecq, I., & Greca, I. M. (2002). Algunas consideraciones históricas, epistemológicas y didácticas para el abordaje de la teoría de la relatividad especial en el nivel medio y polimodal. *Ciência & Educação, 8*(1), 55–69.

Camacho, J., & Quintanilla, M. (2008). Resolución de problemas científicos desde la historia de la ciencia: Retos y desafíos para promover competencias cognitivo-lingüísticas en la química escolar. *Ciência & Educação, 14*(2), 197–212.

Colombo de Cudmani, L. (2003). ¿Qué puede aportar la epistemología a los diseños curriculares en física? *Ciência & Educação, 9*(1), 83–91.

Díaz, M., & Jesús, M. (2002). Enseñanza de las ciencias, ¿para qué? *Revista Electrónica de Enseñanza de las Ciencias, 1*(2), 57–63.

Duschl, R., & Grandy, R. (2013). Two views about explicitly teaching nature of science. *Science & Education, 22*(9), 2109–2139. https://doi.org/10.1007/s11191-012-9539-4.

Eflin, J., Glennan, S., & Reisch, G. (1999). The nature of science: A perspective from the philosophy of science. *Journal of Research in Science Teaching, 36*(1), 107–116.

Fernández, I., Pérez, D. G., Carrascosa, J., Cachapuz, A., & Praia, J. (2002). Visiones deformadas de la ciencia transmitidas por la enseñanza. *Enseñanza de las Ciencias, 20*(3), 477–488.

Fernández-González, M. (2008). Ciencias para el mundo contemporáneo: Algunas reflexiones didácticas. *Revista Eureka sobre Enseñanza y Divulgación de las Ciencias, 5*(2), 185–199.

Flick, L., & Lederman, N. G. (2004). Introduction. In L. Flick & N. G. Lederman (Eds.), *Scientific inquiry and nature of science: Implications for teaching, learning, and teacher education*. Dordrecht: Kluwer Academic Publishers.

Guisasola, J., & Morentin, M. (2007). ¿Comprenden la naturaleza de la ciencia los futuros maestros y maestras de educación primaria? *Revista Electrónica de Enseñanza de las Ciencias, 6*(2), 246–262.

Guridi, V., & Arriassecq, I. (2004). Historia y filosofía de las ciencias en la educación polimodal: Propuesta para su incorporación al aula. *Ciência & Educação, 10*(3), 307–316.

McComas, W. (Ed.). (1998). *The nature of science in science education: Rationales and strategies*. Dordrecht: Kluwer Academic Publishers.

Osborne, J., Collins, S., Ratcliffe, M., Millar, R., & Duschl, R. (2002). What "ideas-about-science" should be taught in school science?: A Delphi study of the expert community. *Journal of Research in Science Teaching, 40*(7), 692–720. https://doi.org/10.1002/tea.10105.

Praia, J., Pérez, D. G., & Vilches, A. (2007). O papel da natureza da ciência na educação para a cidadania. *Ciência & Educação, 13*(2), 141–156.

Vázquez Alonso, Á., Acevedo Díaz, J. A., Mas, M. A. M., & Romero, P. A. (2001). Cuatro paradigmas básicos sobre la naturaleza de la ciencia. *Argumentos de Razón Técnica, 4*, 135–176.

Viau, J., Zamorano, R. O., Gibbs, H., & Moro, L. (2006). Ciencia y pseudociencia en el aula: El caso del "bosque energético". *Revista Electrónica de Enseñanza de las Ciencias, 5*(3), 451–465.

Corpus of Papers Analyzed in This Study

Acevedo Díaz, J. A. (2008). El estado actual de la naturaleza de la ciencia en la didáctica de las ciencias. *Revista Eureka sobre Enseñanza y Divulgación de las Ciencias, 5*(2), 134–169.

Acevedo Díaz, J. A., Alonso, Á. V., Martín, M., Oliva, J. M., Romero, P. A., Paixão, M. F., & Mas, M. A. M. (2005). Naturaleza de la ciencia y educación científica para la participación ciudadana: Una revisión crítica. *Revista Eureka sobre Enseñanza y Divulgación de las Ciencias, 2*(2), 121–140.

Acevedo Díaz, J. A., Alonso, Á. V., Mas, M. A. M., & Romero, P. A. (2007a). Consensos sobre la naturaleza de la ciencia: Fundamentos de una investigación empírica. *Revista Eureka sobre Enseñanza y Divulgación de las Ciencias, 4*(1), 42–66.

Acevedo Díaz, J. A., Alonso, Á. V., Mas, M. A. M., & Romero, P. A. (2007b). Consensos sobre la naturaleza de la ciencia: Aspectos epistemológicos. *Revista Eureka sobre Enseñanza y Divulgación de las Ciencias, 4*(2), 202–225.

Adúriz-Bravo, A. (2005a). ¿Qué naturaleza de la ciencia hemos de saber los profesores de ciencias?: Una cuestión actual de la investigación didáctica. *Tecné, Episteme y Didaxis* Número extra: 23–33.

Arriassecq, I., & Greca, I. M. (2002). Algunas consideraciones históricas, epistemológicas y didácticas para el abordaje de la teoría de la relatividad especial en el nivel medio y polimodal. *Ciência & Educação, 8*(1), 55–69.

Colombo de Cudmani, L. (2003). ¿Qué puede aportar la epistemología a los diseños curriculares en física? *Ciência & Educação, 9*(1), 83–91.

Martín Díaz, M. J. (2002). Enseñanza de las ciencias, ¿para qué? *Revista Electrónica de Enseñanza de las Ciencias, 1*(2), 57–63.

Fernández-González, M. (2008). Ciencias para el mundo contemporáneo: Algunas reflexiones didácticas. *Revista Eureka sobre Enseñanza y Divulgación de las Ciencias, 5*(2), 185–199.

Guisasola, J., & Morentin, M. (2007). ¿Comprenden la naturaleza de la ciencia los futuros maestros y maestras de educación primaria? *Revista Electrónica de Enseñanza de las Ciencias, 6*(2), 246–262.

Guridi, V., & Arriassecq, I. (2004). Historia y filosofía de las ciencias en la educación polimodal: Propuesta para su incorporación al aula. *Ciência & Educação, 10*(3), 307–316.

Praia, J., Pérez, D. G., & Vilches, A. (2007). O papel da natureza da ciência na educação para a cidadania. *Ciência & Educação, 13*(2), 141–156.

Viau, J., Zamorano, R. O., Gibbs, H., & Moro, L. (2006). Ciencia y pseudociencia en el aula: El caso del "bosque energético". *Revista Electrónica de Enseñanza de las Ciencias, 5*(3), 451–465.

Chapter 4
Scientific Skills in Secondary Education: A Study of Curriculum Expectations and Teachers' Thinking

María Teresa Guerra-Ramos and José Baltazar García-Horta

4.1 Introduction and Antecedents

Teachers inevitably communicate messages about the world of science in the classroom. Even more, when according to science curriculum, such messages should be explicit, teachers' ideas (for instance, about how scientific knowledge is elaborated and validated, how scientists proceed in their inquiries) are displayed in their discourse and actions and can influence how students find science interesting, stimulating, and understandable (Zeidler and Lederman 1989). Teachers' ideas about science have pedagogical relevance because they have the important function of introducing students to several aspects of the world of science as a major human intellectual achievement.

A view of natural sciences, their areas of interest, procedures, practices and values go along with any formal or implemented science curriculum. In the context of secondary education in Mexico, the curriculum reform of 2006 introduced the intention to communicate a particular image of science in the subjects science I, II and III. Apart from the study of traditional themes, the programs of study incorporated the innovative idea that teachers should teach *about* science: their methods, values, and how scientists communicate the results of their work. This type of curriculum innovation makes relevant the study of teachers' representations of science and scientific research. There is still a place to understand how science teachers perceive and describe science and how they adapt and display such ideas in the classroom.

M. T. Guerra-Ramos (✉)
Center for Research and Advanced Studies of the IPN, Monterrey Unit, Apodaca, NL, Mexico
e-mail: tguerra@cinvestav.mx

J. B. García-Horta
Autonomous University of Nuevo León, San Nicolás de los Garza, NL, Mexico
e-mail: jose.garciaht@uanl.edu.mx

M. E. B. Prestes, C. C. Silva (eds.), *Teaching Science with Context*,
Science: Philosophy, History and Education,
https://doi.org/10.1007/978-3-319-74036-2_4

Studies aimed at exploring teachers' representations of several aspects of science have adopted different methodological perspectives.[1] However, qualitative approaches have been scarce, and those intended to identify, inform, and develop a wider understanding of natural sciences, and to be prepared to reflect on this in the classroom are even rarer.

Many previous studies lacked a relevant pedagogical context or have adopted criteria or pre-elaborated categories to judge the validity of teachers' conceptions related to the world of science. Some studies have generated scales and questionnaires aimed at quantifying what teachers know and to verify whether such conceptions coincide with normative criteria.[2] This type of study adopts some form of a *deficit* view when they suggest that teachers know little or nothing in comparison with what they "should know" and recommend training programs on methodology and philosophy. Unfortunately, this perspective provides no account of the relationship between teachers' ideas about science and pedagogical practice and have obstacled the possibility of characterizing the diversity and complexity of such ideas.

Other studies have assumed that what teachers know about science is a type of articulated stable and general knowledge (Irez 2006; Liu and Lederman 2007), when exploring teachers' ideas using open-ended general questions requiring articulated and de-contextualized responses. Certainly, this type of study has progressed in describing that teachers' conceptions of science and scientific procedures are eclectic and multifaceted, but they have not contributed to determining their pedagogical relevance in concrete teaching practices in the science classroom.

More recently, some studies have adopted a situated cognition perspective on teachers' conceptions in specific themes related to scientific procedures as evidence in measurement and experimentation situations (Taylor and Dana 2003), science inquiry processes, and their application in the science classroom (Windschitl 2004), or wider aspects of the nature of science (NOS) (Guerra-Ramos et al. 2010). This last type of study has provided rich descriptions about the complex understanding that teachers display in the design of experiments, the identification of variables, and scientific procedures as observation and data analysis. They have pointed out as many affordances as weaknesses in teachers' ideas and have identified several concrete aspects in which teachers can develop knowledge and pedagogical skills.

The research study reported here adopted the view of Driver et al. (1996), when they suggest that teachers and students tend to develop a repertoire of representations of sciences that emerge from their exposure to images of science and scientists in their cultural contexts and from messages, both implicit and explicit, in formal educational settings. The diversity or uniformity of this repertoire of ideas in the case of teachers can constitute, respectively, an affordance or an obstruction when they attempt to introduce young people to the world of science, Considering that teachers' knowledge is highly situated, that is, depends a great deal on the context

[1] For a review of studies in this area, see Lederman (1992), Schwartz and Lederman (2002), and Guerra-Ramos (2012).

[2] Such as Kimball (1968), Chen (2006), and Buaraphan (2011).

in which it is used (Brown et al. 1989), this study explored the representations or conceptions of secondary science teachers of *scientific skills*, from qualitative and exploratory perspectives. The questions guiding this research study were:

1. What aspects or elements of *scientific skills* have been introduced into the Mexican secondary science curriculum?
2. In the context of a hypothetical pedagogical situation, what repertoire of representations of *scientific skills* do secondary science teachers display?
3. How can those representations be qualitatively characterized?

4.2 Methodology and Participants

First, we focused on a content analysis of the official science curriculum documents, which are elaborated by the Ministry of Education in Mexico and contain the description of a particular teaching approach and the programs of study for secondary science. Two documents associated with curriculum reforms in 2006 (SEP 2006) and 2011 (SEP 2013) were considered for comparative purposes. A new curriculum proposal is being elaborated in 2017, but the document in question was not considered because of its preliminary nature. The content analysis allowed us to identify several brief texts and elements expressing curriculum expectations related to scientific skills.

Then, we conducted a study that combined a paper and pencil questionnaire and a semi-structured interview on 22 in-service secondary science teachers working in schools located in Monterrey, Nuevo Leon, an industrial city located in the north of Mexico. All teachers taught the subject science I in the first year of secondary education, which in Mexico consists of 3 years after 6 years of primary education. Therefore, these teachers mainly taught students aged 12–13; but seven teachers also taught science II or science III, respectively, in the other 2 years of the secondary education.

Concerning teachers' training and experience, nine teachers graduated from teacher schools (denominated "*escuelas normales*") and had a degree in secondary education specializing in biology. Six teachers hold university degrees in scientific disciplines such as biology, odontology, and chemistry–pharmacobiology. None of them received specialized courses or training related to scientific inquiry or science education as part of their pre-service or in-service training. Participants' teaching experience ranged from 4 to 18 years. All teachers were familiar with the official curriculum documents, the themes in the subjects sciences (I, II, and III) and the pedagogical perspective, including scientific skills.

Six schools, suggested by the local educational authorities, were finally considered to be part of this study. Then, all teachers from those schools were invited to participate. The final sample included only those teachers accepting the invitation.

To offer a common and relevant context to all participants, we used a questionnaire designed to explore teachers' ideas about scientific skills (Guerra-Ramos et al.

2010). It included a "pedagogical scenario," consisting of the description of a hypothetical situation as a reference context to teachers' responses and two open-ended questions. Eight different actions were described and prompted the teacher to select those actions he/she may use as examples of scientific skills to consider in his/her class and the reasons for doing so (Table 4.1).

The reading of this scenario was followed by the opportunity to write an initial answer to prompting questions and an individual interview focusing on the arguments that the teacher made to justify whether the descriptions involved scientific skills, did not involve scientific skills or why he/she was unsure about it. Teachers were also asked whether they would use the descriptions as examples of scientific inquiry in the classroom, and why.

The antecedents of this methodological approach can be found in studies evaluating students' thinking (Driver et al. 1996) and teachers' thinking (Nott and Wellington 1996; Taylor and Dana 2003; Windschitl 2004) about different aspects of the world of science. The description of the pedagogical scenario, initial written questions and a semi-structured interview agenda, were designed and refined in a process that considers trials with teachers and an external review by two researchers not directly involved in this study.

During the data collection there were three moments: the presentation of pedagogical scenarios, the immediate request to answer in writing the initial associated questions and the individual interviews to explore in detail the ideas of each teacher building on the initial written answers.

Table 4.1 Pedagogical scenario

Preparing a class on scientific skills
Imagine that you are preparing a lesson for a class of students in first grade in which you would like to exemplify procedures or skills that could be regarded as scientific. You find the following descriptions of actions/activities in a resource book:
I. This person is carefully observing different insects and arachnids with a magnifier
II. This person is giving painkillers to a group of people suffering backache and sugar tablets to a similar group to see which group reports any improvement
III. This person is adding slices of potato to a very salty soup to see if it is true that the potato absorbs the salt
IV. This person is comparing the size of coffee beans from this year's harvest with that from the previous year
V. This person is watching a television program about famous volcanoes and their activity
VI. This person is claiming that a plant became yellow because the lack of sunlight impeded the development of chlorophyll
VII. This person is doing a test to see if ice cream melts faster in a metallic container than in a plastic one
VIII. This person is suggesting, based on his/her experience, that ulcers are caused by a microorganism and not by stress
Discuss the following questions:
1. In your view, which of these descriptions involves the use of a scientific skill or procedure? Why?
2. Which of these descriptions would you present in your class to exemplify the use of a scientific skill or procedure? Why?

The interviews were recorded and transcribed for further analysis. In a first stage, data were focused with an ideographic approach to deriving categories of analysis from the same content as the speech of teachers. Such an approach was consistent with the situated cognition perspective. Attention was also paid to the relationship between the teachers' representations of scientific procedures and the intentions of the official curriculum that was determined in the previous document analysis.

Qualitative analysis of the interview transcripts consisted of obtaining an initial set of categories, derived from the same answers of teachers. This was done by the systematic comparison of the responses of different teachers to identify differences and similarities of content. The analysis units were sentences or groups of sentences that were coded and recoded several times using NVivo 7.0, software designed to manage and support the process of qualitative analysis.

An analyst developed the initial categories through a reiterative process based on an analysis line by line. A second analyst was involved in a process of "blind coding" to establish the validity and communicability of the coding scheme. The second analyst had access to descriptions of the analysis codes and three transcribed interviews, but did not know which codes had been applied in each case by the first coder. A rate of consistency between encoders of 0.79 was obtained by dividing the number of agreements between the number of agreements plus disagreements of coding, as suggested by Miles and Michal Huberman (1994, p. 64). This rate of consistency indicated that both analysts working independently applied *roughly* the same categories to the same segments of the transcripts and attributed similar meanings to the responses of teachers. Subsequently, disagreements between analysts were used to improve the ambiguous definitions of certain categories of analysis, some were refined, and others originated new categories. This process enabled the final coding scheme to have higher internal consistency. The descriptive analysis allowed the identification of some regularities and coincidences in the ideas expressed in teachers' responses.

4.3 Results

The outcomes presented here build upon a previous work (Guerra-Ramos 2013), presenting preliminary results with a documentary content analysis of one document and a qualitative analysis of the responses of a smaller teacher sample.

Results are presented in two sections: the first one deals with the analysis of curriculum documents, as discussed earlier; then, we describe the main categories and types of responses found in the analysis of teacher's interviews. Segments of interviews are used to exemplify some of the main categories and trends of responses.

4.3.1 Outcomes of Documentary Analysis

The content analysis of curricular documents included the official program of Science for Secondary Education (SEP 2006, 2013), with special attention paid to sections such as *Introduction*, *Foundations*, and *Pedagogical approach to scientific training*. To identify text related to scientific skills or procedures, we traced those mentioning skills to develop in science education, how students are supposed to be engaged in scientific processes, the incorporation of projects and scientific methods, connecting sciences and scientific skills, and explicit ideas about how scientific skills should be taught. We identified a total of ten texts that suggested teachers must support their students in recognizing a range of procedures or scientific skills (see Table 4.2).

We found that in the curriculum documents the terms *procedures* and *skills* were used interchangeably. From the identified texts, it is important to underline the emphasis on the practice of scientific procedures in school science as part of the scientific training of students and also as an image of scientific practice.

Although the presence of these texts in the official curriculum deserves a more extensive discussion, here, we merely point them out as clear evidence for the explicit inclusion of certain aspects of the NOS. In particular, the presence of methodological aspects seeking to reflect what scientific communities do to create, validate, systematize, communicate, and agree on new knowledge is remarkable. In our interpretation, it is assumed that a version of such processes instilled in students contributes to their scientific training.

A lasting presence of curriculum expectations regarding scientific skills is evident in the analyzed documents. Although the intention was to compare potential differences, we found very similar points in the pedagogical rhetorical discourse in documents of both curriculum reforms. More detailed concern regarding scientific skills was noticed in the curriculum document of 2011, as the extension of texts and emphasis suggest. The listing of specific skills was extended and detailed. As Hodson (2014) points out, several aspects of the NOS have adopted different shifting emphasis in science curricula around the world. To date, the current curriculum document is the second edition of SEP (2013), with no substantial changes and it must be considered by secondary teachers as a guideline for their teaching practice. Although the document of the curriculum reform in 2011 substitutes the previous version; the documentary analysis reveals a relatively recent emphasis on scientific skills in Mexican science curriculum documents that seems to be consistent and reiterative since 2006.

Table 4.2 Texts with educational purpose related to scientific skills in Mexican curriculum documents

Curriculum reform 2006	Curriculum reform 2011
Skills to develop in science education	
Procedures (to develop in science education)	A basic science education implies that children and teenagers (…) develop the following skills:
Search, selection, interpretation, and analysis of data (observation, comparison, measurement)	Use and elaboration of models
Research (prediction, hypotheses, relationship among variables, experimental design, classification, seriation, use of models, drawing conclusions)	Formulation of questions and hypotheses
Construction and use of materials (manipulation of instruments of observation and measurement)	Data analysis and interpretation
Oral and written communication (SEP 2006, p. 10)	Observation, measurement, and recording
	Comparison, contrast, and classification
	Relationship among data, causes, effects, and variables
	Formulation of inferences, deductions, predictions, and conclusions
	Experimental design, planning, implementation, and evaluation of research
	Identification of problems and alternative solutions
	Manipulation of materials and elaboration of artifacts (SEP 2013, pp. 21–22)
Students and scientific processes	
(...) Students will have to give answers to questions raised by themselves, using rigorous scientific procedures and reflecting on scientific attitudes, to develop personal attitudes as part of their basic scientific training (SEP 2006, p. 13)	The study of sciences in secondary education is aimed at teenagers making progress in the development of skills to represent, interpret, predict, explain, and communicate biological, physical, and chemical phenomena (SEP 2013, p. 14)
	Skills associated with science (that students should develop):
	Design scientific research in which social context is considered
	Apply skills for scientific research: ask questions, identify themes or problems, gather data through observation or experiments, make and test hypotheses, analyze and communicate results, and elaborate explanations
	Plan and develop experiments requiring analysis, control, and quantification of variables

(continued)

Table 4.2 (continued)

Curriculum reform 2006	Curriculum reform 2011
	Use technological instruments to enhance senses and gather information of natural phenomena with detail and precision
	Make interpretations, deductions, conclusions, predictions. and representations of natural phenomena and processes, from the analysis of data and evidence, and express how he/she gets to them
	Develop and apply models to interpret, describe, explain or predict natural phenomena and processes as an essential part scientific knowledge
	Apply interpersonal skills in teamwork while conducting scientific research
	Communicate the outcomes of observations and inquiries using resources such as diagrams, tables, graphics presentations, graphs and other symbolic representations, along with information and communication technologies (ICT), and justify their use (SEP 2013, pp. 18–19)
Projects and "scientific method"	
In these (scientific) projects, students have the opportunity to develop activities related to the formal scientific work when they describe, explain, and predict the investigation of natural phenomena and processes that occur in their environment. In this type of project, the promotion of empiricist, inductivist, and simplified views of research should be avoided, as are those reduced to follow a unique and inflexible "scientific method" that starts invariably with observation (SEP 2006, p. 13)	Students can make activities related to formal scientific knowledge in which they describe, explain, and predict natural phenomena and processes occurring in the surrounding environment
	This process promotes the wish to know, investigate, and discover, perseverance, honesty, meticulosity, informed skepticism, openness to new ideas, creativity, participation, confidence in themselves, respect, and compromise. In these projects, (efforts should be made) to avoid the promotion of empiricist, inductivist, and simplified views on research, such as those limited to following a unique and inflexible "scientific method" that invariably starts with observation (SEP 2013, pp. 25–26)
Sciences and scientific skills	
Science is mainly devoted to constructing plausible explanations of natural phenomena, to predicting their behavior and effects, and to building theories that give sense and meaning to observations and discoveries. This field refers to the skills and attitudes toward obtaining information, the use of all senses – directly or indirectly, the use of instruments and reasoning, the formulation of personal explanations and hypotheses, identification of relationships and patterns, and the drawing, assessment, and communication of conclusions (SEP 2006, p. 16)	As biology, physics, and chemistry are experimental sciences that use theoretical models for explanation, the use of material, digital and representational models is necessary, in addition to several types of tools and experimental strategies supporting the development of scientific thinking skills: asking questions, searching for answers, reflection and argumentation based on gathered information in experiments or documental research with ICT tools (SEP 2013, p. 111)

(continued)

Table 4.2 (continued)

Curriculum reform 2006	Curriculum reform 2011
Teaching and learning scientific skills	
The teaching of procedures should proceed from three essential perspectives: that students know about them, apply them in the appropriate context, and obtain more knowledge with them. The learning of skills and procedures must follow a similar pattern to meaningful learning of concepts; in their acquisition, reorganization or extension, students' previous knowledge should be considered. The main skills and procedures to be strengthened in this subject are related mainly to the development of practical activities, experimentation, and research (SEP 2006, p. 24)	

4.3.2 Themes, Categories, and Trends in Teachers' Responses

Participant teachers offered different arguments on whether they considered the description to be related to scientific skills or procedures. At the beginning of the interview, they replied to the direct question (Do you think this description involves the use of a scientific skill or procedure?). Figure 4.1 presents the frequencies of teachers' responses.

Teachers tended to consistently consider that most of the descriptions were associated with some form of scientific skills or procedures. Descriptions I (magnifier) and VII (ice cream) obtained affirmative responses more frequently. Only descriptions V (TV program) and VIII (ulcers) obtained 50% affirmative responses or less. When asked the second direct question (Do you think this description involves the use of a scientific skill or procedure?), teachers tended to consider that they would present those that they considered to be related to scientific skills.

In terms of the qualitative analysis of transcripts of interviews, a total of 21 different categories were identified (Table 4.3). The identified categories suggested that teachers consistently applied the criteria of perceived domain (scientific or non-scientific) and perceived purpose (investigative, noninvestigative) to select the actions that, from their point of view, involved a scientific procedure.

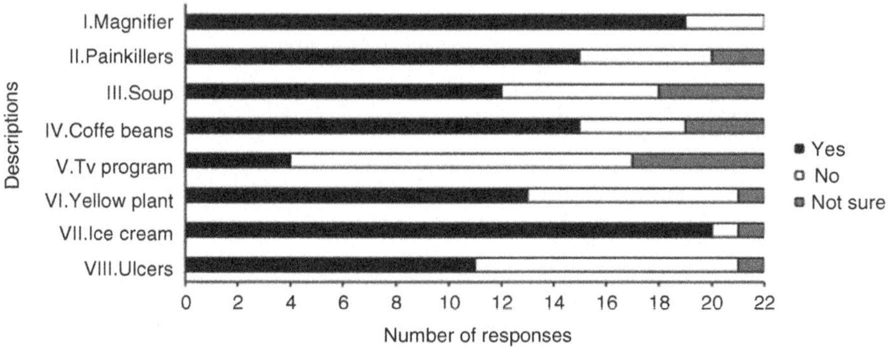

Fig. 4.1 Responses to the interview question: "Do you think this description involves the use of a scientific skill or procedure?"

To illustrate the categories of Table 4.3, here follows some examples and types of responses given by the teachers.

When the responses and their justifications alluded to a scientific domain, teachers made reference to the use of scientific procedures in everyday, impersonal or professional contexts, but predominantly in the classroom, as shown below:

> [It is a scientific procedure because...] If I brought a yellow plant to the classroom, I would ask my students to observe and say why they think the plant became yellow. They might have no idea, or somebody could say that it is because it had not enough soil or water, another one could say that it is because it had too much water. They may not mention sunlight because it is taken for granted. [Q3a-9-27]

> I think is about a scientific procedure because I have something like that [Ice cream]. I ask students to check how long it takes for ice cubes to melt depending on their size and container. These kinds of experiments are about variables like size and isolating materials, observation and recording. [Q3a-21-34]

When teachers judged that a description did not correspond to a scientific procedure, responses suggested that it was because scientific knowledge or skills were not in place. The following are examples of this category of response:

> [It does not necessarily imply a scientific procedure because...] It says that a person observes insects and arachnids, but I was wondering what would be the skill involved? I might observe them just for curiosity or contemplation with no specific ability to identify a structure or function. Some children do that spontaneously. Nothing connected to science, just "seeing" for natural curiosity. …. [Q2a-11-3]

> Watching a program is a doing something passively. Maybe the program contains scientific information but if you do nothing with it… Information is not used for an explanation or to solve a problem. No, there is no scientific skill or procedure. [Q4a-16-45]

However, there was a tendency among teachers to consider the descriptions to be connected to some scientific procedure. When teachers grasped the purpose of the activity to argue whether a scientific procedure was involved, they made a clear distinction between research and nonresearch purposes. At the collective level, they

Table 4.3 An initial characterization of teachers' ideas about scientific skills

Coding categories		Number of teachers providing relevant responses	Coded segments
Perceived domain and persons (actors?) involved	*Scientific domain*		
	School science	11	14
	Professional science	7	8
	Science in everyday contexts	4	7
	Science in impersonal action	3	4
	Nonscientific domain		
	Not applying knowledge	9	11
	Not applying scientific procedure	4	16
Purpose of activity	*Investigative purpose*		
	Obtaining useful information	9	10
	Evaluating an idea	5	14
	Knowing something new	3	3
	Non-investigative purpose		
	Outcome already known	4	6
	Matter of opinion or individual experience	5	10
	Lack of clear purpose	4	6
Procedural aspects	*Actions involved in scientific enquiry*		
	Observing in detail	13	18
	Comparing different conditions	7	7
	Looking for a mechanism	9	9
	Confirming–probing	11	8
	Applying knowledge	3	4
	Features of scientific enquiry		
	Standard scientific method	12	13
	Methodological diversity	5	5
	Integral process	7	7
	Independent steps	11	13

tended to emphasize that obtaining useable and specific information was connected to a research purpose, and to a lesser extent they mentioned assessing the validity of ideas or getting to know something new.

Teachers' responses suggested that they perceived at least five different procedures in the descriptions provided (observation in detail, comparison of something under different conditions, identification of a mechanism or process, testing an idea, and application of knowledge), but *observation* was the most frequently mentioned.

Similarly, half of the participants referred to the scientific method as a form of application of scientific procedures that everyone must follow, including scientists.

> They [scientists] should use the steps of the method, because without these steps they could not reach what they pursue. Steps to follow, one has to follow them. [Q6-3-38]

> In any case, we have to apply the scientific method if we want to solve a scientific question. If we want to know the effects of a medicine or a diet, how cells incorporate nutrients, why plants grow better under certain conditions… It is always the same, in the school laboratory or in a scientific laboratory. [Q7-19-57]

The use of disciplinary knowledge to illustrate their responses and their justifications was infrequent and limited in range. As a common aspect among participant teachers, this reveals the lack of a basic repertoire of examples, themes, and contexts associated with the scientific procedures, perhaps because many do not have scientific training before becoming teachers, which would probably be of great support to addressing them explicitly in the classroom. In this sense, the use of historical episodes and current research could provide examples of scientific procedures that support teachers. Such historical episodes or current research could be incorporated into the strategies of teacher training and educational materials.

References to the *steps* of the *scientific method* were very common in the transcripts of the interviews. Teachers often had difficulty enlisting the *steps* of the method they claimed to have learned at school or during their initial training as teachers. However, they mentioned different prototypical actions involved in scientific work (observing, comparing, making experiments, etc.). Some descriptions included in the pedagogical scenario were perceived as if they represented one or two of such actions and were therefore considered as examples of scientific procedures.

Scientific procedures were conceived by participant teachers as motion actions rather than as intellectual activity. Procedures were also seen as piecemeal and disaggregated actions that are not necessarily guided by theoretical aspects, frames of reference or specific purposes. This shared representation contrasts with the perspective adopted in the official curriculum documents related to the teaching of sciences in secondary education. Such documents propose to present scientific activity as a job that involves the practice of different complementary intellectual and procedural skills, which are interrelated with particular purposes. The documents also insist on avoiding the idea of a rigid, sequential or standard scientific method. The expectation of communicating such messages in secondary education science classes is, to say the least, idealistic given the kind of representations that teachers share on scientific procedures and the few opportunities they usually have to develop them during their initial and in-service training.

Taking into account the representations characterized in this study, it is tempting to speculate that teachers tend to address scientific procedures as a series of mechanical actions, independent of context or theme to investigate. An outstanding aspect was also that teachers did not acknowledge the presence of *scientific procedures* as an element of the pedagogical concern of science as a subject. The repertoire of representations of scientific procedures that have been developed by secondary education science teachers participating in this study tends to reflect some over-simplifications that probably constrain a socialization and genuine practice of scientific procedures in the school context.

In summary, at the collective level, the responses of teachers were multifaceted and although responses that alluded to the "scientific method" were prominent as a stereotypical image, they also presented a range of responses that varied in development (extension, vocabulary, and argumentation) and contextualization (inclusion of examples and references to people and specific spaces).

4.4 Discussion and Implications

This study only focused on teachers' representations of scientific procedures. The intention was to adopt a qualitative perspective to explore what teachers say and how they say it; that is to capture their discourse as a means of accessing content and ways of thinking. It was also an attempt to complement the information provided by other studies exploring teachers conceptions associated with the NOS using scales and questionnaires with a closed-ended format (Kimball 1968; Buaraphan 2011) that indicate what teachers respond, but not why they do so in a specific way, nor do they suggest hints about the pedagogical relevance of the ideas. A more productive perspective seems to be the recognition of the situated nature of teachers' representations of scientific procedures adopted by authors such as Taylor and Dana (2003) and Windschitl (2004). Therefore, this study adopted a situated perspective and can only provide an initial characterization of teachers' representations from which it is possible to advance some hypothesis on teachers reasoning and pedagogical practice. The obtained characterization can inform further research and suggest that some representations shared among teachers might have the potential to interfere with expectations of the science curriculum and its rhetorical pedagogical discourse; or at least to become elements of this interpretation. For instance, García et al. (2013), explored different conceptions of scientific knowledge and learning and their implications on the teaching process, and found that university teachers have different conceptions depending on the context, i.e., performing science may involve one particular epistemological stance whereas teaching science may refer to a different one. Among other consequences of such an effect is that teachers may have a philosophically rich conception of science, but this does not necessarily mean that these conceptions may be transferred to their teaching classroom.

The analysis of official curriculum documents confirms the presence of messages related to the introduction of scientific procedures and their study as a theme to review in an explicit manner. This implies that teachers find scientific procedures as a theme in daily teaching practice, and that they have several possibilities: teaching, re-interpreting or avoiding them. Further studies may demonstrate that these representations of scientific procedures prevail on a larger scale. If that were the case, a possible implication might be that secondary teacher training programs would benefit from promoting the recognition of diversity in scientific procedures and in the strategies used in professional areas of experimental sciences, historical sciences, and in school science. They would also benefit from the study of specific purposes, reference frameworks and methodological approaches as a characteristic feature of scientific practice. The systematic incorporation of historical episodes of science could support solid teacher training in this direction.

An official curriculum only brings about broad recommendations for teaching. In practice, the implemented curriculum depends mainly, although not exclusively, on pedagogical knowledge and practices. Attitudes and other contextual factors (e.g., teaching resources, assessment processes) are also important, but they were not the focus in this study. However, we offered here a description and initial analysis of a base of ideas about scientific procedures held by teachers based on empirical data and provides some clues on the ideas that teachers use to respond to related educational innovations and the messages that they probably communicate in the classroom in this respect.

A serious effort to introduce aspects of the NOS in science education requires at least a recognition of the conceptual and methodological differences among scientific disciplines and a more multifaceted and less totalitarian view of the NOS as a cognitive and human labor (Jenkins 1996). It is also important to acknowledge that introducing themes regarding the history and NOS into initial teacher training requires a great deal of awareness and, consequently, "negotiation" of meanings, which could be a long, slow-paced and gradual process (Ferreira 2013). The prospect may not be as clear as it may seem: it is desirable to introduce themes related to the history and the NOS, but as some researchers have pointed out (Amador and Adúriz 2013), it is not sufficiently clear which epistemological stance is the most suitable for each educational level (realism, positivism, relativism, constructivism, functionalism, or a combination of any of them). Amador and Adúriz (2013) suggest that thoughtful reflection is necessary, considering not only the NOS, but the history and sociology of science as well, to choose the best conceptual and theoretical frames to carry out the teaching process and the selection of strategies according to each educational level. For that reason, it is important to recognize the existence of restricted repertoires of ideas about scientists and how they conduct their work, but also the need to clarify and make more explicit the targets in the science curriculum in these themes. A realist objective in this area should consider the excessive curricular load and the lack of a robust teacher education in the Mexican context. A clear expression of targets and a better definition of learning outcomes are much needed. The emergence of a concern to communicate ideas about the world of science through school science implies that teachers already face these themes. The way they respond to teaching demands concerning scientific procedures in secondary science education is still not sufficiently researched in Mexico and in other contexts. There is still a place to reflect and explore how teaching and learning about the NOS can take place in secondary science classrooms, and the possible effects and challenges.

References

Amador Rodríguez, R. Y., & Adúriz Bravo, A. (2013). Naturaleza de la ciencia: acuerdos teóricos en la comunidad iberoamericana de didáctica de las ciencias. In C. C. Silva & M. E. B. Prestes (Eds.), *Aprendendo ciência e sobre sua natureza: abordagens históricas e filosóficas* (pp. 303–312). São Carlos: Tipographia.

Brown, J., Collins, A., & Duguid, P. (1989). Situated cognition and the culture of learning. *Educational Researcher, 18*, 32–42.

Buaraphan, K. (2011). Pre-service physics teachers' conceptions of the nature of science. *US-China Education Review, 8*, 137–148.

Chen, S. (2006). Development of an instrument to assess views on nature of science and attitudes toward teaching science. *Science Education, 90*, 803–819.

Driver, R., Leach, J., Millar, R., & Scott, P. (1996). *Young people images of science*. Buckingham: Open University Press.

Ferreira, J. M. H. (2013). Abordando a natureza da ciência na formação de professores de física: desafios e perspectivas. In C. C. Silva & M. E. B. Prestes (Eds.), *Aprendendo ciência e sobre sua natureza: abordagens históricas e filosóficas* (pp. 245–257). São Carlos: Tipographia.

Guerra-Ramos, M. T. (2012). Teachers' ideas about the nature of science: A critical analysis of research approaches and their contribution to pedagogical practice. *Science & Education, 21*, 631–655.

Guerra-Ramos, M. T. (2013). Representaciones sobre los procedimientos científicos en el discurso de los docentes de ciencias de secundaria. In C. C. Silva & M. E. B. Prestes (Eds.), *Aprendendo ciência e sobre sua natureza: abordagens históricas e filosóficas* (pp. 297–308). São Carlos: Tipographia.

Guerra-Ramos, M. T., Ryder, J., & Leach, J. (2010). Ideas about the nature of science in pedagogically relevant contexts: Insights from a situated perspective of primary teachers' knowledge. *Science Education, 94*, 282–307.

García, M., Vilanova, S. & Sol, S. (2013). Concepciones epistemológicas y enseñanza de las ciencias. In C. C. Silva & M. E. B. Prestes (Eds.), *Aprendendo ciência e sobre sua natureza: abordagens históricas e filosóficas* (pp. 271–289). São Carlos: Tipographia.

Hodson, D. (2014). Nature of science in the science curriculum: Origin, development, implications and shifting emphasis. In M. Matthews (Ed.), *International handbook of research in history, philosophy and science teaching* (pp. 911–970). Dordrecht: Springer.

Irez, S. (2006). Are we prepared? An assessment of preservice science teacher educators' beliefs about the nature of science. *Science Education, 90*, 1113–1143.

Jenkins, E. (1996). The 'nature of science' as a curriculum component. *Journal of Curriculum Studies, 28*, 137–150.

Kimball, M. E. (1968). Understanding the nature of science: A comparison of scientists and science teachers. *Journal of Research in Science Teaching, 5*, 110–120.

Lederman, N. (1992). Students' and teachers' conceptions of the nature of science: A review of the research. *Journal of Research in Science Teaching, 29*, 331–359.

Liu, S.-Y., & Lederman, N. (2007). Exploring prospective teachers' worldviews and conceptions of nature of science. *International Journal of Science Education, 29*, 1281–1307.

Miles, M., & Michal Huberman, A. (1994). *Qualitative data analysis: An expanded sourcebook.* Thousand Oaks: Sage.

Nott, M., & Wellington, J. (1996). Probing teachers' views of the nature of science: How should we do it and where should we be looking? In G. Welford, J. Osborne, & P. Scott (Eds.), *Research in science education in Europe: Current issues and themes* (pp. 283–293). London: Falmer Press.

Schwartz, R., & Lederman, N. (2002). "It's the nature of the beast": The influence of knowledge and intentions on learning and teaching nature of science. *Journal of Research in Science Teaching, 39*, 205–236.
SEP. (2006). *Educación Básica. Secundaria. Ciencias. Programas de Estudio 2006*. Mexico: Secretaría de Educación Pública.
SEP. (2013). *Programa de estudio 2011. Guía para el maestro. Educación básica secundaria. Ciencias*. México: Secretaría de Educación Pública.
Taylor, J., & Dana, T. (2003). Secondary school physics teachers' conceptions of scientific evidence: An exploratory study. *Journal of Research in Science Teaching, 40*, 721–736.
Windschitl, M. (2004). Folk theories of "inquiry": How pre-service teachers reproduce the discourse and practices of an atheorical scientific method. *Journal of Research in Science Teaching, 41*, 481–512.
Zeidler, D., & Lederman, N. (1989). The effect of teachers' language on students' conceptions of the nature of science. *Journal of Research in Science Teaching, 26*, 771–783.

Chapter 5
Theory, Evidence, and Examples of Teaching the Nature of Science and Biology Using the History of Science: A Chilean Experience

Hernán Cofré, Paola Núñez, David Santibáñez, José Pavez, and Claudia Vergara

5.1 Introduction

There is currently a consensus that to obtain scientific literacy, the main objective of science education, science teachers have to teach not only scientific content, but also how this scientific knowledge is created, and how this can have an impact on society (Osborne et al. 2003; Lederman and Lederman 2014). This curricular issue, in which students learn how scientific knowledge is generated and tested and how scientists do what they do, is called the nature of science (NOS) (Lederman and Lederman 2014; McComas and Kampourakis 2015). In this context, the history of science (HOS) can play a relevant role because it is a bridge between the subject matter that a science teacher must teach and the scientific, social, and cultural context in which this knowledge is elaborated.[1] For example, historical episodes of controversy in the sciences can provide a social context that can help students understand how discoveries and scientific theories emerged, and can also be a useful tool for developing students' argumentation skills (Clary and Wandersee 2013).

In this context, the aim of this chapter is twofold. First, we discuss, in general terms, the role of HOS in the context of international works, assessing how much

[1] Abd-El-Khalick and Lederman (2000), Allchin (2012), Rudge and Howe (2009), and McComas (2013).

H. Cofré (✉) · P. Núñez · J. Pavez
Pontifical Catholic University of Valparaíso, Valparaíso, Chile
e-mail: hernan.cofre@pucv.cl

D. Santibáñez
Catholic University Silva Henríquez, Santiago, Chile
e-mail: dsantibanezg@ucsh.cl

C. Vergara
University Alberto Hurtado, Santiago, Chile
e-mail: clvergara@uahurtado.cl

M. E. B. Prestes, C. C. Silva (eds.), *Teaching Science with Context*, Science: Philosophy, History and Education,
https://doi.org/10.1007/978-3-319-74036-2_5

scientific evidence supports some claims about the benefits and efficiency of using different models for teaching NOS or subject matter. Second, we show some empirical evidence about our research using HOS to teach NOS to students and biology teachers, including lesson materials and activities that could be useful for in-service and pre-service science teachers.

5.2 The Role of the History of Science in Science Education

For a long time, HOS has been an important component to incorporate into science classes, both for teaching content and for understanding how this knowledge is generated (Wang and Marsh 2002; McComas 2013). For example, some scholars argue that reflection on the evolution of scientific thought, fostered by HOS, can help students to achieve critical thinking. It is also mentioned that science teaching, incorporating elements of HOS, can promote student motivation for learning and reflection on science (McComas 2013). For Alvarez (2006), the interest in the teaching of HOS is twofold: it helps with understanding of science concepts that are difficult to learn, and it acts on the representations of science that students hold (NOS). On the other hand, Chang (2010) also states that HOS, specifically historical experiments, can be used to encourage students to question nature, and therefore, include an additional element of motivation in the lesson. According to Izquierdo et al. (2007), HOS can be used to model and understand scientific concepts. According to Allchin (2012), scientific errors may also be useful for guiding more effective teaching and a deeper understanding of the content. HOS may also be beneficial to students in their awareness of conceptual obstacles and to teachers in identifying and directing their actions toward the reorganization of students' thinking (Monk and Osborne 1997). Some authors suggest that there might be a parallel between the conceptions of students and the conceptions that scientists have held throughout history (Galili and Hazan 2001). In this sense, scientific errors can be very useful for understanding students' misconceptions (Allchin 2012). Monk and Osborne (1997) argue that the study of scientific ideas in their original contexts helps pupils to understand why their thinking may be parallel to that of a scientist or similar to ideas they present because of the opposing historical knowledge. Finally, HOS has also been widely proposed and used to improve understanding of NOS, both by students and by teachers.[2]

[2]Clary and Wandersee (2013), Allchin (2012), McComas (2013), McComas and Kampourakis (2015), and Pavez et al. (2016).

5.3 Teaching NOS and Subject Matter Through HOS: Looking for Empirical Evidence

Recently, science educators have recognized different ways of including HOS to teach science and NOS: the use of original papers, case studies on controversies, biographies and autobiographies, HOS in textbooks, dramatic performances, activities, role-play, and replication of classic experiments (Clary and Wandersee 2013; McComas 2013). Moreover, scholars have also proposed different teaching strategies or models to incorporate HOS, with Monk and Osborne (1997) and Rudge and Howe (2009) being the most widely used (see Pavez et al. 2016 for a review).

There are multiple examples of interventions to promote understanding of NOS through HOS (Niaz 2007; Kim and Irving 2010; Paraskevopoulou and Koliopoulos 2010; Pavez et al. 2016; Williams and Rudge 2016). Table 5.1 shows some of these studies and their main results. Such interventions exhibit a variety of approaches, strategies, and implemented activities, in addition to a diversity in selected historical elements. For example, the use of scientific disputes appears to be a contribution to the construction of an informed view of NOS. Using the dispute between Millikan and Ehrenhaft about the existence of an elementary electrical charge for the teaching of NOS, Paraskevopoulou and Koliopoulos (2010) concluded that this teaching strategy served to improve high school students' understanding of five NOS aspects (the empirical, subjective, and creative nature of science, and the differences between observation and inference).

In another experience with secondary students, Kim and Irving (2010) used the development of the theory of heredity to illustrate the differences between theories and scientific laws, the nature of experiments and observations, and the tentativeness and social effect of science. Through reflections and group discussions, guided by questions that explicitly directed attention to the aspects of NOS mentioned, the experimental group showed a better understanding of NOS than participants in the control group at the end of the study. Furthermore, these authors note that using classroom time for HOS was not a waste of time, but was valuable for providing students with insights into how science works while maintaining significant achievement in the science content area.

On this last issue, little effort has been made to explore the effectiveness of HOS on students' learning of content knowledge. In Table 5.2, we can see that there is some evidence that HOS can improve problem-solving abilities in chemistry (Lin et al. 2002) or promote students' attitudes toward science (Lin et al. 2010). However, it is also evidence that lessons including HOS do not result in better outcomes than control groups (Seker and Welsh 2006; Kim and Irving 2010) (Figs. 5.4 and 5.5). Therefore, further research is needed for a clearer conclusion on this issue.

Table 5.1 Summary of empirical research about the role of history of science (HOS) in learning about the nature of science (NOS)

Author/year	Characteristics of the intervention	Activities	Effect
Solomon et al. (1992)	8 sessions, 94 high school students on a course with scientific content. HOS was used including an historical discovery	Practical research laboratories, lectures, exhibition posters, modeling, cartoons, role playing, text analysis	Qualitative and quantitative data show that HOS contributed to the student understanding of NOS
Abd-El-Khalick and Lederman (2000)	Three HOS courses lasted 10 weeks. A 50- to 80-min session each week. 166 undergraduate and graduate students and 15 pre-service secondary science teachers. HOS was used, including scientific controversies	Mainly implicit approach, with some explicit references, discussion-oriented and lecture-oriented courses; content-embedded activities and generic activities, mostly of the "black-box" variety	Very few and limited changes in participants' NOS views were evident at the conclusion of the course
Lin and Chen (2002)	One semester, 63 chemistry pre-service teachers. Course of methods of teaching chemistry and the current development of science education. HOS was used, including debates and original discussions	Discussion groups, debates, demonstrations, projects, experiments	Positive, the experimental group showed better progression in several aspects
Dass (2005)	One course for four semesters (15 weeks each). 52 students. HOS was used, including a review of the history of science since antiquity until the mid-twentieth century	Debate and discussion of historical events class to class, according to the textbook chapter. Essay construction. Final examination	Minimum improvement in students about NOS understanding
Howe and Rudge (2005)	One unit at the end of the course, eight lessons of 2.5 h each. 24 elementary educator students. HOS was used, including the development of the research	Explicit and reflective approach, constructivist approaches, group work, discussion, analysis of real data	Explicit and reflective methods can positively affect (change or enrich) students' NOS views for some of the aspects addressed
Niaz (2007)	11 weeks (62 h.) on a chemistry method course. 17 in-service teachers, methodological course. HOS was used, including scientific controversy	Written reports, class discussions based on student presentations, written examinations	The use of several controversial episodes is directly related to the improved understanding of NOS by teachers
Kim and Irving (2010)	7 activities on a high school biology course. 33 students of 10th grade. HOS was used, including the development of a theory (theory of heredity)	Reflections, discussions, and learning groups. Activities include explicit questions related to NOS	The tests show that the experimental group showed a greater understanding of NOS after intervention

Paraskevopoulou and Koliopoulos (2010)	6 units of instruction on a physics course. 24 students of 10th grade high school (15/16 years). HOS was used, including scientific controversy	HOS (4 small stories followed of questions that focus students on the NOS aspect of this story)	There were improvements into view of the participants in the 5 aspects of NOS worked
Forato et al. (2012)	20 h of lessons (2 h daily for 2 weeks) in an experimental course on the history of optics. 38 high school students. HOS was used, including scientific controversy	Using a reflexive, explicit, and implicit approach to NOS. Games, guides, discussions and debates	Evidence for positive change around NOS in some aspects (only description data not shown)
Rudge et al. (2014)	3 days of an introductory biology course. 130 students in elementary school teachers (pre-service). HOS was used, including the development of research (industrial melanism)	Guided discussions, individual and group work, review of a film ("Evolution in Progress")	There are significant improvements in some aspects of NOS
Kampourakis and Gripiotis (2015)	Five lessons, 45 min each. Fifty 17-year-old students. HOS was used, including short stories about the social, cultural, and scientific factors that influenced the development of Darwin's theory	Historical case study, presentations of the historical events, analysis, and discussion of original writings	It is concluded that this kind of highly contextualized NOS instruction can provide students with a more authentic view of science
Pavez et al. (2016)	Three sessions of 2 h in a professional development program about NOS and teaching evolution. 8 biology teachers. HOS was used, including 3 historical episodes of research in biology	They used an integrated biological explicit content reflective approach. Activities for discussion, debate and reflection. Planning a class that included NOS through HOS	significant improvements were found in understanding of 8 aspects of NOS
Williams and Rudge (2016)	One unit. Three 2.5-h lessons in an introductory biology course for pre-service elementary teachers. 11 teachers between 18 and 26 years old. HOS was used including the history of Mendel's classic work	Explicit and thoughtful approach, lecture-format laboratory course, mini-lectures, work in small groups, discussion questions, and research "silver box" activity	The data collected indicate that students showed improvements related observations, inferences, and the influence of culture on science

Table 5.2 Summary of investigations into the role of HOS on other different aspects instead of learning of NOS, for example, science learning or science attitude

Author and year	Characteristics of intervention	Learning activities	Object to include HOS	Outcomes
Tsatsarelis et al. (2000)	1 lesson, 2 pupils, 12 year-old in year 7. elementary science c0lass	Narrative based on a metaphor, observation of cells under the microscope, elaboration of their own analogy and metaphor, historical pictures of cells, drawings of cells	To shape pupils' views about the concept of cells	The rhetorical way that entities were constructed is similar for early scientists and young pupils
Dedes and Ravanis (2009)	Two stages, with three and two tasks respectively. 48 students from different schools, 12- to 16-year-olds	Recreation of Kepler's historic experiment, making predictions, immediate validation, discussions, guidance, and interpretations	To investigate the effectiveness of a teaching conflict experimental procedure in the transition from a spontaneously formatted pupil's representation to one compatible with the scientific view	The main goal was accomplished. Most of the subjects accepted the model of geometrical optics
	109 13-year-old pupils and 148 student-teachers participated in the questionnaire	Galdano and Giblert's experiment, Faraday's experiment, electric, magnetic, thermal and gravity experimental tasks	To search for pupils' alternative ideas in the HOS. To overcome pupil's alternative ideas by using early experiments carried out by scientists in the past	Most of the student-teachers and pupils relate electrostatic to magnetic phenomena, in the same way the scientists did, thereby helping pupils and teachers to overcome their alternative ideas
Dibattista and Morgese (2013)	5–18 h, depending on the teacher's unit plan. Twenty-four teachers and classes of 9–65 students (10–18 years old), depending on the school	The 15 teaching units built include various types of products ranging from written tests to scientific instruments built as models, multimedia products, laboratory activities	To increase the appeal of science and the learning of science to students, and to promote a more complete perception of science	According to the students' responses to the questionnaire, most objectives were achieved

Lin et al. (2010)	The class met for four periods a week and the experiment extended over a period of 1.5 months. 329 students in grade 7 junior high school class	Control group was taught using the textbook only, whereas the experimental group was taught using the textbook plus HEM materials and associated discussion	Understanding the effect of inclusion of the HOS in science teaching on promoting an understanding of NOS in addition to the attitudes toward science	The findings reveal that the exposure of students to HEM materials did promote the students' attitudes toward science
Lin et al. (2002)	One year of teaching. Two classes of 8th graders (*n* = 74). Physics 8th grade class	Historical-rich supplementary materials, including simulations of previous scientists' experiments, debates, and discussions of their ideas, demonstrations or hands-on activities, small group discussions and student presentations	To investigate the efficacy of promoting students' problem-solving ability through the HOS teaching	The experimental group students outperformed their counterparts in their chemistry conceptual problem- solving ability
Kim and Irving (2010)	7 activities. 33 students in 10th grade high school. Science content course (honors biology)	Reflections, discussions, and learning groups	To promote genetics content knowledge and learning of NOS	Experimental and control group shows similar genetics knowledge in the post-test
Seker and Welsh (2006)	4 months. 91 8th grade students randomly assigned to four classes Science class in an urban school	HOS cases, historical ideas of scientific phenomena, scientists' opposing ideas, discussions, stories from famous scientists' lives	To promote learning science, understanding NOS, and students' interest in science	Data showed a positive effect, in the change of perceptions of NOS, and of interest in science. But, the changes between classes in meaningful learning did not show any differences

HEM historical episodes map

5.4 Teaching NOS and Evolution Using the History of Science

In Chile, the understanding of NOS has been an objective of official curricula since the end of the last century (Ministry of Education of Chile 2014). Nevertheless, what students should know about NOS or HOS is not clearly explained. Furthermore, NOS and HOS have also been included in the new standard documents for science teacher education (Cofré 2012; Cofré et al. 2015), but most of the country's science teacher education programs do not yet include NOS or HOS as topics to be covered (Pavez et al. 2016), and the Chilean government has still not made NOS or HOS a priority for teacher professional development (Cofré et al. 2010, 2015). Therefore, it would not be surprising if many science teachers in Chile hold uninformed views of NOS (Cofré et al. 2014; Pavez et al. 2016).

In the following sections, we show some preliminary findings about using HOS for teaching NOS and evolutionary theory. The study with in-service biology teachers used a pretest/post-test single-group research design to understand the impact of HOS intervention on biology teachers' understanding of NOS. Qualitative and quantitative methods were used to characterize changes in teachers' views of NOS. The study involving high school students used a pretest/post-test quasi-experimental design with one control group and two treatments using HOS instruction to teach NOS and evolution.

5.4.1 Teaching NOS Using HOS with In-Service Biology Teachers

Since 2013, we have been developing research into teaching NOS and evolution to high school students (Cofré et al. 2017) and biology teachers (Cofré et al. 2016; Pavez et al. 2016; Bravo and Cofré 2016). The results shown here come from an intervention administered within the context of a professional development program (PDP) about NOS and teaching evolution at a private university in Santiago, Chile. The PDP lasted for 6 months, and the participants met for 5 h each Saturday morning between August and November 2015, and for one intensive week in January 2016, for a total class time of 120 h. Some of the HOS activities used during the intervention are available in Spanish at: https://goo.gl/TWdCGX (for more detail about the intervention see also Pavez et al. 2016).

The sample comprised seven secondary biology teachers (five females and two males) who voluntarily participated in the study. They ranged in age from 24 to 45 years old, and they had a wide range of teaching experience (between 1 and 18 years). At the beginning of the first module, the teachers completed the VNOS-D+ questionnaire (pre-test), and then all participants were interviewed to clarify their answers (Lederman and Khishfe 2002). Next, the intervention was performed, which included first decontextualized activities and then lessons that included HOS to teach NOS. Among other strategies, "ticket out of class" was used (Russel and Bullock 2010). At the end of the lessons, including decontextualized activities, each teacher individually answered a written test that was used to assess their views of four aspects of NOS. At the end of the second module, after the lessons including HOS instruction, teachers again completed the VNOS-D+ (post-test). Then, a final semi-structured interview was conducted. Data analysis involved the assignation of an informed, mixed or naïve understanding of the eight targeted aspects of NOS according to a rubric developed based on previous works (Cofré et al. 2014; Pavez et al. 2016). The authors then met and discussed each score to achieve consensus. In total, 90% of the initial scores were identical, and all disagreements were resolved in conversation (pre-test). At the post-test, most of the authors' coding was identical (95%). To generate a quantitative analysis in addition to the qualitative description of the profile of each teacher, a variable "understanding of NOS" was created by the sum of the values obtained for each aspect for each student. For the current analysis, only four aspects were studied, and because of the range of values for understanding NOS being between 0 and 2, the quantitative profile of each teacher could fluctuate between 0 and 8 points.

The results of the intervention are shown in Fig. 5.1. After the first part of the intervention, based on mainly decontextualized activities, most of the teachers increased their NOS understanding significantly ($t\,[6] = -3.44$; $p = 0.014$; $r = 0.66$). After this part of the intervention, we implemented a group of lessons focused on teaching NOS using HOS. The instruction in these lessons was conducted using mainly the model proposed by Rudge and Howe (2009), in which participants (in this case biology teachers) face the data, questions, and/or the observations made by

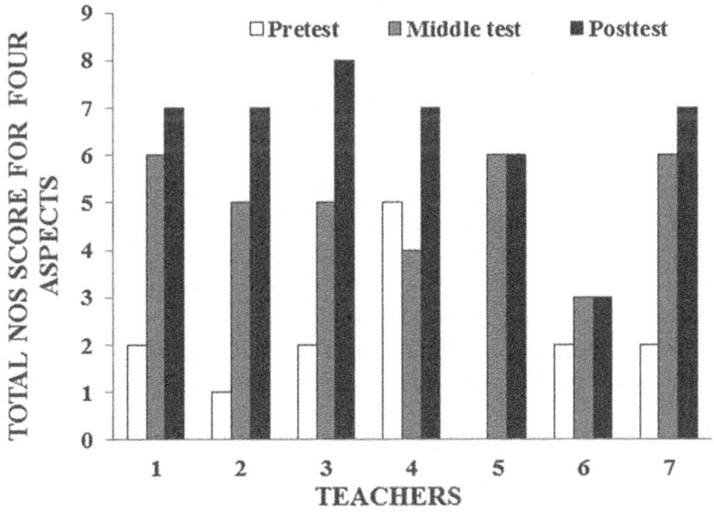

Fig. 5.1 Total nature of science (NOS) score for four aspects by teachers pre-test, mid-test, and post-test. Because only four aspects of NOS were studied, and the informed view was scored as 2, the mixed view as 1, and the naïve view as 0, the maximum value reached by one teacher is 8

the scientist (for more details about the HOS lessons, see Pavez et al. 2016). Similar to the first part of the intervention, significant improvements were observed in teachers' understanding of NOS after the HOS lessons (t [6] = −3.0; p = 0.03; r = 0.60). On the other hand, as in our first study (Pavez et al. 2016), most of the biology teachers participating in this new program valued the HOS activities because they allowed them to understand the context in which scientific research is conducted and because HOS activities were easier to incorporate into the school science curriculum than decontextualized ones:

> The first, the tube activity was very significant, but the activities that I really enjoyed were the historical ones… the VIH activity or the activity about chromosomal theory, everything, as a story was very enjoyable. I think that when you tell a history, and you stop and you ask questions to students, and they make predictions or propose explanations it is very exciting for them. They want to know what happened at the end: Am I right? What did the scientist find? I think it is much more dynamic… (Teacher 1)

> I think they are not mutually exclusive, and must be worked in a continuous way, first without context and then with biological context … I think considering HS is a good element, especially to capture this other group of students not so interested in science, but who like history … that student that may be interested in the context and data as the scientist lived, and indeed when one highlights the aspects of sociocultural science, that kind of student is put on alert … (Teacher 2).

5.4.2 Teaching Evolution and NOS Using History of Science for High School Students

In this intervention, four lessons of 90 min were applied in three classes of 11 grades by a biology teacher who finished our teacher development program (TDP). In this program, she knew HOS resources, literature on teaching NOS with HOS and about teaching evolution. The intervention included one class as a control (n = 25) and two classes as treatments (n = 50). The control class was exposed to four lessons about natural selection and evolution without any explicit instruction of NOS and without any reference to HOS. The four lessons used in treatment classes included a first lesson about NOS (reviewing concepts of theory, hypothesis, and law in the context of the theory of evolution by Darwin and Wallace); a second lesson that included the historical review of T. Dobzhansky's research (see Appendix); a third lesson about the field work of Rosemary and Peter Grant, who collected empirical data for more than 40 years about evolution in Galapagos finches; and a last lesson about the historical development of the evolutionary hypothesis about lactose intolerance in humans.

We used the Assessment of Contextual Reasoning about Natural Selection (ACORNS) instrument to assess understanding of the natural selection mechanism (Nehm et al. 2012), because it is one of the most widespread questionnaires used in the current literature (Ha et al. 2015) and because a detailed study of validity and reliability had been reported (Nehm et al. 2012). From the open questionnaire responses, ACORNS could determine both the correct items in the answers (ten) and the presence of misconceptions (six). Key concept (KC) scores for each item could range from 0 to 10, and alternative conception scores could range from 0 to −6. The analysis of the responses was conducted by two authors, and a rate of 92% consistency was obtained. To perform a quantitative analysis of change in students' understanding of natural selection, for each question (4 in total) we counted the number of correct concepts and the number of preconceptions included in each answer.

After the intervention, it was shown that HOS has a different effect on the learning of NOS and of evolution. On the one hand, we found that on average, students learned significantly more NOS in treatment classes (t [50] = −4.11; $p < 0.001$; $r = 0.5$) than students in control classes without explicit NOS instruction (t [25] = −1.75; $p = 0.1$; $r = 0.3$). Figure 5.2 shows the distribution values of NOS scores for control classes (pre- and post-) and Fig. 5.3 shows the distribution values of NOS scores for treatment classes (pre- and post-).

On the other hand, in relation to the learning of evolution (specifically about the natural selection mechanism), the data showed that there is no significant difference between treatments (t [43] = −10.91; $p < 0.001$; $r = 0.74$) and the control (t [19] = −5.55; $p < 0.001$; $r = 0.63$). Figure 5.4 show that students in the control group increased their knowledge of natural selection, in a similar way to the students in the treatment classes (Fig. 5.5). Therefore, the effect of including NOS and HOS was not statistically significant. It is important to highlight that the control treatment was not a didactic or traditional instruction, but a student-centered instruction without any explicit references to NOS and HOS.

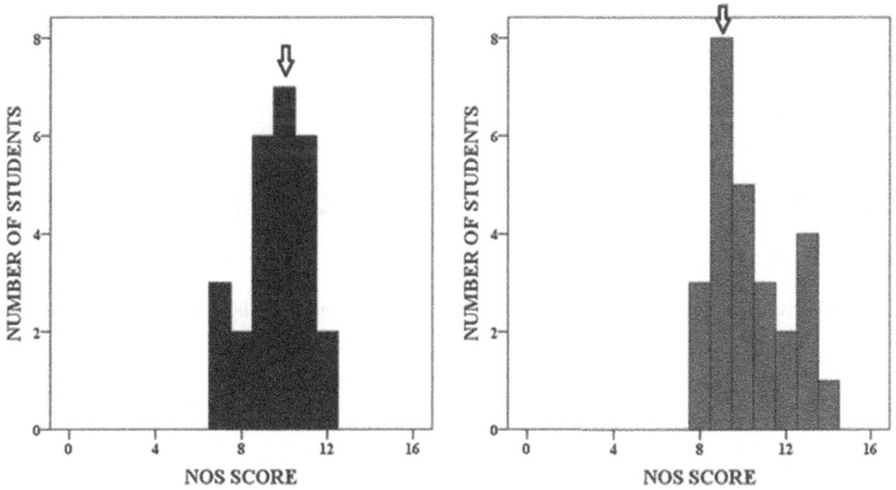

Fig. 5.2 Pre-test and post-test in class without NOS and history of science (HOS). The mean value of the NOS score before instruction was 9.7 and after instruction it was 10.3. Because seven aspects of NOS were studied, and the informed view was scored as 3, the maximum possible value achieved by the student was 21

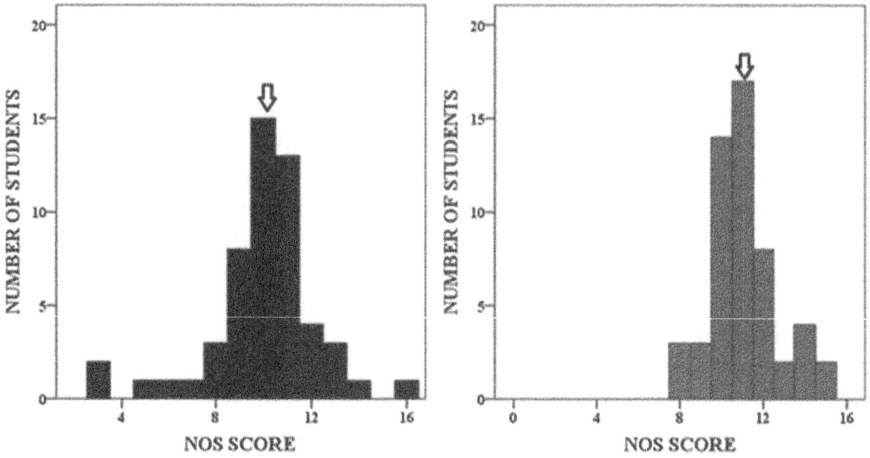

Fig. 5.3 Pre-test and post-test in classes with NOS and HOS. The mean value of the NOS score before the intervention was 9.8 and after the intervention it was 11.1. Because we studied seven aspects of NOS and the informed view was scored as 3, the maximum score was 21 points

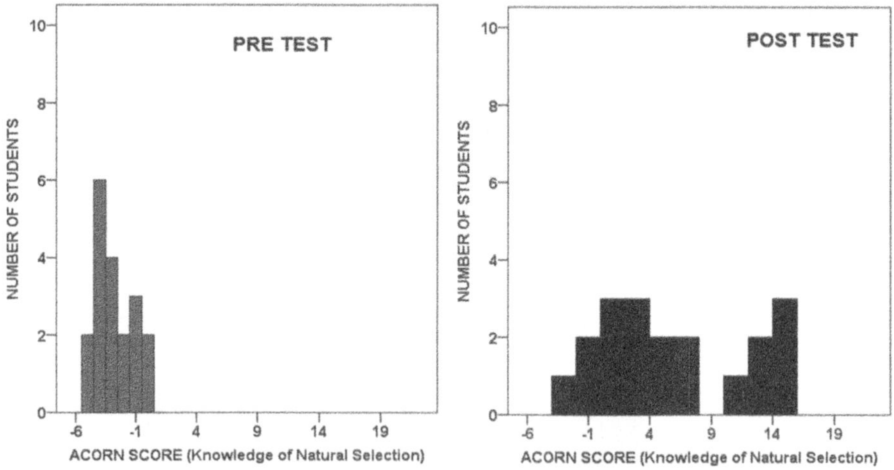

Fig. 5.4 Histograms of the ACORN score of students' responses in the pre-test and post-test in class without NOS and HOS. The mean value of the ACORN score before instruction was −2.8 and after instruction, it was 5.6

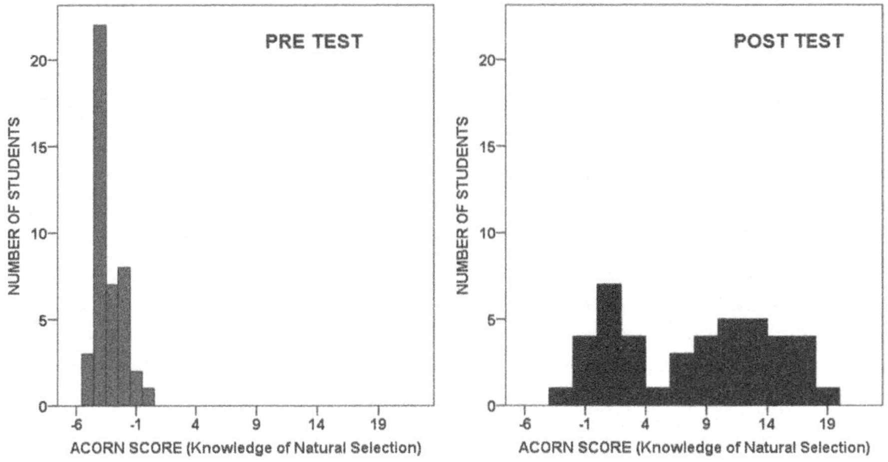

Fig. 5.5 Histograms of the ACORN score of students' responses in the pre-test and post-test in class with NOS and HOS. The mean value of the ACORN score before instruction was −3.3 and after instruction, it was 7.4

5.5 Some Examples of Lessons Using HOS for Teaching NOS and Biology

In the following activities, the history of science was utilized as an instructional intervention to promote both NOS and scientific content knowledge. The activities are based on the proposed models of Rudge and Howe (2009), and Paraskevopoulou and Koliopoulos (2010). The first activities were used to teach NOS in the study with in-service biology teachers, and the second was used to teach evolution and NOS in high school students.

5.5.1 Teaching Cellular Theory and the Nature of Science

The session: "The untold story of the cell theory", was based on the articles "The invention of the cell" (Vial 1999) and "The history of science in teaching the cell", (Carrillo et al. 2011). The first is a detailed account of the events in the laboratory of Johannes Müller in Berlin during the second half of the 1830s, when Theodor Schwann and Matthias Schleiden agreed to work together, compare notes, and determine patterns in the microscopic organization of plant and animal tissues (see also: https://goo.gl/TWdCGX). Comments on the cell reproduction of Rudolf Virchow, another pupil of Müller, are also described. The author emphasizes the series of errors and random events that gave rise to what we now know as the statements of the cell theory. The second paper introduces the story of Vial (1999) and ends with the discoveries of the twentieth century that allowed a more complex model cell to be shaped. The aim of the activity knows biographical aspects about the origin of the cell theory, to discuss the effect of incorporating elements of HOS into teaching biology.

The emphasized aspects of NOS were the subjectivity, the distinction between theory and law, and how science is influenced by society and culture. In the first stage, teachers were presented with background on the role of teaching HOS in a biology class and some working modalities. Later, they were asked about their own experience as biology teachers, about how they had been taught cell theory and about how they teach it to their students. There was a consensus that everyone had learned and taught more or less the same: the statements of the cell theory, without any context or background that would allow them to give meaning to this content. They were immediately asked to read a passage by Vial (1999) and discuss to what extent this reading modified their perception of the cell theory (see: https://goo.gl/TWdCGX) to obtain the summary of the article used in the activity). The specific questions in the activity were:

1. To what extent does this reading change your perception about cell theory?
2. A teacher believes that such information impairs the teaching of biology. Another teacher believes that the information could benefit it. Who agrees? Why?
3. Review the curriculum learning objectives. Which of them would be relevant for addressing elements of the history of science? Why?

Most teachers answered that they had changed their perception of this subject, but that it also made them think about the ease with which they have accepted a large amount of biological knowledge, such as dogmas, during the training process. They were especially struck by the absence of intent in the investigation that led to the cell theory. Subsequently, they are shown the results of the implementation of a similar activity, but with 9th-grade students. The results of subsequent evaluations after the intervention showed reasonable difficulty in understanding the margins of the validity of a theory concerning a law and accepting that science is influenced by the socio-cultural context. After giving time to discuss the possible causes of such difficulties, biology teachers relate to whether to deliver this type of "additional content information." After a brief discussion, the teachers agreed together that showing the obstacles and circumstances that have developed theories underlying the discipline they teach cannot be considered a detrimental effect on the validity of the content. Based on reading Carrillo et al. (2011) they recognize a benefit to the extent that is able to approach the figure of the scientist to the student and unfold aspects of the construction of scientific knowledge.

5.5.2 Teaching T. Dobzhansky's Contribution to Evolutionary Theory and NOS

The session "Discovering evidence of evolutionary change in wild populations" was based on the articles "Genetics of natural populations IX. Temporal changes in the composition of population of *Drosophila pseudoobscura*" (Dobzhansky 1943) and

"Variations in the gene arrangement in the chromosomes of *Drosophila pseudoobscura*" (Dobzhansky and Sturtevant 1938). In the lesson, work was conducted on preconceptions associated with NOS and evolution concerning that the "theory of evolution only has fossil evidence" and "evolutionary theory is just what Darwin proposed." Thus, students worked explicitly on NOS aspects of tentativeness and an empirical basis.

The lesson begins by briefly contextualizing students about the academic historical context in the development of synthetic theory of evolution. Based on this context, the activity is presented as a possible answer to the question of the existence of evolutionary change in natural populations. The activity consists of three parts (see Appendix). In the first part, students observe a picture of the natural environment where Dobzhansky conducted his investigation. According to this observation, the students are asked to propose some fly adaptations. After that, information on the methods of collecting samples in the field by Dobzhansky is delivered. In the second part of the activity, students work with a second picture, which identifies the different mutants of chromosome 3 found by Dobzhansky. They are asked to observe and describe one of the mutants (Chiricahua), as the researcher did. In the third part, they are faced with Dobzhansky's question in relation to the change of allele frequencies in *Drosophila,* and they worked with a third picture that included data in a table (see Appendix) that includes the frequency of the described chromosomal mutation (Chiricahua) in the population. Students are asked to plot the frequency and observe the pattern to answer the question.

Finally, we discussed questions with the students that led to reflection on obtaining empirical evidence for evolution (unlike fossils) by T. Dobzhansky and reflection on the contribution of this and other scientific developments to the theory of evolution.

5.6 Discussion and Conclusions

In this chapter, we showed that in the current literature and in our own research, there is empirical evidence that suggests that HOS serves as a rich instructional context for teaching different aspects of NOS (Table 5.1, Figs. 5.1 and 5.3) (Pavez et al. 2016). There is also empirical evidence that HOS is more efficient when lessons are deeply contextualized into research-specific situations, e.g., the development of inheritance theory (Kim and Irving 2010), the development of the synthetic theory of evolution and when students have an opportunity to reflect and discuss both NOS and content within specific historical and research contexts.

On the other hand, some studies reviewed here (see Table 5.2), and our own data, found that students in experimental groups, with HOS as an instructional context, neither helped nor hindered students' content learning. Therefore, science teachers need to realize that using classroom time for HOS is not a waste of time, but is an effective starting point for teaching about NOS and for addressing students' preconceptions about science (Kampourakis 2016), while maintaining significant achievement in the science content area (Kim and Irving 2010).

It is also important to note that lessons in control groups in many investigations (Kim and Irving 2010) are not usually didactic or traditional interventions. The control lessons have usually been student-centered practices including inquiry activities that only differ with treatment groups in the lack of explicit and reflective work with HOS and NOS.

In conclusion, we can state that including HOS in teaching NOS is an effective instructional context, but more research is needed to state how or to what extent HOS is a better context for teaching subject matter in comparison with the socio-scientific or inquiry contexts. These results are very important for the TDP and pre-service science teacher training in South America, because we know that in most of the country, teachers do not have much experience or training in HOS.

Appendix: Student Handout for the Dobzhansky NOS and Evolution Activity

Introduction

Theodosius Dobzhansky was one of the most influential scientists in the development of the modern synthesis. Thanks to his work and that of other researchers, evolution was re-defined as "changes in the frequency of genes that occur in populations over time." However, is there evidence of changes over time in the frequency of genes in natural populations?

Activities

1. Look at picture 1 (Mount San Jacinto, California) and describe the environment where *Drosophila pseudoobscura* lives and propose one adaptation of the species to this habitat.
2. Look at chromosome 3 of *Drosophila pseudoobscura* in picture 2 (Plate I in Dobzhansky and Sturtevant 1938). As known, some flies naturally had mutated versions of this chromosome. For example, mutations Arrowhead, Santa Cruz and Chiricahua. Now look at Chiricahua and describe what is distinctive about it (take into account the inversions or deletions).
3. When T. Dobzhansky found these different mutants in natural populations he wondered, **Will the proportion of mutant flies within populations change over time?** In Table 1 you can find the gene frequency data obtained by Dobzhansky, for Chiricahua mutant. Now, make a graph with these data and then answer Dobzhansky's question.

Table 1 Monthly data during 4 years of the frequencies of Chiricahua gene arrangements in the third chromosome in the populations of Andreas Canyon

Year	Month/day	Freq.	n	Year	Month/day	Freq.	n	Year	Month/day	Freq.	n
1939	Apr/24	6.6	106	1940	Apr/20	26.0	100	1941	Oct/04	15.4	26
1939	May/13	11.3	62	1940	May/19	28.3	60	1941	Nov/08	11.0	100
1939	Jun/04	36.8	38	1940	Oct/19	6.0	100	1941	Dec/06	15.0	100
1939	Sep/21	20.6	102	1940	Nov/20	9.1	37	1942	Jan/11	0.6	146
1939	Oct/28	10.8	102	1940	Dec/31	6.2	32	1942	Feb/02	15.0	100
1939	Dec/9	19.2	104	1941	Feb/10	13.8	93	1942	Mar/14	6.8	44
1940	Jan/13	14.4	104	1941	Mar/08	14.0	114	1942	Apr/02	9.6	104
1940	Feb/10	6.1	114	1941	Apr/19	12.0	100	1942	May/02	20.7	116
1940	Mar/28	12	126	1941	Sep/06	14.8	122	1942	Jun/12	38.5	104

Modified from Dobzhansky 1943

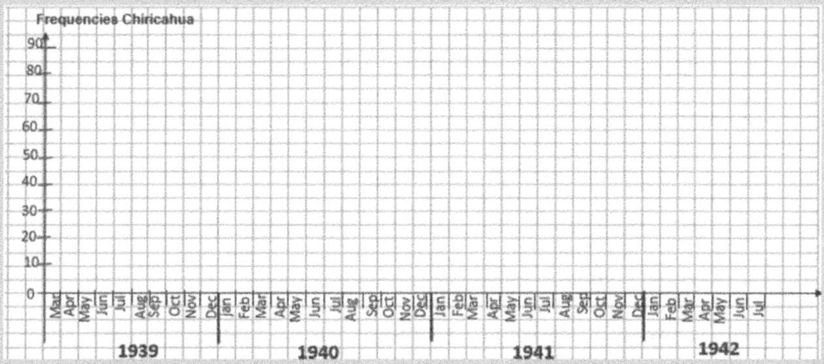

Fig. 1 Graph template of the frequencies of Chiricahua gene arrangements in the populations of Andreas Canyon

4. According to the graph, do you think that it is possible to say that there is empirical evidence for evolution in natural populations? Explain.
5. Read the following statement and explain whether you agree with it or not, basing your answer with what worked in today's session. "**The theory of evolution, proposed by Darwin over 100 years ago, has remained as a knowledge unchanged until today**".
6. If due to the **current climate warming**, the summer temperature in the 40s in Andreas Canyon currently can be observed **all year round.** Explain, by the mechanism of natural selection, how this new **selection pressure** could result in a new subspecies of *D. pseudoobscura*.

References

Abd-El-Khalick, F., & Lederman, N. G. (2000). The influence of history of science courses on students' views of nature of science. *Journal of Research in Science Teaching, 37*(10), 1057–1095.

Allchin, D. (2012). The Minnesota case study collection: New historical inquiry case studies for nature of science education. *Science & Education, 21*(9), 1263–1281.

Alvarez, M. (2006). La historia de la ciencia en la formación del professorado de ciencias naturales. In M. Quintanilla & A. Adúriz-Bravo (Eds.), *Enseñar ciencias en el nuevo milenio: retos y propuestas*. Santiago: Ediciones Universidad Católica de Chile.

Bravo, P., & Cofré, H. (2016). Developing biology teachers' pedagogical content knowledge through learning studies: The case of teaching human evolution. *International Journal of Science Education*, 38(16), 2500–2857.

Carrillo, L., Morales, C., Pezoa, V., & Camacho, J. (2011). La historia de la ciencia en la enseñanza de la célula. *Tecné, Episteme y Didaxis, 29*, 112–127.

Chang, H. (2010). How historical experiments can improve scientific knowledge and science education: The cases of boiling water and electrochemistry. *Science & Education, 20*(3–4), 317–341.

Clary, R., & Wandersee, J. (2013). Arguing history. *The Science Teacher*, 39–43.

Cofré, H. (2012). La enseñanza de la naturaleza de la ciencia en Chile: del currículo a la sala de clases [Teaching nature of science in Chile: From the curriculum to the classroom]. *Revista Chilena de Educación Científica, 11*, 12–21.

Cofré, H., Camacho, J., Galaz, A., Jiménez, J., Santibáñez, D., & Vergara, C. (2010). La educación científica en Chile: debilidades de la enseñanza y futuros desafíos de la educación de profesores de ciencia. [Science education in Chile: Teaching weaknesses and future challenges of science teacher education]. *Revista Estudios Pedagógicos, 36*, 279–293.

Cofré, H., Vergara, C., Lederman, N., Lederman, J., Santibáñez, D., Jiménez, J., & Yancovic, M. (2014). Improving Chilean in-service elementary teachers' understanding of nature of science using self-contained NOS and content embedded min-courses. *Journal of Science Teacher Education, 25*, 759–783.

Cofré, H., Gonzalez-Weil, C., Vergara, C., Santibáñez, D., Ahumada, G., Furman, M., Podesta, M. E., Camacho, J., Gallego, R., & Pérez, R. (2015). Science teacher education in South America: The case of Argentina, Chile, and Colombia. *Journal of Science Teacher Education, 26*(1), 45–63.

Cofré, H., Jiménez, J., Santibáñez, D., & Vergara, C. (2016). Chilean pre-service and In-service teachers and undergraduate students' understandings of evolutionary theory. *Journal of Biological Education, 50*(1), 10–23.

Cofré, H., Santibáñez, D., Jiménez, J., Spotorno, A., Carmona, F., Navarrete, K., & Vergara, C. (2017). The effect of teaching the nature of science on students' acceptance and understanding of evolution: myth or reality? *Journal of Biological Education*. https://doi.org/10.1080/00219266.2017.1326968

Dass, P. M. (2005). Understanding the nature of scientific enterprise (NOSE) through a discourse with its history: The influence of an undergraduate 'history of science'course. *International Journal of Science and Mathematics Education, 3*(1), 87–115.

Dedes, C., & Ravanis, K. (2009). History of science and conceptual change: The formation of shadows by extended light sources. *Science & Education, 18*, 1135–1151.

Dibattista, L., & Morgese, F. (2013). Introducing history (and philosophy) of science in the classroom: A field research experience in Italy. *Science & Education, 22*, 543–576.

Dobzhansky, T. (1943). Genetics of natural populations IX: Temporal changes in the composition of population of *Drosophila pseudoobscura*. *Genetics, 28*, 162–186.

Dobzhansky, T., & Sturtevant, A. (1938). Variations in the gene arrangement in the chromosomes of *Drosophila pseudoobscura*. *Genetics, 23*, 28–64.

Forato, T. C. M., Martins, R. A., & Pietrocola, M. (2012). History and nature of science in high school: Building up parameters to guide educational materials and strategies. *Science & Education, 21*, 657–682.

Galili, I., & Hazan, A. (2001). The effect of a history-based course in optics on students' views about science. *Science & Education, 10*(1–2), 7–32.

Ha, M., Baldwin, B., & Nehm, R. (2015). The long-term impacts of short-term professional development: Science teachers and evolution. *Evolution: Education and Outreach, 8*, 11.

Howe, E., & Rudge, D. (2005). Recapitulating the history of sickle-cell anemia research improving students' NOS views explicitly and reflectively. *Science & Education, 14*, 423–441.

Izquierdo, M., Quintanilla, M., Vallverdú, J., & Merino, C. (2007). Una nueva reflexión sobre la historia y filosofía de las ciencias y la enseñanza de las ciencias. In M. Quintanilla (Ed.), *Historia de la ciencia: Aportes para la formación del profesorado* (pp. 30–51). Santiago: Arrayan Editores.

Kampourakis, K. (2016). The "general aspects" conceptualization as a pragmatic and effective means to introducing students to nature of science. *Journal of Research in Science Teaching, 53*, 667–682.

Kampourakis, K., & Gripiotis, C. (2015). Darwinism in context: An interdisciplinary, highly contextualized course on nature of science. *Perspectives in Science, 5*, 25–35.

Kim, S., & Irving, K. (2010). History of science as an instructional context: Student learning in genetics and nature of science. *Science & Education, 19*(2), 187–215.

Lederman, N., & Khishfe, R. (2002). *Views of nature of science, Form D*. Unpublished paper. Chicago: Illinois Institute of Technology.

Lederman, N., & Lederman, J. (2014). Research on teaching and learning of nature of science. In N. Lederman & S. Abell (Eds.), *Handbook of research on science education* (Vol. II, pp. 600–620). New York: Routledge.

Lin, H.-S., Hung, Y.-J & Hung, S.-C. (2002). Using the history of science to promote students' problem-solving ability. *International Journal of Science Education, 24*(5), 453–464.

Lin, C.-Y., Cheng, Y.-J., & Chang, W.-H. (2010). Making science vivid: Using a historical episodes map. *International Journal of Science Education, 32*(18), 2521–2531.

McComas, W. (2013). Uma proposta de classificação para os tipos de aplicação da história da ciência na formação científica: implicações para a pesquisa e desenvolvimento. In C. C. Silva & M. E. B. Prestes (Eds.), *Aprendendo ciência e sobre sua natureza: abordagens históricas e filosóficas* (pp. 419–441). São Carlos: Tipographia.

McComas, W., & Kampourakis, K. (2015). Using the history of biology, chemistry, geology, and physics to illustrate general aspects of nature of science. *Review of Science, Mathematics and ICT Education, 9*(1), 47–76.

Ministry of Education of Chile. (2014). *Bases curriculares 78 básico a 28 medio ciencias naturales*. Santiago: Ministerio de Educación, Chile.

Monk, M., & Osborne, J. (1997). Placing the history and philosophy of science on the curriculum: A model for the development of pedagogy. *Science Education, 81*(4), 405–424.

Nehm, R., Beggrow, E., Opfer, J., & Ha, M. (2012). Reasoning about natural selection: diagnosing contextual competency using the ACORNS instrument. *American Biology Teacher, 74*(2), 92–98.

Niaz, M. (2007). Progressive transitions in chemistry teachers' understanding of nature of science based on historical controversies. *Science & Education, 18*(1), 43–65.

Osborne, J., Collins, S., Ratcliffe, M., Millar, R., & Duschl, R. (2003). What "ideas-about-science" should be taught in school science? A Delphi study of the expert community. *Journal of Research in Science Teaching, 40*, 692–720.

Paraskevopoulou, E., & Koliopoulos, D. (2010). Teaching the nature of science through the Millikan-Ehrenhaft dispute. *Science & Education, 20*(10), 943–960.

Pavez, J. M., Vergara, C., Santibañez, D., & Cofré, H. (2016). Using a professional development program for enhancing Chilean biology teachers' understanding of nature of science (NOS) and their perceptions about using history of science to teach NOS. *Science & Education, 25*, 383–405.

Rudge, D., & Howe, E. (2009). An explicit and reflective approach to the use of history to promote understanding of the nature of science. *Science & Education, 18*(5), 561–580.

Rudge, D., Cassidy, D., Fulford, J., & Howe, E. (2014). Changes observed in views of nature of science during a historically based unit. *Science & Education, 23*(9), 1879–1909.

Russel, T., & Bullock, S. M. (2010). From talk to experience: Transforming the preservice physics methods course. *Brock Education, 20*(1), 19–33.

Seker, H., & Welsh, L. (2006). The use of history of mechanics in teaching motion and force units. *Science & Education, 15*, 55–59.

Solomon, J., Duveen, J., Scot, L., & McCarthy, S. (1992). Teaching about the nature of science through history: Action research in the classroom. *Journal of Research in Science Teaching, 29*(4), 409–421.

Tsatsarelis, C., Ogbom, J., Jewitt, C., & Kress, G. (2000). Rhetorical construction of cells in science and in a science classroom. *Research in Science Education, 30*(4), 451–463.

Vial, J. (1999). La invención de la célula. *Biological Research, 32*(4). https://doi.org/10.4067/S0716-97601999000400004. Accessed 20 Jul 2016.

Wang, H., & Marsh, D. (2002). Science instruction with a humanistic twist: Teachers' perception and practice in using the history of science in their classrooms. *Science & Education, 11*, 169–189.

Williams, C., & Rudge, D. (2016). Emphasizing the history of genetics in an explicit and reflective approach to teaching the nature of science: A pilot study. *Science & Education, 25*(3), 407–427.

Chapter 6
Epistemological Conceptions of University Teachers and Students of Science

María B. García, Silvia Vilanova, and Sofía Sol Martín

6.1 Introduction

Even though the study of epistemological conceptions has been comprehensibly addressed – as important revision works on the topic reveal[1] – some aspects have been surveyed in only a limited way. The works that have been carried out evaluate secondary school students' and teachers' conceptions, with the study of epistemological conceptions coming from the university field less well explored. This explains the brief reference to these works – especially by students – in the latest revision made by Lederman and Abell (2014). Furthermore, the studies conducted in this area have dealt with the problem from theoretical frameworks that portray the conceptions as ideas explicitly held by subjects, instead of undertaking studies to explore their implicit beliefs, a recommendation made by Schraw (2010) in her revision of the research field.

An earlier work on epistemological conceptions of faculty members of Exact and Natural Sciences courses of the National University of Mar del Plata from the perspective of implicit theories carried out by this group served as a contribution to the topic (García et al. 2013). In the present work, the university teachers' conceptions presented in the previous work are compared with those of their students, i.e., the undergraduate students of Exact and Natural Sciences courses from the same university. When looking at responses, we found that both groups have conceptions that reflect current visions on the essence and possibility of scientific knowledge;

[1] Revisions such as in: (Koulaidis and Ogborn 1995; Abd-El-Khalic and Lederman 2000; Hofer and Pintrich 2001; Scrhaw and Olafson 2002; Audi 2003; Schraw 2010; Lederman and Abbel 2014).

M. B. García (✉) · S. Vilanova · S. S. Martín
Science Education Department, Faculty of Exact and Natural Sciences, Universidad Nacional de Mar del Plata, Mar del Plata, Argentina
e-mail: bagarcia@mdp.edu.ar; svilano@mdp.edu.ar

M. E. B. Prestes, C. C. Silva (eds.), *Teaching Science with Context*, Science: Philosophy, History and Education,
https://doi.org/10.1007/978-3-319-74036-2_6

half the subjects have conceptions associated with relativist visions and another substantial portion of both groups have conceptions close to critical realism.

6.1.1 Conceptual Framework

In general terms, the objective of science is the exploration of the world that surrounds us, the search for fundamental and unifying principles that make up the basis from which that exploration continues. These principles, however, have been interpreted in different ways throughout history. The first positions were dogmatic, based on naïve realism and an exaggerated faith in its principles (Heraclitus of Ephesus epitomized this stand, as he asserted that the *arche* was the fire and that the senses showed us reality as it truly was). Intermediate stands understand that principles do not rule over science, but result from theories that, although constructed with a realistic intention, are provisional, as they represent successive approaches to reality (as the ideas mooted by Kant, Hume, and Popper exemplify to different degrees). Last, the most relativist positions believe that theory – always being provisional – does not give a single final explanation. Instead, theory is understood as an interpretative image, a source of new discoveries, generalizations, and relations between symbolic entities (as expressed in the epistemologies developed by Lakatos and Khun, amongst others). These three positions and their different versions are present in the research conducted throughout the history of science and are also scattered, as epistemological conceptions, to researchers and to university teachers and students of the sciences (Vosniadou 2013).

As the conceptions surveyed in this work are understood to be implicit theories, i.e., as intuitive ideas or beliefs resulting from each individual's experience with the world (Hodson 1994), and not as a product of formal instruction on the subject, it is assumed that:

- The conceptions to be surveyed do not necessarily have a lineal correspondence with a specific formal epistemological theory – e.g., critical realism, apriorism – but may share certain assumptions with some of them. In this sense, Fourez (1994) calls them "spontaneous philosophies of the sciences," as they are ingrained in the cultural context, revealing their slightly formal and intuitive nature
- Conceptions make up a system – as the last investigations show – and are, thus, a product of the interaction between external representations, social processes and inner mental representations. This unveils the need to carry out their study with methods that can give access to those ideas that are not easily verbalized
- As conceptions originate in shared sociocultural scenarios, there is a chance that students and professors from the same field share the same type of conceptions. In this work, this assumption was the initial hypothesis.

6.1.2 Background Research on the Topic

An analysis of the most relevant studies conducted on university teachers' conceptions (Lederman and Abell 2014) show discrepancies in the results obtained concerning the content of epistemological conceptions. For instance, works such as those by Acevedo (1994) point out that although university teachers bear empiricist traces, they should not be simplistically framed in a radical inductivism or naïve realism. The results obtained by Pomeroy (1993) and Raúl (1994) also follow this direction; Porlán Ariza and collaborators Raúl et al. (1998) concluded that the greater the teaching experience, the more absolute the vision of science. Conversely, other authors, such as Abd-El-Khalic and Lederman (2000), Chai (2010), and Enna and del Rocío Gómez Vallarta (2002), suggest that experienced high-level university teachers have more relativist and elaborate conceptions. A third stand comprises those works that state that university teachers' conceptions about the nature of scientific knowledge cannot be framed in a particular philosophical trend, as the subjects respond from different perspectives, depending on the aspect that they are being asked (Manassero and Alonso 2000).

Recent studies of students reflect previous ones: without a formal education on epistemological themes, students do not have updated conceptions not only as regards the nature of scientific knowledge, but also of its origin and progress. It is known, from the first investigations on the topic, that schools portray science as the "rhetoric of conclusions" (Schwab 1964), offering students few opportunities to develop more sophisticated epistemological conceptions. In fact, it is common for middle and high school students to think of science as "problem-free knowledge" (Carey and Smith 1993), in which the most sophisticated epistemological beliefs simply emerge. This also seems to be the belief – though not always – among late adolescents and freshmen (see Hofer and Pintrich 1997 for a review). There are studies showing that students tend to explain scientific phenomena with simple descriptions, instead of providing deeper explanations (Carey et al. 1989; Driver et al. 1996). Furthermore, they believe that the aim of experiments is to find out *how* something works instead of interpreting them as proofs of competing explanatory hypotheses (Carey et al. 1989; Schauble et al. 1991).There are also works pointing that students believe that the models are concrete replicas instead of tools for the development and revision of theories (Grosslight et al. 1991).

It is noteworthy that most of the studies carried out with student participants have focused on the analysis of epistemological conceptions in connection with metacognition; addressing general dimensions, such as their beliefs about the justification of knowledge, structure, source, and boundaries, etc. (Hofer and Pintrich 1997; Sandoval 2005). There are fewer investigations addressing students' conceptions as they work on specific scientific topics, and those that have been conducted have basically dealt with physics, focusing on topics such as "power and movement" (see review by Lederman and Abell 2014). Because of this, the study on conceptions streaming from different disciplinary domains (chemistry, physics, and biology) is paramount for coming up with comprehensive approaches to the subject. It is in this sense that this work is aimed at making a contribution.

6.2 Objective

The objective of the present work is to describe, discuss, and compare university science teachers' conceptions of the nature of scientific knowledge and its acquisition (previously examined) with those conceptions of their students, i.e., undergraduate students of exact and natural science courses (presently examined).

6.3 Method

One of the major difficulties in accumulating evidence about epistemological conceptions is the selection of an adequate methodology. So far, the research conducted has used two alternative methods of collecting data. We collected data using interviews and questionnaires. The instruments gave us access to participants' explicit knowledge consistent with the research carried out by Aldridge et al. (1997), Hammer (1994), McGinnis et al. (1997), and Schommer (1990). We also used an ethnographic approach to developing an understanding of participants' conceptions within a more natural setting, similar to Baena (2000).

Even though ethnographic approaches produce richer information and solve problems such as those connected with context, they present difficulties in connection with the size and representativeness of the sample that are not found in quantitative approaches. A choice has, thus, to be made between collecting in depth or substantial information.

To take into account the theoretical perspective adopted in this work and, in turn, to consult a representative number of teachers and students, a Choice Dilemma Questionnaire (CDQ) was used. The dilemmas, as opposed to other types of tools such as the Likert scale, force the subject to decant the choices and finally lean toward one: one that, although it does not reflect all the subject's beliefs, which could otherwise be obtained from an answer produced by the subject, better approximates his/her idea of the problem being posed. Furthermore, the CDQ is of an argumentative nature, i.e., the same stand can be defended by different reasons. Arguments for each alternative are offered in the questionnaire items (see Appendix). Finally, the dilemmas enable the answers to be contextualized and the topics to be varied, raising issues in different disciplines and in line with the conceptual framework of the implicit theories.

6.3.1 Participants

The participants of the research were the following:

- Group 1: 100 university teachers of Sciences at Universidad Nacional de Mar del Plata, Argentina
- Group 2: 96 university students taking Exact and Natural Science courses at Universidad Nacional de Mar del Plata, Argentina.

6.3.2 Variables

6.3.2.1 Variable 1 (V1): Conceptions of Scientific Knowledge

Definition: ideas and beliefs that people have about two-dimensional analysis: what scientific knowledge is (nature of knowledge) and how scientific knowledge is produced (acquisition of knowledge).

Categories of the variable and level of measurement: the following categories were set a priori, based on the classical epistemological categories about the principles behind the different types of representations of knowledge. They are related to the results from the main studies conducted so far.[2] The variables, in addition to its dimensions, are categorical and its level of measurement is nominal, as shown in Table 6.1:

6.3.2.2 Variable 2 (V2): Grouping Variable

This variable takes two values: university teachers and students. The level of measurement of the variable is also nominal.

Table 6.1 Categories assigned to the "Conceptions of scientific knowledge" variable

	Theories with which they share assumptions	
Category	Nature of knowledge	Acquisition of knowledge
Position I	Objectivism, naïve realism	Empiricism
Position II	Objectivism, critical realism	Intellectualism
Position III	Subjectivism, phenomenalism	Apriorism

[2] As in: (Hessen 1973; Koulaidis and Ogborn 1989; Porlán 1994; Porlán et al. 1998; Aldridge et al. 1997; Abd-El-Khalick and Lederman 2000; Sander et al. 2000; Pecharromán and Pozo 2006).

6.3.3 *Materials*

For the reasons mentioned above, a Choice Dilemma Questionnaire was adopted to collect data (see Appendix for examples). To design it, a bibliographic search was conducted, and different types of questionnaires were found. Even though these instruments were taken into account in the design, they were not used as they were for different reasons (because they were oriented to educators of different educational levels, or examined explicit conceptions, or because they did not take into account the context of inquiry, etc.). Finally, a new Dilemma Questionnaire about scientific knowledge was drawn up and validated. To design it, the dimensions of the variable and the characteristics of the population that comprises the study were taken into account. The final instrument consisted of ten dilemmas – each one intended to evaluate one of the two dimensions of the variable, having three possible answers corresponding to the three assigned categories (Positions I, II, and III). Before its implementation, the questionnaire was subjected to a validation process (Alpha de Cronbach: 0.8342). For more details on the construction and validation of this instrument, see García et al. (2007), and García and Vilanova (2008). Each of *the ten dilemmas with their three categories of answers, were distributed according to the dimensions* assigned to the variable as shown in Table 6.2.

The dilemmas are, in turn, contextualized in topics from different disciplines, i.e., dilemmas 3, 4, 10, and 9 deal with physics, 1 and 6 with chemistry, and 7 and 8 with biology. Dilemmas 2 and 5 are not contextualized in a specific discipline.

6.4 Results

To examine the content of the conceptions, a calculation was made of the distribution of relative frequencies for each alternative presented in the questionnaire for both groups: university teachers and students. The results obtained are shown in Figs. 6.1a, b and 6.2a, b.

Both groups show a variation in the most adopted position depending on the aspect being evaluated. Although "Position III" was mostly taken concerning the "Nature of Knowledge," "Position II" predominates in connection with the "Acquisition of Knowledge". This means that, on general grounds, the epistemological conceptions of most of the subjects that are part of both groups could be described as follows:

Table 6.2 Distribution of dilemmas in the questionnaire to evaluate conceptions of scientific knowledge

	Nature of knowledge	Acquisition of knowledge
Dilemmas	D1, D2, D3, D6, D8, D9	D4, D5, D7, D10

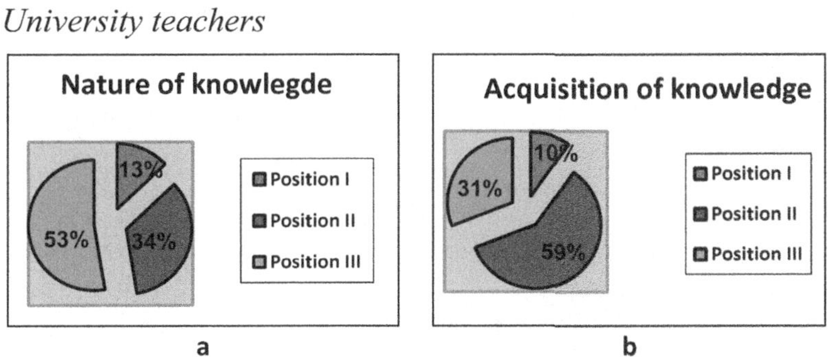

Fig. 6.1 (**a**) Nature of knowledge by teachers. (**b**) Acquisition of knowledge by teachers

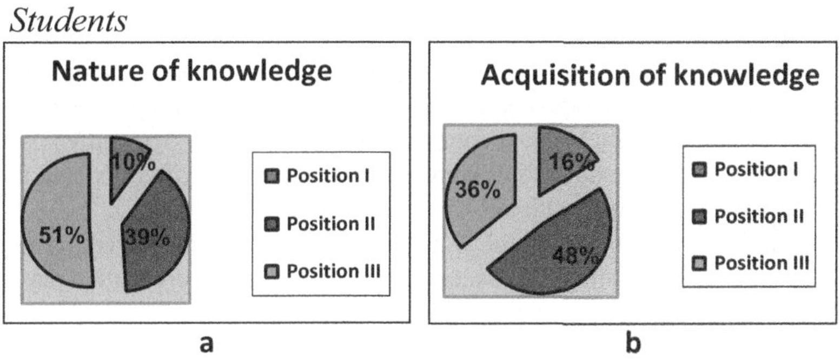

Fig. 6.2 (**a**) Nature of knowledge by students. (**b**) Acquisition of knowledge by students

1. Regarding the nature of knowledge, they consider that the phenomenon is paramount, without considering the autonomous existence of physical entities. Things are not known for what they are, but for how they appear
2. Concerning the acquisition of knowledge, they consider that experience and reason intervene in knowledge. Sensible images are received from concrete things and, then, thought removes their general concepts – what Hessen (1973) calls "intellectualism."

It should be noted that there is also a significant percentage of individuals whose conceptions about the nature of knowledge are closer to critical realism (Position II) and others whose conceptions are closer to apriorism concerning the acquisition of knowledge (Position III). A detail of this analysis with the complete instrument can be seen in García, et al. (accepted for publication).

Table 6.3 Association between Dilemma 4 and grouping (university teachers and students)

Cross-tabulation					
			VAR00011		Total
			University teachers	Students	
VAR00004	1	Count	2	19	21
		Standardized residue	−2.9	3.3	
	2	Count	16	25	41
		Standardized residue	−1.5	1.7	
	3	Count	52	10	62
		Standardized residue	2.9	−3.3	
Total	Count	70	54	124	

6.4.1 *Group Comparison*

To investigate if, beyond the general similarities observed in the descriptive analysis, there were differences between the groups when comparing each dilemma, possible associations between V_1 Conceptions of scientific knowledge and V_2 Grouping were evaluated. Contingency tables were elaborated for each dilemma and the Chi-squared test was calculated with a 5% significance level.

Once the test was concluded, the Conceptions and Grouping variables in dilemmas 4, 5, and 10 pertaining to Acquisition of Knowledge resulted associated ($p < 0.05$.). That is, they are related, they are no independent.

So as to examine the reasons for these associations, standardized residues were calculated in each of the three tables. In these cases, a residue larger than the one expected (I1.96I) was observed in the boxes from the crosstab between Positions I and III and the grouping; thus, a hypothesis of independence between the two variables cannot be accepted. By way of example, in Table 6.3, Dilemma 4 is shown.

Similar tables resulted from Dilemmas 5 and 10, in which an association between Position I and the students' group and Position III and the teachers' group was observed: when dealing with the acquisition of scientific knowledge, students answer from Position I, whereas university teachers answer from Position III.

In the rest of the dilemmas, the p value showed independence between both variables.

6.5 Conclusions

As regards the description of university teachers' and students' conceptions, the results obtained show that in the Nature of Knowledge dimension, both groups have conceptions that reflect current visions of the essence and possibility of scientific knowledge. Half the subjects have conceptions associated with relativist visions and another substantial portion of both groups has conceptions close to critical realism. These results are encouraging, as scientific knowledge should not be treated, in university classrooms, as full knowledge or a sole paradigm capable of explaining all scientific problems. To portray science as static goes against the dynamic nature of scientific knowledge – such as showing students different points of view about a particular phenomenon so that they can critically analyze and verify scientific concepts. Therefore, it is preferable that the teacher presents a *conceptual profile* (Mortimer 2001)[3] from which the dynamic nature of knowledge is conformed to and that can favor the construction of an updated notion of science by students.

Even though professors and students have conceptions that bear conceptions about scientific knowledge – as a provisional and a richer and more reflective philosophical and sociological vision of science – this does not guarantee that their vision will be fully transferred to classroom practice. An issue that raises further questions is the fact that the results obtained from the analysis of the *Acquisition of knowledge* dimension show differences in relation to those obtained from the *Nature of knowledge* dimension, showing Position II to be the majority in both groups (university teachers and students). This dimension, more connected to the procedural, could be showing a more implicit character of the conceptions (Pozo and Scheuer 2000).

As regards the comparison between the conceptions in both groups, the results show that in the *Nature of knowledge* dimension, the belonging group does not have an influence, i.e., there is no association between variables because both university teachers and students have similar conceptions. This may be because the two participant groups are teachers and students of the same class. In the *Acquisition of knowledge* dimension, however, an association between the grouping variable (V_2) and this dimension of V_1 is observed, as in four of the five dilemmas connected to it, teachers "conceptions go against those of their students." Although, as regards *Acquisition of knowledge* (Position III), teachers show conceptions closer to apriorism, their students have conceptions that are more connected to empiricism. A possible explanation could be that university teachers (most of whom are also researchers), have a particular view about the acquisition of scientific knowledge, but, when they teach about it in university classrooms, they present it as something complete, done by others and fixed (Ravanal et al. 2014; Chai 2010).

This leads to an examination of the influence that university teachers' epistemological conceptions and the strategies that they use, i.e., the way in which knowledge is presented to the students, may have on their teaching of sciences.

Some factors related to teaching practice at university – as the structure of the syllabi, the rigid division between theory and practice, the assessment methods that mainly focus on results and the traditional activities to confirm and practice theory – do not foster the idea of science as a human product in continuous construction, but rather as a full knowledge. Furthermore, these factors hinder the possibility of coherence between a personal perspective on science – as developed though it might be – and the teaching practice.

---

[3] "The basic presupposition is that, in any culture or person, there is no single homogeneous way of thinking but different types of verbal thinking. I have tried to characterize this heterogeneity of verbal thinking in terms of a *conceptual profile* that recognizes the coexistence, in an individual, of two or more meanings for a single word or concept, which are correctly used in different contexts. This coexistence is also possible for a scientific concept in which the classical and modern visions of the same phenomenon are not always comparable" (Mortimer 2001, p. 488).

The difficulty in transferring the epistemological vision developed during the teacher training and the professional development of teachers and researchers could be partly explained by the teaching approach that the university has: rationalist (theory is more important than practice), traditional (the teacher has knowledge that he/she "transmits" to the students by being a mediator between science and the student), and encyclopedic (the sum of the transmitted knowledge must be evaluated as the accumulation of this knowledge). This approach is replicated in classrooms and accepted as the proper way of teaching at university. For this reason, it is imperative to work on teachers' training through the creation of reflection spaces and incorporate epistemological issues into the university curriculum.

Finally, it should be pointed out that in this work, as in all the research, a cross-section of reality has been made and, consequently, the results obtained are limited and can only be analyzed within certain limits, coming from:

1. *The problems associated with the methodology and data collection instruments.* Although posing dilemmas allows for indirect and contextualized inquiry, it is necessary to address the problem from more qualitative methodologies to obtain in-depth information. This is being currently developed by studies based on the direct observation of the teaching practice and in-depth interviews
2. *The size and representativeness of the sample*, which makes it hard to generalize the results. Studies – which were not included in the first stage of the investigation – are being currently conducted with teachers at the Faculty of Social Studies from the same university, and with professors of exact sciences from other universities
3. *The mainly descriptive rather than explicative nature of the work*, which justifies the need to investigate the creation, interaction, and changing mechanisms of the conceptions, as Limón (2006) and Hofer (2006) suggest.

Appendix

Some examples of dilemmas to examine implicit conceptions of scientific knowledge (To see the complete instrument, consult García and Mateos, 2013 or García et al. (accepted for publication).

Below are situations and answers/comments made by teachers.

For each scenario, mark the one you agree with:

Dilemma 6-A (Translated from Spanish)

Below is a dialogue between people who are discussing about whether the theory of evolution is scientific. Which person better reflects your ideas?

Esteban: I believe that the theory of evolution is not scientific. It explains too much and it is difficult to test. Darwinism describes singular events, something that is not repeatable and is, thus, not available for experiments. It cannot be scientific. (Position I)

Carlos: Following your criteria, it would be impossible to prove that the world existed yesterday. Even though experimentation is a fundamental starting point, theories are obtained from a process that goes beyond data. Besides, what is important and what makes it truly scientific is its predictive power. For example: if a hypothesis that polar bears with thicker fur survive the icy Arctic winter is formulated, the hypothesis can be contrasted and it can be stated if theory can or cannot explain reality. (Position II)

Pedro: I agree with Carlos in that what makes a theory scientific is, as regards Darwin's theory of evolution, its predictive power; but I do not agree in that experimentation is the starting point. I think that Darwin already had a theory in mind when he started to experiment. Any kind of knowledge stems from reason and then comes the experimentation. (Position III)

(a) Esteban
(b) Carlos
(c) Pedro

Dilemma 8

The image presented in Fig. 1 corresponds to a piece of graphite (a substance made of carbon atoms) seen through a scanning tunneling microscope with a 1×10-to 12-m resolution (approximately the size of an atom). Look at the picture: what do you see?

Fig. 1 Picture of a piece of graphite seen through a scanning tunneling microscope

(a) It can only be said that it is the picture of a piece of graphite obtained through a microscope. I would not assert that atoms can be seen. This would be a personal interpretation influenced by the wording of this dilemma. (Position III)
(b) If the picture is interpreted from a quantum theory stance, carbon atoms that make up the structure of graphite – as described by science – are observed. (Position II)
(c) The empirical confirmation of carbon atoms arranged for graphite as science states. (Position I)

References

Abd-El-Khalic, F., & Lederman, N. (2000). Improving science teachers' conceptions of the nature of science: A critical review of the literature. *International Journal of Science Education, 22*, 665–701.

Acevedo, J. (1994). Los futuros profesores de enseñanza secundaria ante la sociología y la epistemología de las ciencias. *Revista Interuniversitaria de Formación del Profesorado, 19*, 111–125.

Aldridge, J., Peter, T., & Chung-Chi, C. (1997). *Development, validation and use of the belief about science and school science questionnaire* http://www.chem.arizona.edu/tpp/basssq.pdf. Accessed 20 Jun 2016.

Audi, R. (Ed.). (2003). *Diccionario Akal de filosofía*. Madrid: Akal.

Baena, M. D. (2000). Pensamiento y acción en la enseñanza de las ciencias. *Enseñanza de las Ciencias, 18*(2), 217–226.

Carey, S., Evans, R., Honda, M., Jay, E., & Unger, C. (1989). An experiment is when you try it and see if it works: A study of grade 7 students' understanding of the construction of scientific knowledge. *International Journal of Science Education, 11*, 514–529.

Carey, S., & Smith, D. C. (1993). On understanding the nature of scientific knowledge. *Educational Psychologist, 28*, 235–252.

Enna, C. C., & del Rocío Gómez Vallarta, M. (2002). Concepciones y representaciones de los maestros de secundaria y bachillerato sobre la naturaleza, el aprendizaje y la enseñanza de las ciencias. *Revista Mexicana de InvestigaciónEducativa, 7*(16), 577–602.

Chai, C. S. (2010). Teachers' epistemic beliefs and their pedagogical beliefs: A qualitative case study among Singaporean teachers in the context of ICT-supported reforms. *TOJET: The Turkish Online Journal of Educational Technology, 9*(4), 128–139.

Driver, R., Leach, J., Millar, R., & Scott, P. (1996). *Young people's images of science*. Buckingham: Open University Press.

Fourez, G. (1994). *La construcción del conocimiento científico: filosofía y ética de la ciencia*. Madrid: Narcea.

García, M., Mateos, M., & Vilanova, S. (2007). Las concepciones epistemológicas de los profesores universitarios de química. *Educación Química, 18*(2), 133–138.

García, M., & Vilanova, S. (2008). Diseño y validación de un instrumento para analizar concepciones implícitas sobre el aprendizaje en profesores de matemática en formación. *Revista Electrónica de Investigación en Educación en Ciencias, 3*(2), 27–34.

García, M., & Mateos, M. (2013). Las cuestiones de dominio, intersujeto e intrasujeto, en el contenido de las concepciones epistemológicas en estudiantes universitarios. *Avances en Piscología Latinoamericana, 31*(3), 586–619. Universidad del Rosario, Colombia. ISSN: 1794-4724.

García, M., Vilanova, S., & Martín, S. (2013). Relaciones entre las concepciones epistemológicas de los docentes y la enseñanza de las ciencias en el nivel universitario. In C. C. Silva & M. E. B. Prestes (Eds.), *Aprendendo ciência e sobre sua natureza: abordagens históricas e filosóficas* (pp. 277–295). São Carlos: Tipographia.

García, M., Mateos, M., & Vilanova S. (2017). ¿Qué concepciones sobre el conocimiento científico tienen los docentes universitarios de ciencias? Diseño, validación y aplicación de un cuestionario de dilemas para evaluar concepciones implícitas. *Revista Docencia Universitaria,* CEDEDUIS, ISSN: 2145–8537, Colombia. Accepted, in press.

Grosslight, L., Unger, C., Jay, E., & Smith, C. L. (1991). Understanding models and their use in science: Conceptions of middle and high school students and experts. *Journal of Research in Science Teaching, 28*(9), 799–822.

Hammer, D. (1994). Epistemological beliefs in introductory physics. *Cognition and Instruction, 12*(2), 151–183.

Hessen, J. (1973). *Teoría del conocimiento.* Madrid: Espasa-Calpe.

Hodson, D. (1994). In search of a meaningful relationship: An exploration of some issues relating to integration in science and science education. *International Journal of Science Education, 14*, 541–562.

Hofer, B. (2006). Domain specificity of personal epistemology: Resolved questions, persistent issues, new models. *International Journal of Educational Research, 45*(1–2), 85–95.

Hofer, B. K., & Pintrich, P. R. (1997). The development of epistemological theories: Beliefs about knowledge and knowing and their relation to learning. *Review of Educational Research, 67*(1), 88–140.

Hofer, B. K., & Pintrich, P. R. (2001). *Personal epistemology: The psychology of belief about knowledge and knowing.* London: LEA.

Koulaidis, V., & Ogborn, J. (1989). Philosophy of science: An empirical study of teachers'views. *International Journal of Science Education, 11*(2), 173–184.

Koulaidis, V., & Ogborn, J. (1995). Science teachers' philosophical assumptions: How well do we understand them*? International Journal of Science Education, 17*(3), 273–283.

Lederman, N. G., & Abell, S. K. (Eds.). (2014). *Handbook of research on science education* (Vol. II). New York: Routledge.

Limón, M. (2006). The domain generality specificity of epistemological belief: A theoretical problem, a methodological problem or both? *International Journal of Educational Research, 45*(1–2), 7–27.

Manassero, M. A., & Alonso, Á. V. (2000). Creencias del profesorado sobre la naturaleza de la ciencia. *Revista Interuniversitaria de Formación del Profesorado, 37*, 187–208.

McGinnis, R., Greber, A., & Watanabe, T. (1997). Development on an instrument to measure teachers candidates' attitudes and beliefs about the nature of and the teaching of mathematics and science. In *Annual Meeting of the National Association for Research in Science Teaching*, January 9th–12th, Cincinnati, OH, USA.

Mortimer, E. (2001). Perfil Conceptual: formas de pensar y hablar en las clases de ciencias. *Infancia y Aprendizaje, 24*(4), 475–490.

Pomeroy, D. (1993). Implications of teachers' beliefs about the nature of science: Comparison of the beliefs of scientists, secondary science teachers, and elementary teachers. *Science Education, 77*(3), 261–278.

Raúl, P. A. (1994). Las concepciones epistemológicas de los profesores: El caso de los estudiantes de magisterio. *Investigación en la Escuela, 22*, 67–84.

Raúl, P. A., García, A. R., & Del Pozo, R. M. (1998). Conocimiento profesional y epistemología de los profesores II: Estudios empíricos y conclusiones. *Enseñanza de las Ciencias, 16*(2), 271–288.

Pecharromán, I., & Pozo, J. I. (2006). ¿Cómo sé que es verdad?: epistemologías intuitivas de los estudiantes sobre el conocimiento científico. *Investigações em Ensino de Ciências, 11*(2), 153–187.

Pozo, J. I., & Scheuer, N. (2000). Las concepciones sobre el aprendizaje como teorías implícitas. In J. I. Pozo & C. Monereo (Eds.), *El aprendizaje estratégico: Enseñar a aprender desde el currículo.* Madrid: Santillana.

Ravanal, E., Camacho, J., Escobar, L., & Jara, Y. N. (2014). ¿Qué dicen los profesores universitarios de ciencias sobre el contenido, metodología y evaluación? Análisis desde la acción educativa. *REDU Revista de Docencia Universitaria, 12*(1), 307–335.

Sandoval, W. A. (2005). Understanding students' practical epistemologies and their influence on learning through inquiry. *Science Education, 89*(4), 634–656.
Sander, P., Stevenson, K., King, M., & Coates, D. (2000). University students' expectations of teaching. *Studies in Higher Education, 25*, 309–323.
Schauble, L., Klopfer, L., & Raghavan, K. (1991). Student's transition from an engineering model to a science model of experimentation. *Journal of Research of Science Teaching, 28*(9), 859–882.
Schwab, J. J. (1964). Structure of the disciplines: Meanings and significances. In G. W. Ford & L. Pugno (Eds.), *The structure of knowledge and the curriculum* (pp. 1–31). New York: Rand McNally.
Schommer, M. (1990). Effects of beliefs about the nature of knowledge on comprehension. *Journal of Educational Psychology, 82*(3), 498–504.
Schraw, G. (2010). Teachers' epistemological stances and citizenship education journal of peace. *Education and Social Justice, 4*(1), 78–107.
Schraw, G., & Olafson, L. J. (2002). Teachers' epistemological world views and educational practices. *Issues in Education, 8*, 99–148.
Vosniadou, S. (Ed.). (2013). *International handbook of research on conceptual change* (2nd ed.). New York: Routledge.

Chapter 7
Is the Electron Real? Who Discovered the Expanding Universe? Debating Nonconsensus Topics of Nature of Science in Science Classrooms

André Noronha, Alexandre Bagdonas, and Ivã Gurgel

7.1 Considerations on Nature of Science Approach Proposals

A long and particular pathway, that happened in recent years, more inside academic discussions than outside, led to a consensus among science education researchers that an adequate scientific education needs not only conceptual teaching, but also teaching *about* science. The nature of science (NOS) comprises metascientific general questions (historical, sociological, and especially philosophical) about what science is (Lederman 1992). In recent decades, there has been quite a large predominance of studies that employ the history of science (HOS) to discuss topics of NOS in the classroom. Even though it is recognized that there is much disagreement and many controversies among historians, sociologists, and philosophers of science about NOS, a general pedagogical "consensus view" has been constructed as the most relevant aspects of these discussions for basic education. Studies about students' and teachers' conceptions about NOS that adopted this "consensus view" usually concluded that it is important to change science classes to transform and improve their NOS conceptions. Despite the many caveats concerning the validity and reliability of research methods, most of the research regarding the assessment of students' and teachers' conceptions about NOS found mostly confused, naive or incomplete views about science (Hodson 2014).

A. Noronha (✉)
Federal Institute of Education, Science and Technology of São Paulo, São Paulo, SP, Brazil
e-mail: noronha@ifsp.edu.br

A. Bagdonas
Federal University of Lavras, Lavras, MG, Brazil
e-mail: alexandre.bagdonas@dex.ufla.br

I. Gurgel
University of São Paulo, São Paulo, SP, Brazil
e-mail: gurgel@if.usp.br

M. E. B. Prestes, C. C. Silva (eds.), *Teaching Science with Context*,
Science: Philosophy, History and Education,
https://doi.org/10.1007/978-3-319-74036-2_7

The importance attributed to history and philosophy of dcience (HPS) and NOS by science education researchers did not prevent criticisms of the attempts to bring this discussion to basic education in the last decades. Recently, many authors have criticized the NOS consensus view, for philosophical, historical, and pedagogical reasons. During the last few decades, philosophers interested in science education argued against an essentialist view of NOS, according to which would be "one essence of nature or set of criteria that describes all and only the activities and inquiries that count as science" (Eflin et al. 1999, p. 108). Instead, it would be better to treat science as a "family resemblance" concept, an idea developed by the philosopher Ludwig Wittgenstein (1889–1951), as shown here:

> "Science" is not a sharply circumscribed concept, but denotes rather a series of paradigmatic examples and – importantly – a rider such as "and other closely similar activities." Any science uses methods and has values belonging to this series. However, no science need exhibit all of them. (Eflin et al. 1999, p. 108)

Irzik and Nola developed this argument, criticizing explicitly the "consensus view" approach to NOS "because it portrays a too narrow and monolithic picture of science, blind to the differences among scientific disciplines" (Irzik and Nola 2011, p. 592), proposing the family resemblance approach as a better alternative to NOS teaching.

Michael Matthews recommended to researchers in science education "a change of terminology and research focus from the essentialist and epistemologically focused 'Nature of Science' (NOS) to a more relaxed, contextual and heterogeneous 'Features of Science' (FOS)" (Matthews 2012, p. 4). Instead of teaching NOS tenets in a dogmatic and acritical way, proposing them as phrases to be memorized and evaluated in examinations, he argued that it would be more interesting to discuss, investigate, and formulate features of science, including in these discussions controversial aspects of science regarding studies published by philosophers, historians, and sociologists of science. In fact, since the 1990s, Mathews and other science educators have been publishing studies arguing controversial aspects of science. Many of them became special issues of the journal *Science & Education*, such as "Values and Socio-scientific Issues," "Mathematisation," "Technology Explanation," "Worldviews and Religion," "Theory Choice and Rationality," "Feminism, Realism, and Constructivism," and others (Matthews 1994, 2012, 2015).

Douglas Allchin also criticized NOS teaching by the presentation of abstract principles as new phrases to be memorized. Like Eflin and collaborators (1999), and Irzik and Nola (2011), Allchin considered that "the very phrase 'nature of science' (especially with its reference to "nature") tends to connote some inherent, universal essence" (Allchin 2011). He argued that there is a lot to unlearn about NOS (Allchin 2013, p. x). Although he recognizes the utility of this concept in some situations, he emphasizes its problems, as ideas about science need concrete contexts and examples. This involves highlighting social aspects of science, such as science funding, aims, inherent cognitive processes, peer review, frauds, and new method validation. One way to do it would be to allow students to evaluate evidence by themselves, making judgments as scientists do. After the predominance of constructivism among science education researches in the 1990s, several authors interested in contributions of history, philosophy, and sociology of science presented critical evaluations of social constructivism both at the Science Studies and Science Education research communities (Mathews 1998). Some authors, such as Adúriz-Bravo have highlighted the "urgent need to recover temperate versions of realism and rationalism, which are more compatible than relativist philosophies with the aims of a liberal science education for all" (Adúriz-Bravo 2004, p. 717).

In a literature review of research into NOS, Bagdonas et al. (2015a) concluded that most of the articles published in Ibero-American journals showed a strong tendency to adopt passively and without criticisms the NOS "consensus view" proposed by the Lederman, McCommas, and others. This is a sign, on one hand, of the strength of the consensus on NOS among researchers. However, on the other hand, there is also a risk that these NOS tenets become accepted dogmatically. Some authors who highlighted controversial discussions regarding NOS in Ibero-American publications were Angel Vásquez, Maria Manassero, and Acevedo Diaz. They published several papers (see, for example, Vázquez Alonso et al. 2007) in which they debated disagreements with the Lederman group regarding the role of sociological studies on NOS. Although, Lederman and collaborators restrict the use of the term NOS to *epistemological* reflections on scientific knowledge, excluding from this concept considerations regarding what scientists *do,* or what has been called *scientific inquiry* (Abd-el-Khalick 2012, p. 367), other authors (such as Allchin 2013; Hodson 2014) have used NOS as a broader concept, which includes scientific inquiry and social influences on science. On the other hand, as an answer to recent criticisms of social constructivism as a dangerous influence for science education researchers, Vázquez and Manassero (1999) argued that moderate views about the social influences of science would be more adequate to show NOS as a historical and social dialectic construct that emerges from many tensions. Some of these tensions, such as those regarding realism, constructivism, and world views, were also debated in teacher training courses by Bagdonas and Silva (2010, 2013).

Martins (2015) also argued that even though the NOS tenets, which are simple and categorical statements, may be interesting for teachers as they may contribute to more concrete and directed discussions about NOS in science classes, they may also cause much confusion, such as the tenets' poor interpretations caused by the lack of context and deepening of the discussion. According to the author, Lederman's NOS tenets approach can lead to an exacerbated form of relativism, considering the misinterpretation of its statements. Thus, Martins proposes guiding principles (organized by three axes, sociological, historical, and epistemological) that may change the form and the content of the work perspective on NOS.

Rozentalski et al. (2012) have also argued that instead of avoiding controversies, science education could provide a richer view of the features of science by presenting a plurality of views about nonconsensual issues that are important for teachers who want to contribute to developing students' intellectual autonomy and critical and reflective thinking. The diversity of philosophical positions that would possibly be manifested in discussions on controversial NOS issues is extremely interesting to the teaching of physics, for at least two reasons: it reflects, in its own way, the diversity of the philosophical positions of scholars of science and scientists, and allows the reflective and undoctrinaire education about what scientific knowledge and enterprise can be.

Bagdonas et al. (2014) argued that even though the proposal of the NOS consensus view and its tenets may allow clearer directions to debates and evaluation in science classes, according to curriculum theories this would be one more example of missing an opportunity for critical reflection, choosing instead just the apprehension of a group of ideas as a product and not a process. Noronha (2014) incorporated the criticism of Irzik and Nola (2011) of the NOS tenets approach, suggesting, however, a reinterpretation of their thesis on the family resemblance approach. Noronha and Gurgel (2012) points out more general problems associated with the consensus view of NOS approaches by tenets. The presence of NOS tenets in national assessments and evaluation standards may contribute to a new victim of the "teaching to test" phenomena in schools. Also, the authors argue that it feeds political ideologies that are responsible for the process of commodification of education worldwide. In this direction, Noronha and Gurgel (2015) propose a political–curricular argument for a nonconsensus view of NOS as relevant approach against consensus ideologies regarding both the society and the nature of scientific knowledge and endeavor.

Despite all these recent critiques, the possibilities of elaborating didactic proposals of nonconsensus topics on NOS are still in the first stages. In particular, how to assess student learning on NOS is a question that raises controversies. Defenders of the consensus view of NOS approaches argue that to work with tenets is a didactical exigency (Lederman 2007). We show that this exigency can be relaxed, and that it is possible to work in science classrooms with controversial NOS subjects, dismissing any tenet approach or kernel list of features of NOS. Mainly, this work is aimed at presenting two different examples of proposals that sought to address nonconsensus topics of NOS in science classrooms. We show how pedagogical practices based on new educational perspectives allow the teaching and evaluation of controversial issues of NOS.

7.2 COSMIC: An Educational Game on Cosmology

We present a brief analysis of an educational game designed to teach cosmology as an example of how controversies about NOS might be discussed in basic education.[1] Basically, in this game, the students and the teacher play roles as members of an agency for science funding, with a focus on cosmology. The students are invited to carry out research about HOS in the period 1914–1931, and to create arguments

[1] A complete analysis of the game can be found, in Portuguese, in Bagdonas (2015). A description of the game, and summary of students' conceptions of science investigated during the game will be presented in English in Bagdonas et al. (2017). A previous version of this text, with a focus on the risk for fostering relativism, was also presented at the IHPST conference in 2015 (Bagdonas et al. 2015b).

about which countries and scientists should receive funding. This allowed investigations about the value attributed by students to science, including not only descriptions of its history, but also debating prescriptive questions about how science should or could be. We aim to problematize the neutrality of science and social determinism, showing science as a human construction influenced by its social and historical context.

The game COSMIC was developed collectively by science educators at the Institute of Physics of the University of São Paulo. All of them were physics high school teachers and graduate students, doing research into physics education. In this game (part of a teaching learning sequence), the general objectives are to problematize the vision of a neutral scientist, free of ideological influences, and a naive relativism, the vision of scientists completely dominated by ideological influences. It is considered important for the student to understand science development as a human production influenced by its social and historical contexts, but not determined by these contexts.

The teaching sequence includes an introduction to the concepts of cosmology and their development, and three main phases of the game, which involves the search for information related to science and cosmology and its appreciation and debate. The main activities of the three phases of the game (detailed in Bagdonas et al. 2017) are the investigation of relevant historical events in the period 1914–1939, which includes the history of cosmology and its cultural, political, and social context.

In the game, the teacher and students play roles: the teacher presents himself as the president of a science funding agency focused on studies about cosmology. The students are divided into groups, and all of them except one (the Jury), act like investigators that work for this agency. The investigators do research about the recent developments on cosmology "traveling" around the world, in a fictional scenario that allows them to choose countries where they will receive information on the form of cards.[2] After studying the cards and discussing them within each group, they should formulate arguments and vote for a country that should receive funding in the round of the game.

While the investigators study and formulate arguments, the Jury must evaluate the arguments and their decision makes one of the groups of investigators the winner of the round. The formulation of arguments is constantly tied to the tensions between the value of internal aspects of scientific community (were the models well formulated? Are there testable predictions? Does astronomical evidence support these models?) and the social context for the production of these theories (are the economic, social, and political conditions in this country favorable to the development of science and cosmology?). It is evident that all these questions are highly controversial. Therefore, we have not adopted the so-called "consensual view of

[2] It is important to highlight that this game is inspired by HOS, but it includes several fictional elements that were introduced by the group to create a playful scenario. For example, in real funding agencies, the scientists apply for funding and there are no world travels.

NOS" and have investigated the possibility of debating controversial issues about science when using the game in high school physics classes.

The game has both a regional division and a historical evolution, in which each round simulates one time period (first round 1914–1924; second round 1925–1931, and third round 1932–1939). Each phase of the game is divided into three moments, which resemble the "three pedagogical moments" proposed by Delizoicov in light of the pedagogy of Paulo Freire.[3] They are:

1. Investigation (search for information) and choice of a country to be funded
2. Organization: summary of main events of the period (presented by the teacher)
3. Application: the choice of the "Cosmology Award" for a scientist.

In the Table 7.1, we present a summary of the activities proposed, such as the main issues that were debated in each pedagogical moment.

In the investigation moment, the teacher is mainly an organizer of students' research, so that they can choose the issues that were seen to be more interesting. They are asked to vote on a country, because this involves less scientific knowledge than when voting for a scientist. In the organization moment, the teacher seeks to dialogue with students' interests stated in the previous moment, but also has the opportunity to present some issues that were considered important, but did not receive much attention from the students. In this moment, the most complicated scientific concepts are explained, while always trying to relate them to the students' interests. Finally, in the application moment, students are supposed to use what they learned in the previous moments to vote for a scientist.

The scenario of the game starts with the first Great War (1914). The teacher presents himself as a representative of the division of Agriculture and Natural Sciences

Table 7.1 Overview of the structure of the game COSMIC

Investigation	Organization	Application
Activity 3	Activity 4	Activity 5
1st phase	Static and expanding models	Einstein x Friedmann
(1914–1924)	Boycott on German Science	
Activity 6	Activities 7 and 8	Activity 9
2nd phase	Redshift – distance relation, priority disputes	Lemaître x Hubble
(1925–1931)	De Sitter effect	
	1929 economic crisis	
Activity 11	Activities 11 and 12	Activity 13
3rd phase	Tired light model	Global evaluation of the game
(1932–1939)	Soviet cosmology	
	Expansion of the universe	

[3]The three pedagogical moments, inspired by the works of Paulo Freire, were presented as investigated by Delizoicov et al. (2002) since the 1980s. For an analysis of the Freirean perspective of science education in English, see Santos (2009).

of the Cosmology Foundation. He tells the students that they were invited to become investigators of a new subdivision: studies of the universe (cosmology). At this time, there were no professional cosmologists, as most scientists who studied this area were astronomers, mathematicians or physicists.

In the first activity, the groups should vote on a *country* to receive financial help from the Foundation. This leads to a more contextual debate, with focus on political and economic influences on science. In the third activity, the groups should vote on a *scientist*, which leads the debate from context to content.

The game allowed the students to discuss two types of controversies: the scientific controversies, regarding disputes among different theories about the origin and development of the universe, and controversies regarding NOS, including descriptive statements about how science *is* and prescriptive statements about how science *should* be.

In the first type of controversy, some examples were the discussions among supporters of static universes (such as Einstein and De Sitter) and expanding universes (such as Friedmann and Lemaître), or regarding different interpretations of astronomical data, especially about the redshift-distance law and the expansion of the universe. The doubts of Edwin Hubble regarding the expansion of the universe were central to the discussions, remaining skeptical regarding cosmological interpretations of his studies on redshifts and measurements of nebula distances.[4] In the second type of controversy, we find discussions on values of the scientific community that were debated by the students while playing the role of science funding investigators. The fictional aspects of the game allowed them not only to learn about descriptive aspects of how the cosmological theories were constructed, by studying historical sources such as scientists' papers, correspondence and debates, but they were also encouraged to think about prescriptive NOS statements, imagining how science *should* be.

Some interesting examples were the discussion regarding the *role of errors* in science, debating Einstein's initial rejection of Friedmann's paper submitted to *Zeitschrift fur Physik* (Physics Magazine). Einstein, who had previously published a paper in which he argued that the universe is static, believed that Friedmann would have presented a new model of an expanding universe based on a mathematical mistake. However, Friedmann was able to convince him, later, that this proposal was correct. Einstein accepted it, but only as a mathematical possibility, and kept his preference for a static universe as a model for the real universe (Tropp et al. 1993; Bagdonas 2015, pp. 97–99).

This game was played for 2 months by high school students of a State School in São Paulo, Brazil, in 2013. On the one hand, Einstein's static model for the universe was criticized by students who supported the expansion of the universe; on the other, supporters of this static model considered Einstein's mistake to be a normal and acceptable part of the process of creation of theories, as it would be important for scientific progress for new theories to be evaluated cautiously. Einstein had behaved adequately because he accepted later that he was wrong.

---

[4] Despite being often presented as the "discover of the expansion of the universe," Hubble remained skeptical about the interpretation of redshifts as radial velocities and showed sympathy for alternative approaches to explaining them, such as Zwicky's tired light model (Assis et al. 2009).

Other values that were present in most students' arguments were typical of the Mertonian "ethos of science": universalism, communism of intellectual property, disinterestedness, organized skepticism, humility, and others (Merton 1973). Some examples among high school students' arguments that appeared in the game were:

- Priority, debated regrading who would be the first scientist to receive credit for Hubble's law[5]
- Fecundity, in the case of Einstein and the De Sitter static models, that despite being considered wrong nowadays by most contemporary cosmologists, were important for the discussions and developments that led to the emergence of relativistic expanding universe models
- Impartiality, which appeared in a wrong argument, was caused by some students' confusion while reading the historical sources presented by us. They believed that Arthur Eddington would have hidden Lemaître's paper published in French in 1927 on purpose, and waited for 4 years before he started promoting his contribution and convincing himself that the universe was expanding. In fact, Eddington argued that he just had not paid attention to Lemaître's paper, and not only himself, but almost no-one read Lemaître's paper before 1930. The teacher noted that the students interpreted the historical data with an accusation with no evidence to support it. Therefore, after criticizing the argument, he valued their participation in the game, showing that this "mistake" was an interesting opportunity to show students different views on NOS.

Other controversial questions debated were: should we value scientists such as Einstein, who admitted that he was wrong, or Hubble, who was successfully engaged in promoting his own scientific contributions as being more important than his rival's? Should science be influenced by external aspects, such as religion, politics, and culture? Is it acceptable that scientists' theoretical commitments and beliefs influence their theory choices? Was Lemaître's primeval atom model promoted by him for his own religious beliefs? Despite the complexity of all these concepts and discussions, not only because they require comprehension of difficult scientific concepts, but also of elements of history, philosophy, and sociology of science, we believe that this first presentation on this debate contributed to the development of more critical and less naive NOS conceptions after the course.

7.3 Physics Undergraduate Students' Views on the Scientific Realism Debate

Our main purpose now is to briefly show our analysis of another educational experience dealing with a controversial and nonconsensual topic of NOS: the scientific realism and anti-realism debate. Philosophical investigations on the possibility and justification of knowledge is not a novelty. Specifically in the context of scientific knowledge, the scientific realism and anti-realism debate is connected to almost all

[5] Some of the "candidates" were Lemaître, Silberstein, Lundmark, and De Sitter (Kragh and Smith 2003; Bagdonas 2015).

other debates within the philosophy of science, as it touches on core issues with regard to the nature of scientific knowledge (Chakravartty 2013). Looking back, it is not difficult to identify examples of realist and antirealist interpretations of scientific theories and models that were relevant in several episodes in the HOS (Niiniluoto 1999). This debate, is comprehensible not only to philosophers (and some scientists), as it may appear, but also manifests itself in peculiar forms in other fields, especially in education, either at a research level or in the classrooms. For instance, students often stand before a teacher's discourse about scientific entities and processes that cannot be observed directly in their day-to-day (Matthews 1994) – they are facing the scientific realism and anti-realism debate alone, without any further or previous discussion.

In this section, we briefly present the structure, analysis, and some of the results of an applied survey with students of the Physics Institute of the University of São Paulo, carried out in the first half of 2013. Its objective was to identify the philosophical tendencies of the students in the scientific realism and anti-realism debate. The research had three stages:

1. Application of a questionnaire
2. Analysis of students' papers
3. A focus group.

The target audience comprised physics undergraduate students who were taking the lectures course "Evolution of Physics Concepts".[6] The survey was conducted over the entire course semester (about 4 months). Twenty-one students were engaged in this survey. In this chapter, we present our analysis of only the first stage, which was conducted by means of 14 questions, one question per week.

The questions used five Likert-scale answers. Seven were related to Realism or Anti-realism of Scientific Entities, and seven were related to Realism or Anti-realism of Scientific Theories. Each of the questions followed a basic structure that remained unchanged throughout the course. First, they were contextualized by mentioning historical facts, general knowledge, and often quotes or excerpts from scientists or philosophers. Subsequently, each question had a categorical statement that would require the students' answers. The questions were deliberately constructed in such a way that their statements were explicitly biased to a form of realism or anti-realism (of entities or of theories).

Concerning the analysis of data of the first stage, we began by organizing the qualitative answers for each question of each student: *rE* (for answers related to Realism or Anti-realism of Entities) and *rT* (for answers related to Realism or Anti-realism of Theories). Thus, for each student we had a data set of 14 answers (seven *rE* and seven *rT*). To make a graphic representation of students' answers, all Likert-scale answers were transformed into quantified values.[7] With the quantified data

[6]The course covers different topics of History of Physics, emphasizing nonconsensus issues of NOS.

[7]In our previous work (Noronha 2014), "strongly agree" was considered to be ± 1, "partially agree" ± 0.5, "suspension of judgment" 0, "partially disagree" ∓ 0.5, and "strongly disagree" as ∓ 1. The plus or minus signs vary depending on the bias of each question. For questions whose statement

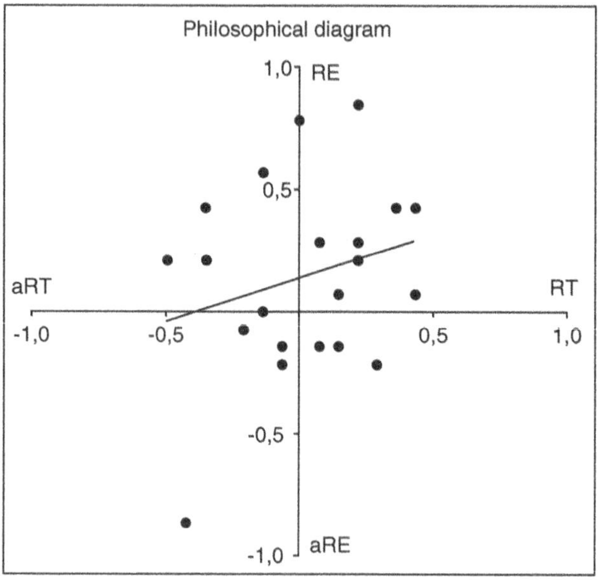

Fig. 7.1 Philosophical trends of all 21 students who engaged fully in the questionnaire survey

sets, we defined the variables *rET* = "Realism or Anti-realism of Entities trend" (given numerically by the average of *rE*), and *rTT* = "Realism or Anti-realism of Theories" trend (given numerically by the average of *rT*) for each student. Hypothetically, *rET* and *rTT* values corresponded in some way to the student's trend to (dis) belief (in the specific context of the survey) about the objective and independent existence of unobservable scientific entities and processes (Realism or Anti-realism of Entities), also about the possibility of knowledge of reality through scientific theories (Realism or Anti-realism of Theories). To be able to "plot" these trends graphically, we associate the *rTT* and *rET* values of each student with ordered pairs: i.e., like a "two-vector" (*rTT*, *rET*). Then, we use a two-dimensional diagram with axes "Realism of Theories" (*RT*) × "Anti-realism of Theories" (*aRT*), and "Realism of Entities" (*RE*) × "Anti-realism of Entities" (*aRE*), perpendicular with each other. The diagram with the trends of the 21 students who answered all 14 questions of the questionnaire is shown in Fig. (7.1).

was biased to some form of realism, the "signature" used was (+, +, 0, −, −). And for questions whose statement was biased to some form of anti-realism, the "signature" used was (−, −, 0, +, +). For instance, if a student answered "strongly disagree" for a question whose statement was biased to Anti-realism of Entities, it would be considered that $rE = 1$. In the case of "partially disagree," $rE = 0.5$. For "strongly agree," $rE = -1$, and in the case of "partially agree," $rE = -0.5$. However, if the statement was biased to Entities Realism, the numeric values would be reversed by the plus or minus sign. We are aware of the problems that come from this quantification process of the responses. In our previous work (ibid.), however, this was not problematic because this research stage was not central owing to all the other research stages involving focal groups, one-to-one interviews, and analysis of students' papers.

It might be expected that a considerable number of students' trends would stay close to the main diagonal of the diagram. On the one hand, this is reasonable as there is a very simplistic view of the scientific realism and anti-realism debate, which "forbids" trends in the second and fourth quadrants with mixed realism and anti-realism postures (Cobern and Loving 2007). To investigate a possible correlation between realism and anti-realism trends, a simple descriptive statistical analysis was performed on quantified data collected by the questionnaire answers of students. We considered only the data sets of those 21 students who answered all 14 questions during the lecture semester.

A working hypothesis is that there is a correlation between the two realism trends and between the two anti-realism trends (i.e., the hypothesis that a realist of entities is also a realist of theories, and that an anti-realist of entities is also an anti-realist of theories). It can be evaluated by calculating the Pearson correlation coefficient ρ for the values of rTT and rET of all data sets.[8] Pearson's coefficient was found to be $\rho \approx 0.25$. In short, the value found for the Pearson correlation coefficient suggests that there was not necessarily a correlation between the Realism and Anti-realism trends, in the context of the survey. Also, the slope coefficient of linear regression strengthens this conclusion. We understand that this all means that the scientific realism and anti-realism debate cannot be simplified to a dispute of "pure realism" against "pure anti-realism" – in fact a significant number of students' trends were located in the quarters of mixed realism and anti-realism.[9]

The plurality of students' philosophical trends that we found was interpreted as reflecting the rich diversity of possible philosophical conceptions. In our previous work, which contains the endeavors of all three research stages (Noronha 2014), we showed how a controversial aspect of NOS can be as rich and interesting from an educational point of view compared with the consensus view approaches. The non-consensus feature of the scientific realism and anti-realism debate was reflected in the heterogeneity of the trends – and this from our point of view should be understood as something positive for science education, the philosophical plurality. From this perspective, there is no "proper" or "inadequate" trend, posture or view.[10] Engaged discussions between the students on the subject were far from any rhetoric of philosophical conclusions, or a maieutic toward a consensus answer. Disagreement between visions, something that is natural in philosophy as well in science, rather urged students to discover new critical reasoning, arguments, and to rethink their own philosophical trends or beliefs.

---

[8] It is an important fact that the correlation itself does not imply that the trends are concentrated around the mean diagonal of the diagram. Thus, beyond ρ calculation, it is interesting to determine the slope coefficient. In particular, if the value of the slope coefficient is close to 1, it means that the trends concentrate around the main diagonal. However, this conclusion is not always valid. Using the least squares method, we found $y = 0{,}35x + 0{,}14$.

[9] First, it is important to pay attention to the fact that any inference of correlation via Pearson's coefficient is a method that works best for large data sets, which is not the case for this survey. Thus, naturally our conclusions are partial and should not be generalized – on the other hand, this does not take away its relevance.

[10] Naturally, one may argue that it could be *insufficiently based* trends. This in fact is an important question to explore, but it goes beyond the scope of this work.

7.4 Final Remarks

Nowadays, it is a consensus that it is important not only to aim for the teaching of science concepts, but also its features, or its "nature" – the NOS. As discussed in the introduction, many surveys in the last 20 years have shown that students have naive conceptions about science. This strengthened some science education researchers' appeal that the goal of NOS is to make students acquire views of science closer to philosophical proposals developed over the past decades. However, authors critical of the so-called NOS consensus view approach claim that it can easily fall back into an education scheme where students simply memorize statements about science that are allegedly appropriate. This "scheme" may feed a dogmatic form of teaching, which we have been trying to fight against for many years (Bagdonas and Silva 2010; Rozentalski et al. 2012; Bagdonas et al. 2014). Instead, we argue that a philosophically biased humanistic education should focus on the development of critical and reflective thinking. To accomplish this, we believe that science education approaches should be based on specific didactic situations and try to seek to analyze its reasons for being. In the case of NOS approaches, we consider it central to contextualize it into historical episodes to prepare the ground for debates about what science is.

The previous examples have sought to show this. First, we presented an educational game in which students are invited to investigate the development of cosmology considering its social and political backgrounds in the context of the early decades of the twentieth century. The students themselves were in charge of judging the work of scientists and were to justify why one should receive support. The conscious decision-making requires the careful consideration of several important features of science, which helps to lead students into an in-depth thinking and critical reasoning with regard to the scientific endeavor. In our second example, we tried to highlight an analysis of a survey developed by a contextualized questionnaire based on a nonconsensus view of NOS, which resulted in a diagram representation of philosophical trends on the scientific realism debate. We considered four possible trends inside the debate and we analyzed how physics undergraduate students argued during half-year lectures on the history of physics. We hope that both examples may contribute to new proposals still to be drawn up by colleagues and science teachers, aiming for more critical and reflective (and controversial, positively speaking) NOS approaches to science education.

References

Abd-El-Khalick, F. (2012). Examining the sources for our understandings about science: Enduring conflations and critical issues in research on nature of science in science education. *International Journal of Science Education, 34*(3), 353–374

Adúriz-Bravo, A. (2004). Methodology and politics: A proposal to teach the structuring ideas of the philosophy of science through the pendulum. *Science & Education, 13*(7–8), 717–731.

Allchin, D. (2011). Evaluating knowledge of the nature of (whole) science. *Science Education, 95*(3), 518–542.

Allchin, D. (2013). *Teaching nature of science: Perspectives & resources*. Saint Paul: SHiPS Education Press.

Assis, A. K. T., Neves, M. C. D., & Soares, D. S. L. (2009). Hubble's cosmology: From a finite expanding universe to a static endless universe. In F. Potter (Ed.), *Proceedings of the conference: 2nd crisis in cosmology conference, ASP Conference Series 2008* (pp. 255–267). San Francisco: Astronomical Society of the Pacific.

Bagdonas, A. (2015). *Controvérsias envolvendo a natureza da ciência em sequências didáticas sobre cosmologia* (Controversies regarding nature of science in teaching and learning sequences about cosmology). Doctoral dissertation in Science Education. Instituto de Física, Instituto de Química, Instituto de Biociências e Faculdade de Educação da Universidade de São Paulo, São Paulo, Brazil.

Bagdonas, A., & Silva, C. C. (2010). Controvérsias sobre a natureza da ciência e a formação de uma visão de mundo científica. In *Atas da 1° Conferência Latino-americana do International History, Philosophy and Science Teaching Group*. Maresias, SP, Brazil.

Bagdonas, A., & Silva, C. C. (2013). Controvérsias sobre a natureza da ciência na educação científica. In C. C. Silva & M. E. B. Prestes (Eds.), *Aprendendo ciência e sobre sua natureza: abordagens históricas e filosóficas* (pp. 209–218). São Carlos: Tipografia.

Bagdonas, A., Gurgel, I., & Zanetic, J. (2014). Controvérsias sobre a natureza da ciência como enfoque curricular para o ensino de física: o ensino de história da cosmologia por meio de um jogo didático. *Revista Brasileira de Histórica da Ciência, 7*(2), 242–260.

Bagdonas, A., Rozentalski, E., & Polati, F. (2015a). Controversial aspects of the construct NOS in the Ibero-American Science Education journals: A literature review. In *Proceedings of the IHPST 13th Biennial International Conference*. Rio de Janeiro, RJ, Brazil.

Bagdonas, A., Gurgel, I., & Zanetic, J. (2015b). The risk of fostering relativism by teaching cosmology through a didactic game that emphasizes the social context of science. In *Proceedings of the IHPST 13th Biennial International Conference*. Rio de Janeiro, RJ, Brazil.

Bagdonas, A., Fabricio, V., Gurgel, I., Noronha, A., & Velasquez, F. (2017). A didactic game about cosmology designed to tackle controversies about science. In M. Pietrocola & I. Gurgel (Eds.), *Crossing the border of traditional science curriculum: Innovative teaching and learning in science basic education*. Boston: Sense Publishers.

Charkravartty, A. (2013). Scientific realism. In E.N. Zalta (Ed.), *The Stanford encyclopedia of philosophy*. https://plato.stanford.edu/archives/win2016/entries/scientific-realism. Accessed 20 July 2016.

Cobern, W., & Loving, C. (2007). An essay for educators: Epistemological realism really is common sense. *Science & Education, 17*, 425–447.

Delizoicov, D., Angotti, J. A., & Pernambuco, M. (2002). *Ensino de Ciências: fundamentos e métodos*. São Paulo: Cortez Editora.

Eflin, J. T., Glennan, S., & Reisch, G. (1999). The nature of science: A perspective from the philosophy of science. *Journal of Research in Science Teaching, 36*(1), 107–116.

Hodson, D. (2014). Nature of science in the science curriculum: Origin, development, implications and shifting emphases. In M. R. Matthews (Ed.), *International handbook of research in history, philosophy and science teaching* (pp. 911–970). Dordrecht: Springer.

Irzik, G., & Nola, R. (2011). A family resemblance approach to the nature of science for science education. *Science & Education, 20*(7–8), 591–607.

Kragh, H., & Smith, R. W. (2003). Who discovered the expanding universe? *History of Science, 41*, 141–162.

Lederman, N. G. (1992). Students' and teachers' conceptions of the nature of science: A review of the research. *Journal of Research in Science Teaching, 29*(4), 331–359.

Lederman, N. (2007). Nature of science: Past, present and future. In S. K. Abell & N. G. Lederman (Eds.). *Handbook of research on science education* (pp. 831–880). Mahwah: Lawrence Erlbaum Associates.

Martins, A. F. (2015). Natureza da ciência no ensino de ciências: uma proposta baseada em "temas" e "questões". *Caderno Brasileiro de Ensino de Física, 32*(3), 703–737.

Matthews, M. R. (1994). *Science teaching: The role of history and philosophy of science*. New York: Routledge.

Matthews, M. R. (Ed.). (1998). *Constructivism in science education: A philosophical examination*. Dordrecht: Springer.
Matthews, M. R. (2012). Changing the focus: From nature of science (NOS) to features of science (FOS). In M. S. Khine (Ed.), *Advances in nature of science research* (pp. 3–26). Dordrecht: Springer.
Matthews, M. R. (2015). Reflections on 25 years of journal editorship. *Science & Education, 24*(5–6), 749–805.
Merton, R. K. (1973). *The sociology of science*. Chicago: University of Chicago Press.
Niiniluoto, I. (1999). *Critical scientific realism*. Oxford: Oxford University Press.
Noronha, A. (2014). *Interpretando a relatividade especial: Discutindo o debate realismo e antirrealismo científicos no ensino de ciências* (Interpreting the special relativity: Discussing the scientific realism debate in science education). Master Thesis in Science Education, Instituto de Física, Instituto de Química, Instituto de Biociências e Faculdade de Educação da Universidade de São Paulo, São Paulo, Brazil.
Noronha, A., & Gurgel, I. (2012). Considerations about the presence of the Nature of Science in official educational documents: A comparison between DCN (Brazil) and NGSS (USA). In *Proceedings of 2nd World Conference on Physics Education*. Universidade de São Paulo, São Paulo, Brazil.
Noronha, A., & Gurgel, I. (2015). On the relevance of a non-consensus view of nature of science in science education: A political-curricular argument. In *Proceedings of 11th Conference of the European Science Education Research Association* (ESERA). Helsinki, Finland.
Rozentalski, E., Bagdonas, A., & Noronha, A. (2012). Realismo e antirrealismo científicos: Pela pluralidade filosófica no ensino de ciências. In *Atas del 2ª Conferência Latinoamericana del International History, Philosophy and Science Teaching Group*. Mendoza, Argentina.
Santos, W. L. (2009). Scientific literacy: A Freirean perspective as a radical view of humanistic science education. *Science Education, 93*(2), 361–382.
Tropp, E. A., Frenkel, V. Y., & Chernin, A. D. (1993). *Alexander A. Friedmann: The man who made the universe expand*. Cambridge, MA: Cambridge University Press.
Vázquez Alonso, Á., & Manassero Mas, M. A. (1999). Características del conocimiento científico. *Enseñanza de las Ciencias, 17*(3), 377–395.
Vázquez Alonso, Á., Manassero Mas, M. A., Diaz, J. A., & Romero, P. A. (2007). Consensos sobre la naturaleza de la ciencia: la comunidad tecnocientífica. *Revista Electrónica de Enseñanza de las Ciencias, 6*(2), 331–363.

Chapter 8
Undergraduate Psychology Students' Conceptions on Scientific Knowledge and Psychology-Specific Epistemological Beliefs

Zuraya Monroy-Nasr, Rigoberto León-Sánchez, Kirareset Barrera-García, and Germán Alvarez-Díaz de León

8.1 Introduction

In August 2008, the academic community began implementing modifications to the curriculum of the License degree in Psychology, at the National Autonomous University of Mexico (UNAM).[1] The curricular changes allowed the inclusion of some new subjects such as History of Psychology and Philosophy of Psychology. The main goal has been to promote the teaching of psychology as a science, based upon comprehension of its history and philosophy. Nonetheless, most of the academics (who taught psychological courses at the licentiate level) argued that History and Philosophy of Psychology were nonpsychological subjects. Despite the opposition of those who saw them as useless courses, we maintained that the understanding of the nature of science (NOS) within a historical and philosophical framework could improve the students' conception of the scientific nature of their discipline. We also view these subjects' potential as integrative forces (Fuchs and Viney 2002) for science teaching in this curriculum.

Pierre Gréco (1972) observed on the scientific nature of psychology, "that is the disgrace of the psychologist: never to be sure of 'making science'. And if he makes it, he is never sure it is psychology" (1972, p. 19). This has been a recurring concern of historians and philosophers of psychology in addition to teachers in these domains.

The American Psychological Association (2007) has curricular guidelines recommending the development of an appreciation for why scientific thinking is necessary. Holmes and Beins (2009) took this into account in their eloquent study "Psychology is a science: At least some students think so." In a large sample of

[1] The previous curriculum was from 1971, with changes added to the programs since then.

Z. Monroy-Nasr (✉) · R. León-Sánchez · K. Barrera-García · G. Alvarez-Díaz de León
Faculty of Psychology, National Autonomous University of Mexico, Mexico, DF, Mexico
e-mail: rigobert@unam.mx; gadl@unam.mx

M. E. B. Prestes, C. C. Silva (eds.), *Teaching Science with Context*,
Science: Philosophy, History and Education,
https://doi.org/10.1007/978-3-319-74036-2_8

psychology majors in a private college in the northeast of the United States, the students reported greater interest in practitioner activities than in scientific ones. Also, their results suggest that the students' scientific interest did not correlate with their scientific literacy at any level. The authors' prediction that students would increasingly come to see psychology as a science as they progressed through the psychology curriculum received no support. Furthermore, there seems to be a pattern showing that as students "take more psychology courses throughout their undergraduate and graduate years, their interests tend to become more polarized and they become interested either in scientific or applied activities – but generally not both" (ibid., p. 9). Gréco's observation finds confirmation in such a polarized pattern.

One important challenge in teaching courses on History of Psychology and Philosophy of Psychology has been to provide the elements that allow our students to understand how and why, before and during the nineteenth century, there was a consensus against the idea of using methods practiced by natural sciences to study psychological phenomena. Nevertheless, a major shift occurred when Wilhelm Wundt launched his research laboratory in Leipzig in 1879. The new experimental psychology readily changed the status of the former philosophy-based discipline. The scientific status of the new discipline has been internally and externally questioned since then, and psychology has developed in the midst of several theoretical controversies. Thus, the new proposal to be implemented is also aimed at articulating knowledge on the nature of science and psychology as a scientific discipline in its diverse traditions of psychological thought. In the previous curriculum of the Faculty of Psychology of UNAM, this was not fully acknowledged and is one of the main transformations intended in 2008.

Our research group, convinced of the relevance of teaching psychology as a science based upon the comprehension of its history and philosophy, thought that the new subjects' contents were important, but that it was also necessary to take into account the students' learning process. They began inquiring about the scientific conceptions and misconceptions of the undergraduate students. One of our goals has been to improve the programs' contents in addition to the teaching strategies to allow the undergraduate students to appreciate the significance of scientific knowledge for psychological applied practices.

A study on epistemological conceptions about science was previously carried out with the first generation of Philosophy of Psychology students who studied under the modified curriculum (February 2010). The results showed that the participants did not identify the scientific characteristics of the main conceptions in the philosophy of science. They did not present a "pure" epistemological conception, but a mixture (Monroy-Nasr et al. 2013a).

According to Schommer-Aikins (2004) students' epistemological beliefs undergo changes in the way that they understand and explain reality. These changes seem to go from a simple and direct "objective comprehension" of reality to a "subjective comprehension" mediated by different theoretical frames of reference. In a third stage, epistemological comprehension would be the result of the coordination of these two dimensions where knowledge is understood to be essentially subjective, although it is dependent on objective evaluations (Kuhn and Weinstock 2002; Leal et al. 2009). Also, it is possible that the change in personal epistemological conceptions might be due to the degree of education of the subjects (Kuhn et al. 2000; Leal 2010): more sophisticated epistemology would correspond to a higher degree of education.

In this research, we subscribed similar evolutionary hypotheses to those held by Kuhn et al. (2000), Leal (2010) and Pecharromán et al. (2009). In this sense, the students of $G1_{Semester}$ (second and fourth semesters) assume "objectivist" conceptions on the nature of psychological knowledge and support the value of this knowledge through the "authority" of who built it: science. On the contrary, students in the $G2_{Semester}$ (sixth and eighth semesters) pay more attention to the context of production of this knowledge, to its "provisional" nature, and coordinates in a better way the objective and subjective aspects of the generation of knowledge. The License degree in Psychology is completed in eight semesters. The curriculum is divided in two main areas: from the first to the fourth semester it is an area of general formation (AFG); from the fifth to the eighth semesters it is an area of professional training (AFP) divided into six main fields: Cognitive and Behavioral Sciences, Psychosocial and Cultural Processes, Psychobiology and Neurosciences, Clinical and Health Psychology, Educational Psychology, and Organizational Psychology.

We present the results of a new study on undergraduate psychology students of different semesters (second, fourth, sixth, and eighth) that aimed to assess their conceptions of scientific knowledge and psychology-specific epistemological beliefs. The research was carried out on 156 undergraduate students of psychology, applying two questionnaires:

1. Psychology-Specific Epistemological Beliefs (Psych-SEBS), designed by McMahan et al. (2015)
2. A modified questionnaire inspired by Raviolo et al. (2010), which explores the students' conceptions of three aspects of the nature of science:
    (a) Scientific knowledge
    (b) Scientific theories
    (c) Scientific models.

Their instrument was built to examine students' conceptions of the nature of knowledge (general knowledge, scientific knowledge, and models). We intend to adapt the questionnaire to the target of our inquiry by focusing mainly on the students' understanding of:

1. The nature of scientific knowledge
2. The concepts of scientific models and explanations
3. The role of social and historical context in scientific production.

8.2 Method

8.2.1 *Participants*

The research was carried out with a nonprobabilistic, intentional sample of 156 undergraduate psychology students (120 women and 36 men, a sex distribution in concordance with the general population attending the Faculty of Psychology). The participants' age ranged from 18 to 31 years (M = 20.75, SD = 2.15). They were organized into two groups according to their academic semester: $G1_{Semester}$ (n = 71) was composed of students in the second and fourth semesters and $G2_{Semester}$ (n = 85) was made up of students in the sixth and eighth semesters. This decision was made based on the structure of the curriculum of the Faculty of Psychology, UNAM. At the time of this study, eight generations had been taught History of Psychology in the first semester and six generations had studied Philosophy of Psychology in the fourth semester.

8.2.2 *Materials*

8.2.2.1 Psychology-Specific Epistemological Beliefs

This questionnaire was designed by McMahan et al. (2015) to evaluate the epistemological beliefs of people about psychology (for instance, the importance of research in psychology, the subjective nature of psychological knowledge and its dependence on the context, in addition to the problem of the predictability of human behavior). An exploratory factor analysis found three factors that were corroborated by a confirmatory factor analysis. However, it is of our opinion that there was no conceptual coherence between some of the items that integrated one factor; thus, another factor analysis was made with the criterion of keeping only those items with weights above 0.300 in Cronbach's alpha test. The items were answered according to a four-point Likert-type scale (1 = strongly disagree, 4 = strongly agree).

A factor analysis, using the main components method and Varimax rotation, showed a fitness index of the sampling KMO = .683, whereas Bartlett's Test of Sphericity showed $X^2(15) = 141.788$, $p < 0.001$. Two factors explained the 56.65% variance (Table 8.1). Factor 1, *social and historical context*, implies the importance for the psychological research of contextual aspects, either individual or social. Factor 2, *explanation and prediction of human behavior*, implies the possibility or impossibility of scientific psychology to explain and predict human behavior.

Table 8.1 Factor structure of the questionnaire Psychology-Specific Epistemological Beliefs (Psych-SEBS)

Item		F1	F2
12	Some psychological knowledge proposed earlier opposes the contemporary knowledge	0.825	
03	Psychologists' research activities are affected by their existing ideas, thoughts, and beliefs	0.681	
13	Currently acceptable psychological knowledge may be changed or totally discarded in the future	0.660	
10	Carefully controlled research is not likely to be useful in solving psychological problems		0.781
05	Cultures around the world are so diverse that explaining and predicting human behavior is impossible for scientific psychology		0.770
11	Our ability as humans to behave in any way we choose makes our attempts to predict behavior ineffective		0.618
Eigenvalue		2.349	1.050
Percent variances		29.10	27.55
Cronbach's alpha		0.601	0.609

8.2.2.2 Questionnaire on Knowledge and Scientific Models (KScienModel-Q)

A modified version of the questionnaire developed by Raviolo et al. (2010) on conceptions of knowledge and scientific models was used. The original questionnaire included 36 statements to be evaluated using a four-point Likert-type scale (from 1 = strongly agree, to 4 = strongly disagree). The items included statements relating to general knowledge, scientific knowledge, and scientific models. This instrument was applied to a sample of 65 students (freshman and tertiary level institute). In the reliability analysis the scale obtained an $\alpha = 0.80$.

In a previous study (Monroy-Nasr et al. 2013b) with 294 undergraduate psychology students, the number of items from the original questionnaire was first reduced from 36 to 32. Four items were eliminated as we considered that they were "precepts" and not a type of knowledge. Removed items include: "Wisdom is not knowing the answers, but knowing how to find them" or "The only certainty is uncertainty itself." Other questions were modified with the purpose of making the question clearer. Thus, the item that said: "Scientific knowledge is artificial or constructed and does not show nature as it really is," was exchanged for the item: "Scientific knowledge is a direct copy of reality." This was carried out based on two criteria: writing concise and specific items and avoiding negative items. The evaluation of the 32 resulting items continued to be pondered according to a four-point Likert-type scale (from strongly agree to strongly disagree).

The 32 items were classified into two epistemological conceptions: naive realism (NR) and scientific realism (SR); a third category was the influence of social context

(SC) on science (Kitcher 1993; Devitt 1997). Before analyzing the results, a reliability analysis of the instrument was performed. In this analysis when 11 items were deleted $\alpha = 0.760$ was achieved. This way, the final instrument included 21 items with factor weights above 0.200. Subsequent factor analysis suggested five factors with a reliability index between 0.752 and 0.481. This was the questionnaire version used in the present study.

Again, the items were evaluated according to a four-point Likert-type scale (from 1 = strongly disagree to 4 = strongly agree). Cronbach's alpha test showed indexes of 0.718 for all 21 items. Given that many of the items were rewritten,[2] a new factor analysis was carried out using the main components method and Varimax rotation. The results showed a fitness index of the sampling KMO = 0.734, while Bartlett's test of sphericity showed $X^2(210) = 879.759$, $p < 0.001$. Seven factors were found, which explained 63.49% of the variance. Nonetheless, and given that two items did not load in any of the factors, although the other factor had only one or two items, these were eliminated. A new analysis showed an index KMO = 0.726, while the test of sphericity showed $X^2(120) = 708.760$, $p < 0.001$. We found that four factors explain 57.30% of the variance (Table 8.2).

Factor 1, *scientific models as true representations of the world*, highlights the role of science as a set of truths. Factor 2, *scientific explanation*, emphasizes the role of explanation and prediction in scientific work. Factor 3, *immutabitiy of scientific knowledge*, implies the assumption that scientific knowledge does not change. Last, Factor 4, *the role of context in scientific production*, involves the function of socio-historic context in the generation of scientific knowledge.

8.2.3 Procedure

One group from the second to the eighth semester was chosen to apply the questionnaire (students in $G2_{Semester}$ belonged to three different fields in psychology: Psychobiology and Neurosciences, Clinical and Health Psychology, and Psychosocial and Cultural Processes). Teachers were asked for authorization to apply the survey at the beginning of their classes. On the designated date the classrooms were visited and the students were told about the purposes of the research. The questionnaires were given to the students that accepted to participate. To control the answers of the students, two different versions of the survey (version A = Psych-SEBS – KSCIENMODEL-Q, version B = KSCIENMODEL-Q – Psych-SEBS) were distributed. The students responded in an average time of 20–25 min. There were no significant differences between the two versions.

[2] We thank Dietmar Höttecke for his useful suggestions that helped to improve the questionnaire.

Table 8.2 Factor structure of the Knowledge and Scientific Models Questionnaire (KSCIENMODEL-Q)

Item		F1	F2	F3	F4
09	An accepted scientific model has the quality of being true	0.734			
21	A scientific model expresses a truth about the world	0.732			
13	Scientific knowledge is always true	0.671			
18	A scientific model corresponds exactly to reality	0.635			
01	Scientific knowledge can exactly represent an entity or phenomenon of the real world	0.632			
04	Unlike what happened in the past, nowadays accepted scientific knowledge is true	0.555			
19	Scientific measurements are reliable and valid	0.421			
16	Scientific knowledge is valid because it is useful to develop explanations and to make predictions		0.872		
17	Scientific knowledge is valid because of its usefulness in explaining and predicting		0.854		
10	A theory is valid because of its power to describe, explain, and predict the analyzed phenomena		0.739		
08	The knowledge expressed by a scientist is irrefutable			0.811	
12	Scientific knowledge never changes			0.799	
02	In science each problem has only one correct solution			0.568	
15	The social environment influences the knowledge content a scientist proposes				0.840
14	All knowledge depends on the context in which it is produced or used				0.774
03	The social environment influences the theories a scientist constructs				0.604
Eigenvalue		3.741	2.498	1.586	1.343
Percent variances		18.34	14.48	12.83	11.69
Cronbach's alpha		0.773	0.786	0.639	0.663

8.2.4 *Results*

8.2.4.1 Psych-SEBS

A *t* test for paired samples found statistically significant differences between **F1** ($M = 2.72$, $SD = 0.599$) and **F2** ($M = 2.14$, $SD = 0.617$), $t(154) = 10.725$, $p < 0.001$, Cohen's $d = 0.953$. The participants are in better agreement with the role of the socio-historic context in psychological research than with the possible prediction of human behavior. On the other hand, a *t* test for paired samples found statistically significant differences, though borderline, between $G1_{Semester}$ ($M = 2.62$, $SD = 0.568$) and $G2_{Semester}$ ($M = 2.81$, $SD = 0.613$), $t(154) = -1.962$, $p = 0.052$, Cohen's $d = -0.321$. That is, to a large extent the participants in G2 tend to ponder more the contextual aspects in the psychological research than the participants in G1. Nevertheless, as can be seen in Table 8.3, over 50% of the participants are in disagreement with the items in factor 2.

8.2.4.2 KScienModel-Q

A one-way ANOVA with repeated measures was carried out to explore whether there were differences among the means of the four factors that comprise the questionnaire. No sphericity was found ($W = 0.797$, $p < 0.05$) and we then report the Greenhouse–Geisser criterion, $F(2.620, 403.441) = 375.228$, $p < 0.001$ (observed power = 1.00; $\eta^2_{partial} = 0.709$).

Table 8.4 shows, with an adjustment of critical levels using Bonferroni correction significant statistical differences among the four factors: **F1** ($M = 2.35$, $SD = 0.463$), **F2** ($M = 3.14$, $SD = 0.492$), **F3** ($M = 1.54$, $SD = 0.474$) y **F4** ($M = 3.08$, $SD = 0.506$).

In the next Table 8.5 we show the means of the two $G_{Semester}$ in each of the factors examined. In F1 the participants in $G1_{Semester}$ score above those of the $G2_{Semester}$. The same is true for F2. Meanwhile, in F4, the participants in $G2_{Semester}$ appraise better the role of the context than the participants in $G1_{Semester}$.

The Pearson test shows that conceiving scientific models or knowledge as an exact representation of reality (F1) has a statistically significant negative correlation when the role of context in the generation of science is better appraised (F4), as seen in Table 8.6.

Table 8.3 Psychometric properties of Psych-SEBS and percentages of agreement–disagreement in each item

Item				Agreement	Disagreement
	A	*M*	*SD*	%	%
12	0.473	2.49	0.791	49.4	50.6
03	0.361	2.74	0.802	66.0	34.0
13	0.397	2.95	0.818	76.9	23.1
10	0.400	2.13	0.843	32.7	67.3
05	0.437	2.28	0.884	38.5	61.5
11	0.423	2.02	0.734	23.7	76.3

Table 8.4 Multiple comparisons of the four factors (with Bonferroni correction) in the subscales that comprise KSCIENMODEL-Q

					CI 95%	
(I)Factor	(J)Factor	Means' differences (I–J)	Standard error	Sig.	LL	UL
F1	F2	−0.787*	0.047	0.000	−0.912	−0.661
	F3	0.813*	0.045	0.000	0.694	0.932
	F4	−0.724*	0.060	0.000	−0.884	−0.565
F2	F3	1.600*	0.056	0.000	1.450	1.750
	F4	0.062	0.055	1.000	−0.084	0.208
F3	F4	−1.538*	0.064	0.000	−1.707	−1.368

Sig. Bonferroni correction for multiple comparisons, *CI* confidence interval, *LL* lower limit, *UL* upper limit

$*p < 0.05$

Table 8.5 Comparison of means of the two groups for each factor in KSCIENMODEL-Q

	$G1_{Semester}$	$G2_{Semester}$			
Factor	*M* (*SD*)	*M* (*SD*)	*t*(154)	*p*	Cohen's *d*
1	2.46 (0.388)	2.26 (0.499)	2.796	0.006	0.447
2	3.25 (0.430)	3.05 (0.522)	2.483	0.014	0.418
3	1.56 (0.434)	1.52 (0.507)	0.495	0.621	0.084
4	2.96 (0.496)	3.18 (0.491)	−2.844	0.005	−0.445

Table 8.6 Correlations among the four factors that comprise KSCIENMODEL-Q

	F1	F2	F3	F4
F1	–	0.249*	0.302*	−0.177**
F2		–	−0.051	0.069
F3			–	−0.302*
F4				–

*The correlation is significant at 0.01 (bilateral) level
**The correlation is significant at 0.05 (bilateral) level

Similarly, there is a negative correlation between this last factor and understanding scientific knowledge as being immutable (F3). There is also a positive correlation between the roles of explanation and prediction (F2) and regarding science as corresponding exactly with reality (F1).

Lastly, an analysis of items (02, 08, 12, 13, and 18) with a mean under 2.00, shows that most of the participants (between 80.2% and 98.0%) *disagree* with those statements that imply an identification of scientific knowledge with an exact description of reality (item 18): "A scientific model corresponds exactly to reality." This also happens with those propositions that state a static view of science, for instance (item 12): "Scientific knowledge never changes". Furthermore, the analysis did not show statistically significant differences in the responses of the two groups for any of these five items. That is, $G1_{Semester}$ and $G2_{Semester}$ express a similar disagreement with these items. On the other hand, in those items with a mean above 3.00 (items 03, 10, 14, 15, 16, and 17), between 85.9% and 93.6% of participants *agreed* with the statements. Namely, the participants agreed with those that emphasize the social context in the generation of scientific knowledge, such as item 3, "The social environment influences the theories a scientist constructs," in addition to those statements that imply the predictive power of scientific theories such as item 10: "A theory is valid because of its power to describe, explain and predict the analyzed phenomena." For these results, it is worth emphasizing that although the participants in $G1_{Semester}$ show their agreement with the items that highlight the predictive value of theories, the participants in $G2_{Semester}$ tend to be in agreement with the items that stress the influence of context in the production of scientific knowledge (Table 8.5).

8.3 Discussion and Conclusions

The results obtained in this research do not support the expected difference between personal epistemologies of the participants in $G1_{Semester}$ and $G2_{Semester}$. That is, the students in the first semesters were expected to express epistemological conceptions in agreement with the statements that identify scientific knowledge with reality or about the immutability of this kind of knowledge. Authors such as Schommer-Aikins (2004) maintain that in terms of the changes the students undergo in the way in which they understand and explain reality, simple and direct "objective comprehension" should move toward a "subjective comprehension," mediated by different reference frames. In this sense, it could be assumed that such a change in personal epistemological beliefs had been propelled by the degree of education (Kuhn et al. 2000; Leal 2010): a higher degree of education would correspond to a more sophisticated epistemology.

Our evidence suggests that this progression might not be so linear. Although the participants in $G1_{Semester}$ tend to value more the functional aspects of scientific theories (explanation and prediction) and those in $G2_{Semester}$ acknowledge more the influence of context in the generation of scientific knowledge; the data suggest that both groups of students might coordinate the objective and subjective aspects of knowledge. As observed, on the one hand, they recognize that contextual aspects influence the generation of psychological scientific knowledge, but, on the other hand, they disagree with the "relativist" approach that denies the possibility of explaining and predicting human behavior, even though they may be situated in different social and cultural contexts. This is corroborated by the result obtained in factor 2 of the KSCIENMODEL-Q. That is, the participants strongly agree that scientific knowledge (or a scientific theory) is valid according to its capacity to predict and explain phenomena. Nevertheless, it does not mean that they assume scientific statements to be an absolute truth. Rather, as can be seen, the mean obtained in factor 3 points to a disagreement with the idea that scientific knowledge is immutable. Both results, in fact, indicate that the participants have an epistemologically and sophisticated coherence, about scientific knowledge in general, and about psychological knowledge specifically.

As several generations have been taught the new subjects (eight in History of Psychology in the first semester and six in Philosophy of Psychology in the fourth semester), we find these results encouraging and we look forward to continuing to evaluate students' conceptions to revise and enrich the contents of the curriculum, in addition to updating teaching strategies. However, our findings should be taken with caution, given that the data obtained in both questionnaires do not allow us to differentiate between the general and specific epistemologies of the participants (Leal 2010; Leal et al. 2009).

Acknowledgements This work has been supported by the research project DGAPA-PAPIIT IN402515/National Autonomous University of Mexico.

References

American Psychological Association. (2007). *APA guidelines for the undergraduate psychology major*. https://doi.org/10.1037/e526232015-001

Devitt, M. (1997). *Realism and truth* (2nd ed.). Princeton: Princeton University Press.

Fuchs, A. H., & Viney, W. A. (2002). The course in the history of psychology: Present status and future concerns. *History of Psychology, 5*(1), 3–15.

Gréco, P. (1972). Epistemología de la psicología. In J. Piaget (Ed.), *Epistemología de las ciencias humanas* (pp. 11–65). Buenos Aires: Proteo.

Holmes, J. D., & Beins, B. C. (2009). Psychology is a science: At least some students think so. *Teaching of Psychology, 36*(1), 5–11. https://doi.org/10.1080/00986280802529350.

Kitcher, P. (1993). *The advancement of science: Science without legend, objectivity without illusions*. Oxford: Oxford University Press.

Kuhn, D., & Weinstock, M. (2002). What is epistemological thinking and why does it matter? In B. K. Hofer & P. R. Pintrich (Eds.), *Personal epistemology. The psychology of beliefs about knowledge and knowing* (pp. 121–144). Mahwah: Lawrence Erlbaum.

Kuhn, D., Cheney, R., & Weinstock, M. (2000). The development of epistemological understanding. *Cognitive Development, 15*, 309–328.

Leal, F. A. (2010). Creencias epistemológicas generales, académicas y disciplinares en relación con el contexto. *Universitas Psychologica, 9*(2), 381–392.

Leal, F. A., Espinoza, C., Iraola, M., & Miranda, M. (2009). El contexto en la epistemología personal: consideraciones teóricas y exploraciones empíricas. *Revista Interamericana de Psicología, 43*(1), 170–180.

McMahan, E. A., Renken, M. D., & Nitkova, M. (2015). Initial validation of an instrument measuring psychology-specific epistemological beliefs. *Teaching of Psychology, 42*(2), 126–136. https://doi.org/10.1177/0098628315569927.

Monroy-Nasr, Z., León-Sánchez, R., de Leon Alvarez-Díaz, G., García, K. B., Flores-Camacho, F., Gallegos-Cázares, L., & Canales, E. C. (2013a). De la filosofía de la ciencia a la filosofía de la psicología: avatares en un cambio curricular. In C. C. Silva & M. E. B. Prestes (Eds.), *Aprendendo ciência e sobre sua natureza: abordagens históricas e filosóficas*. São Carlos: Tipographia.

Monroy-Nasr, Z., León-Sánchez, R., de León, G. Á.-D., & García, K. B. (2013b). An approach to psychology students' conceptions on scientific knowledge, scientific theories and models. In *Conference proceedings 21st*. Pittsburgh: IHPST. https://ihpst.clubexpress.com/content.aspx?page_id=22&club_id=360747&module_id=188194. Accessed 30 June 2016.

Pecharromán, I., Pozo, J. I., Mateos, M., & del Puy Pérez-Echeverría, M. (2009). Psicólogos ante el espejo: las epistemologías intuitivas de los estudiantes de psicología. *Avances en Psicología Latinoamericana, 27*(1), 61–78.

Raviolo, A., Ramírez, P., López, E. A., & Aguilar, A. (2010). Concepciones sobre el conocimiento y los modelos científicos: un estudio preliminar. *Formación Universitaria, 35*(5), 29–36. https://doi.org/10.4067/S0718-50062010000500000.

Schommer-Aikins, M. (2004). Explaining the epistemological belief system: Introducing the embedded systemic model and coordinated research approach. *Educational Psychologist, 39*(19), 19–29. https://doi.org/10.1207/s15326985ep3901_3.

Chapter 9
The "Science" as Portrayed in Documents of the Biological Evolution Versus Intelligent Design Debate

Sílvia Regina Groto and André Ferrer P. Martins

9.1 Introduction

Biological evolution[1] is the scientific explanation currently accepted for the origin of the diversity of life. Its importance to biology and to the teaching of biology is considered undeniable in academic communities that address both areas. Expressions such as "cornerstone," "integrating axis," "central and unifying axis," "central concept," "transversal axis," for example, are used in works that endorse the relevance of biological evolution to biology and/or to its teaching.[2]

On the other hand, the literature on biology education has pointed out several problems related to the teaching of biological evolution. Smith (2010), for example, provides a review of some of the factors that interfere in its teaching and learning. Religious beliefs, particularly those considered most fundamentalist, which have a biblical literalist view, are considered an important factor that would influence both the understanding and the acceptance of evolution.[3]

Accentuating this problem, the organized antievolutionist movement – previously considered a typically United States phenomenon – has been growing in many countries (Numbers 2006). One of the antievolutionist movements that falls within that context is the called Intelligent Design (ID). Historically, ID has its origin in the 1980s in the United States, resulting from several reformulations of the creationist

[1] In this paper, we use the term "biological evolution" to designate the entire scientific theoretical framework that explains, nowadays, the diversity of life on our planet, including the neosynthesis and the extended synthesis (as framed by Jablonka and Lamb 2010).

[2] Bizzo and El-Hani (2009), Meyer and El-Hani (2005), Smith (2010), Silva (2011), Tidon and Vieira (2009).

[3] Almeida (2012), Blancke et al. (2012), Costa et al. (2011), Mazzur (2005), Miller et al. (2006), Oliveira and Bizzo (2011), and Smith (2010).

S. R. Groto (✉) · A. F. P. Martins
Federal University of Rio Grande do Norte, Natal, RN, Brazil

M. E. B. Prestes, C. C. Silva (eds.), *Teaching Science with Context*,
Science: Philosophy, History and Education,
https://doi.org/10.1007/978-3-319-74036-2_9

movement after suffering successive legal defeats in an attempt to be included in that country's science curriculum (Scott 2009; de Souza 2009; Groto 2016). ID followers believe that the so-called "theory" of intelligent design (TID), supported by the group, is an alternative scientific explanation to biological evolution and that it has the structure of a *scientific theory*.

In Brazil, antievolutionist movements – and their ramifications in politics – have also grown over the years. Of particular concern are the recurring attempts to include the teaching of creationism in state schools at municipal, regional, and federal level, through legislative proposals (mostly) presented by groups and politicians linked to the so-called "evangelical bench."[4] ID is one of the antievolutionist movements that has grown in Brazil more recently. It arrived in about 1998, with the formation of the *Brazilian Centre of Intelligent Design* (CBDI),[5] under the coordination of Enézio Eugênio de Oliveira Filho (now president emeritus of TID-Brazil). But it was in 2014, during the *First Brazilian Intelligent Design Congress* (PCBDI), an event in which the *Brazilian Society of Intelligent Design* (TDI-Brasil) was founded, chaired by Marcos Nogueira Eberlin, that the group gained more prominence in the country. The event publicized the *Public Manifesto of the TDI-Brasil on the teaching of evolution and TID in Brazilian state and private schools* (D-ID 1).[6]

The advent of the PCBDI, together with the foundation of the TDI-Brasil, and the presentation of the legislative proposal 8099/2014, at federal level, authored by deputy and minister Marcos Feliciano, which proposed the teaching of creationism in Brazilian state schools, reverberated in some media and incited the current debate among the groups advocating creationism/ID and biological evolution, in Brazil.

Considering this scenario, we conducted a broad piece of research (Groto 2016) that investigated the biological evolution (EVO) versus intelligent design (ID) debate in the light of Ludwik Fleck's (1896–1961) epistemology, aiming to contribute to the teaching of biological evolution (Fleck 2010). The EVO-ID debate was investigated by using *documental analysis* (Holsti 1969; Lüdke and André 1986) and *discursive textual analysis* (Moraes and do Carmo Galiazzi 2011). In this work, we present part of this broader research that describes and analyses elements of the *ideas about science* manifested in the documents investigated.[7] In particular, we aim to answer the following questions: Which ideas about science do the groups involved in the EVO-ID debate express in the documents analyzed? What contribution can the analysis of these ideas about science bring to the teaching of biological evolution? Our discussion has three stages. First, we described the documents reviewed

[4] Congressmen who, regardless of the political parties to which they are linked, grouped around the fact that they are evangelical and defend projects and proposals in common, such as the teaching of creationism.

[5] All acronyms of the Brazilian entities were kept according to their name in Portuguese.

[6] This manifesto is one of the documents examined by us.

[7] The research of Groto (2016) can be found at https://repositorio.ufrn.br/jspui/handle/123456789/21430.

and the methodology used. Second, we present and discuss the ideas about science that are manifest in them. Finally, we bring our conclusions and we reason some of the possible implications that our analysis can offer to the teaching of biological evolution and to the discussions about aspects of nature of science (NOS).[8]

9.2 The Documents: Description and Methodology of Analysis

The documents we analyzed are all in the public domain. They were produced by official entities or individuals linked to the area of biological evolution or intelligent design, who had taken a position in the debate between the two groups (G-EVO and G-ID), in Brazil. For the G-EVO, 11 documents (D-EVO) were analyzed, 6 of them linked to official entities and organized groups that share the evolutionary thought and 5 linked to specific members of these groups (Table 9.1). For the G-ID, 21 documents were analyzed; a significant number of them were lectures given during PCBDI (Table 9.2).

The documents (D-EVO and D-ID) were analyzed using the methodology of discursive textual analysis (DTA) that can be described, in brief,

> [...] as a self-organized process of understanding construction in which new understandings emerge from a recursive sequence of three components: the deconstruction of the texts of the "corpus" – the unitarization; the establishment of relations between unitary elements – the categorization; the capture of the emerging at which the new understanding is communicated and validated. (Moraes and do Carmo Galiazzi 2011, p. 12)

More specifically, the DTA works with two cycles and four foci. The first cycle comprises: focus 1 in which occurs the *disassembly of texts*. It is the *unitarization process* that results in the production of the *units of analysis*; focus 2 involves establishing relations between the units of analysis. It is the *categorization process* that results in the production of *categories (and subcategories) of analysis*; focus 3 involves the capture of the *new emerging*, a process that enables a renewed understanding of the whole, producing a *metatext* as a result of:

> [...] efforts to explain the understanding that presents itself as the product of a new combination of elements built over the previous steps. (Moraes and do Carmo Galiazzi 2011, p. 12)

In the second cycle (focus 4), the *self-organizing process*, the extent of the capture of the new emerging, resulting in final, creative and original understandings,

[8] It is not our intention here to discuss the science practiced (or not) by the two groups. To establish relationships between the ideas about science expressed in the documents we analyzed that effectively practiced by the groups, it would be needed to resort to the theoretical framework that anchors our original research, the epistemology of Ludwik Fleck (2010). This discussion takes place in another publication.

Table 9.1 D-EVO: documents analyzed, produced by G-EVO

D-EVO	Description
1	Manifesto of the Brazilian Society of Genetics on science and creationism (D-EVO 1)
2	Manifesto of the Brazilian Society of Paleontology on the validity of biological evolution and its teaching in schools nationwide (D-EVO 2)
3	News about the repudiation note of the Federal Council of Biology to the Legislative Proposal 8099/2014 (D-EVO 3)
4	Letter of the Brazilian Society for the advancement of science to deputies on the Legislative Proposal 8099/2014 (D-EVO 4)
5	Open letter of the Brazilian Association for Research in Science Education and the Brazilian Association of Biology teaching about the Legislative Proposal 8099/2014 (D-EVO 5)
6	Replica of teachers and Students of the Federal University of Rio Grande do Sul on the Manifesto of the Brazilian Society of Intelligent Design (D-EVO 6)
7	"Intelligent Design: evolutionism or creationism?" – text authored by Mario Alberto Cozzuol[a] published in "Blog of Master Chassot" (D-EVO 7)
8	"The tree of life is not dead, says biologist" – text authored by Diogo Meyer[b] published in the newspaper *O Estado de São Paulo* (D-EVO 8)
9	"Evolution and Religion" – article authored by the geneticist physician Sergio Danilo Pena[c] published in the magazine *Science Today* (D-EVO 9)
10	"Education and Scientific Discourse" – article authored by the biologist Charbel El-Hani[d] published in the newspaper *Folha de São Paulo* (D-EVO 10)
11	"Transforming ignorance in wisdom" – article authored by the biologist Felipe A. P. L. Costa[e] published in the *Press Observatory* (D-EVO 11)

Source: Groto (2016, p. 183)
[a]Biologist with research in paleontology and professor at the Federal University of Minas Gerais
[b]Biologist with research in evolution and professor at the University of São Paulo
[c]Geneticist physician and professor at the Federal University of Minas Gerais
[d]Biologist, educator, professor at the Federal University of Bahia
[e]Biologist and author of popular science book on evolution

with or without the support of a stronger theoretical framework. In this work, we present the results obtained from the use of the first cycle of the DTA, as the second cycle involves the use of Ludwik Fleck's epistemology, which goes beyond the scope and purposes of this chapter. Next, we describe in more detail the procedures adopted in the analysis of the documents during the first cycle of DTA.

In focus 1, the unitarization process, all documents were read or seen and recorded (lectures), aiming at the production of the units of analysis, which can also be designated as *units of meaning*, as we used the *semantic criteria* (Moraes and do Carmo Galiazzi 2011). Such units of analysis have been analyzed to identify ideas/thoughts/arguments linked to G-EVO and G-ID. All units originating in each document have been properly identified by codes (created by us) which refer to the document from which they were originated. The D-EVO 1, for example, generated 15 units of analysis. Focus 2, the categorization process, was carried out in two moments that made possible the creation of categories and their respective subcategories. At first, for each group (G-EVO and G-ID), the entirety of the units of analysis of each document have been printed with their respective identification

Table 9.2 D-ID: documents analyzed, produced by G-ID

D-DI	Description
1	Public manifesto of the Brazilian Society of Intelligent Design (TDI-Brasil) on the teaching of the theory of evolution and the theory of intelligent design in state and private schools and universities (D-ID 1)
2	Rejoinder of TDI-Brasil to the replica of teachers and students of the Federal University of Rio Grande do Sul on the Manifesto of the Brazilian Society of Intelligent Design (D-ID 2)
3	"Fool's gold" – article authored by the president emeritus[a] of the TDI-Brasil published in the *Press Observatory* (D-ID 3)
4	Homepage of the *First Brazilian Congress of Intelligent Design* (D-ID 4)[b]
5	Interview by the CEO[c] of TDI-Brasil to the blog of the *National Association of Evangelical Jurists* (D-ID 5)
6	"The scientific debate that has not yet occurred" – article authored by the president emeritus of TDI-Brasil published in the *Press Observatory* (D-ID 6)
7	"Origin of God is an absurd question" – interview by John Lennox[d] to the newspaper *O Estado de São Paulo* (D-ID 7)
8	The chemistry of life and its evidence at molecular level: spontaneous origin of life or intelligent design? (D-ID 8)
9	The third element of life: irrefutable proof by natural laws of intelligent design (D-ID 9)
10	Scientific freedom: the constitutional right to discuss and investigate intelligent design in academia (D-ID 10)
11	Evidence for ultrafine adjustment in the universe (D-ID 11)
12	Prejudices and frauds in the teaching of evolution and intelligent design in the classroom: a view of an educator (D-ID 12)
13	The idea of intelligent design: from the ancient Greek philosophers to current theorists (D-ID 13)
14	Intelligent design: a fundamental and primary assumption of science (D-ID 14)
15	Darwin and the false dilemma (D-ID 15)
16	An imperfect world with *bad designers*: evidence for evolution or intelligent design (D-ID 16)
17	The fantastic project with mega complex irreducible complexity of cephalopods: evolution or ID (D-ID 17)?
18	Understanding intelligent design: the myths and the reality (D-ID 18)
19	TID and chemistry: thermodynamic bases of the impossibility of the existence of living systems ordered without intelligent design (D-ID 19)
20	Evidence for intelligent design in the chemistry of life: biochemical processes at molecular level (D-ID 20)
21	Evidences for intelligent design in biology: mimicry and camouflage (D-ID 21)

Source: Groto (2016, p. 184)

The D-ID 08–21 are lectures given during the PCBDI

[a]Enézio Eugênio de Oliveira Filho holds a degree in Human Sciences from the Federal University of Amazonas and a Master degree in History of Science from the Catholic University of São Paulo

[b]The official homepage of the PCBDI remained on the World Wide Web in the years 2014–2015 until early 2016. After March 2016, it was withdrawn and remained so until the writing of this work (June 2016)

[c]Marcos Nogueira Eberlin is a chemist and professor at the University of Campinas

[d]American proponent of ID that gave an interview to the newspaper *O Estado de São Paulo*, in 2009, during his participation in the *II International Symposium Darwinism Today*, at the Presbyterian University of Mackenzie

codes. Each unit of analysis was cut to separate it from its original sequence. The random reading of each unit of analysis was performed to identify convergent and divergent elements to form clusters of ideas/thoughts/arguments that shared common meanings: the categories. Once we separated the units of analysis around common meanings, the categories were then identified from the more general sense that allowed each grouping.

For example, one of the categories created was the category *ideas about science*, which has grouped units of analysis that referred to understandings about science, that is, how the G-EVO and the G-ID (official entities or individuals) perceived the science and demonstrated their understanding of it. The second stage involved a new comparative process within each category targeting emerging similar and divergent ideas/thoughts/arguments in the different documents to allow the emergence of subcategories. For example, from the category *ideas about science*, seven subcategories emerged that identified the perception of specific features of science in the case of G-EVO. In focus 3, the new emerging, we designed a metatext aiming to show, briefly, the new understandings resulting from foci 1 and 2.

From the analysis of D-EVO, four categories emerged. They are:

1. *Features/defense of evolution*
2. *Criticism of ID*
3. *Ideas about science*
4. *Teaching.*

From the units of analysis of D-ID, six categories emerged:

1. *Defense of ID*
2. *Criticism of evolution*
3. *Ideas about science*
4. *Teaching*
5. *Religion*
6. *Structural features.*

As already stated, in this work, we only analyze the category ideas about science, which in the D-EVO originated seven subcategories:

1. *The supernatural is not part of science*
2. *Science is an ongoing process*
3. *Science makes use of the scientific method*
4. *Science proves*
5. *Science has its own rules*
6. *The relationship between science and religion*
7. *Science must be respected.*

In the D-ID category, ideas about science originated nine subcategories:

1. *Science as a way to reach the truth*
2. *Science proves*
3. *Science makes use of the scientific method*
4. *Science produces laws*

5. *Naturalism should not be an assumption of science*
6. *Scientific explanations/rules of science may change*
7. *Science is secular*
8. *Refutability is a feature of science*
9. *Science involves freedom of thought.*

9.3 Ideas About Science Manifest in the Documents

In this section, we describe and discuss each of the subcategories listed above. They represent ideas/thoughts/arguments linked to science, used by the groups involved in the EVO-ID debate. They represent ways of understanding science and its functioning. Put in another way, they are elements expressed in the documents reviewed that make up each group's conceptions of science. First, we discuss the groups separately.

9.3.1 Science Portrayed in D-EVO

To argue that *the supernatural is not part of science*, a recurring argument in several documents (D-EVO 1, 2, 4, 6, 7, and 10), it means that only the natural world is investigated by science, which only uses natural causes to explain it. In general, all these documents have a view that science would adopt; therefore, *methodological naturalism* is one of its assumptions. The D-EVO 10, for example, makes this clear. It argues:

> [...] scientific discourse is, in epistemological terms, of empirical character, in the sense that the claims that science makes about the world should be subject to the test of experiment, should be tested against the empirical world. This empirical character means, in turn, that in its ontology, scientific discourse takes on a methodological naturalism. (D-EVO 10)

To argue that *science is an ongoing process* would mean, in principle, to say that it constantly incorporates new knowledge and, therefore, would be something temporary, that is, subject to change. However, there appear to be several understandings of what would be an ongoing process. The following are mentioned:

> [...] the scientific process is continuous, constantly incorporating new discoveries. (D-EVO 1)

> [...] the recurrent discovery of new fossils has led the paleontological community to constantly reshape their understanding of the history of life on Earth, to adapt it to new evidence. (D-EVO 2)

> [...] is by checking its falsity that science progresses. (D-EVO 4)

> Evolutionary biology continuously develops new theoretical approaches. (D-EVO 6)

> [...] for its temporariness allows its constant review, correction and continuous improvement. (D-EVO 7)

> [...] our current scientific knowledge is finite [...] we, scientists, try to push it, increase it, extend it. (D-EVO 9)

It is clear, therefore, a trend in D-EVO toward talking more about expansion than the changing of knowledge. Only D-EVO 7 brings this possibility of understanding, when it uses the term *temporality*. D-EVO 4 also brings another idea, one that links science to progress.

Understanding that *science makes use of the scientific method* is also recurrent in various D-EVOs. In D-EVO 1, 2, and 7 the idea of the existence of a scientific method appears explicitly in different ways. In D-EVO 3, 10, and 11, the scientific method appears (implicitly) to mention steps and procedures of science. There is a defense of empirical, experimental tests, observations, and validation of hypotheses. The documents that explicitly mention the scientific method as a feature of science have different understandings of how this method would be. Thus, in D-EVO 1, prepared by the *Brazilian Society of Genetics* (SBG), the method is linked to the experimental sciences and involves the investigation of phenomena that can be tested experimentally. D-EVO 2, prepared by the *Brazilian Society of Paleontology* (SBP) has a broader view of the method, as it describes two possibilities of testing and validation of hypotheses: one of the empirical sciences and one of the historical sciences. The first would make use of observation and experimentation. The second, which seeks to explain past events, would make use of the comparison of alternative hypotheses. The D-EVO 4, prepared by the *Brazilian Society for the Advancement of Science* (SBPC), presents an understanding of science that is closer to a view that we could call *empirical–inductive, nontheoretical and rigid*, that is, one that means that science develops through a rigid, fixed method, involving observations and neutral experiments without being oriented a priori by theories or hypotheses. According to the SBPC, scientists "observe nature and ask questions about the natural world that can be tested by experiment and new observations, and then build explanations and theories, evidence-based" (D-EVO 4). Another understanding of the scientific method appears in D-EVO 7, authored by a researcher of paleontology. He describes his understanding of the hypothetical–deductive method:

> [...] the scientific method requires that the hypotheses (or theories, which are nothing more than integrated systems of hypotheses) may be testable. Testable means that they can be contrasted with new facts, experiences or observations. In the field of science, never a hypothesis can be considered proven or demonstrated. It just was not (yet) refuted. (D-EVO 7)

The view that *science proves* arises in different ways in D-EVO 1, 2, 3, 4, and 9. It is understood that:

> [...] an enormous amount of data has confirmed and enhanced the original proposal of Darwin and Wallace. (D-EVO 1)

> [...] various aspects of Biological Evolution can be and they are routinely checked in the laboratory. (D-EVO 2)

> The Evolution of species through natural selection is not a theory, but a collection of widely proven facts. (D-EVO 3)

> The evolution [...] has been repeatedly confirmed through observation and experiments. (D-EVO 4)

> It is absolutely undeniable the *fact* of evolution. (D-EVO 9)

However, D-EVO 7, again, is distinguished from the other documents, because in it, it is argued that science is not capable of proof or demonstration – as can be seen in the passage reproduced above.

The comprehension that *science has its own rules* explicitly arises in three documents (D-EVO 2, 3 and 4). It is argued that: science uses the scientific method or steps linked to the method; falsifiability is one of its features; the knowledge produced undergoes peer review before publication in scientific journals and then it is subjected to the scrutiny of the scientific community; despite being a social enterprise, the choices of its explanations do not involve democratic criteria. It is clarified further, for example, that not all knowledge produced by the scientific community comes to schools, as:

> [...] only the most robust ideas, continually tested by the scientific community, will be introduced in textbooks and taught in schools. (D-EVO 2)

Ideas about the *relationship between science and religion* arise in three documents (D-EVO 4, 7, and 9) with different understandings. The first two (4 and 7) comprehend that science and religion would not be conflicting. According to D-EVO 4, science and religion would occupy different areas of human knowledge. D-EVO 7 compromises the two perspectives, by showing the existence of religious scientists:

> [...] perhaps the most notable example is Theodosius Dobzhansky, a practicing Catholic, and one of the foremost proponents of the so-called Synthetic Theory of Evolution. (D-EVO 7)

The document D-EVO 9 shows another understanding. Although it does not explicitly refer to the relationship between science and religion, but rather to the relationship between evolution and religion, it argues that evolution is compatible with belief in God. However, evolution would be incompatible with some biblical literalist fundamentalist views that believe in Young Earth and the special creation of man.

The view that *science must be respected* appears in two documents (D-EVO 1 and 10). This understanding arises, notably, because of the fact that the documents argue against the teaching of creationism/ID and defend the teaching of biological evolution. With this idea (science must be respected), the documents aim to argue against the call for respect for diversity and plurality when defending the teaching of creationist ideas/ID in science classes. It is alleged that:

> [...] when discussing pluralism and respect for diversity, sometimes it is forgotten that also scientific discourse must be respected, should be recognized as legitimate. (D-EVO 10)

We noticed some contradictions in D-EVO. For example, when, on the one hand, science is seen as a continuous process, but temporary, capable of changes and, on the other hand, science is seen as being able to prove, this leads to the vision of the existence of unalterable scientific facts. If science verifies and proves "facts," how may its theories change with time and knowledge be temporary? There is also a contradiction between the idea that science verifies, proves, and at the same time, progresses through falsification (as shown in D-EVO 4).

Table 9.3 Subcategories that emerged from D-EVO

	D-EVO										
Subcategories	1	2	3	4	5	6	7	8	9	10	11
The supernatural is not part of science	X	X		X		X	X			X	
Science is an ongoing process	X	X		X		X	X		X		
Science makes use of the scientific method	X	X	X	X			X			X	X
Science proves	X	X	X	X					X		
Science has its own rules		X	X	X							
Relationship between science and religion				X			X		X		
Science must be respected	X									X	

The category *ideas about science* still allows different understanding. In various D-EVOs, there is a predominance of *Popper's view of science*, that is, centered on Karl Popper's ideas. They mention refutability, falsifiability, and tests, for example. Another feature arises from the understanding of the nature of the scientific method. The method is understood in different ways in the documents. Thus, for example, the document from the SBPC presents a conception of science nearest to an empirical–inductive, nontheoretical and rigid view, whereas the document drawn up by the SBP shows a broader view of science, as it mentions different methods: the method of experimental sciences and the method of historical sciences.

Summarizing, we could say that, in general, the D-EVOs argue that methodological naturalism is an assumption of science; that science is an ongoing process that is notably linked to the idea of expanding knowledge; that science makes use of the scientific method, with different understandings of what this method is (we may say that there is a gradient that goes from a "stricter" view to a "wider" view, which considers the possibility of more than one type of method); that science proves; that science has its own rules elaborated by the community that produces it; that science and religion are not conflicting, offering different types of knowledge, but may conflict, when we consider more fundamentalist religious positions; and that science, as a type of knowledge, also needs to be respected. Table 9.3 highlights which subcategories of the category *ideas about science* have emerged in each of the D-EVO documents.

9.3.2 Science Portrayed in D-ID

The view of *science as a way to reach the truth* appears in D-ID 3, 4, 5, and 18. For example:

> [...] when our ideas do not adequately fit to what the evidence says, we must correct them and/or replace them. Cover our ears or turning our backs to the evidence does not lead us to the knowledge of scientific truth. (D-ID 3)

D-ID 4, aimed at attracting supporters to ID, states:

> [...] if you are like this, a professional or academic who is seeking only the truth about our origins, and is a supporter of the free debate of ideas, join us. (D-ID 4)

The idea that *science proves* also arises in four documents (D-ID 3, 9, 12, and 21). The D-ID 9 discusses scientific laws from the following thought:

> [...] in physics and chemistry is easy to prove, relatively easy to prove or disprove some events using well defined laws. (D-ID 9)

The very title of this lecture given at the PCBDI, "Irrefutable proof by natural laws of intelligent design," shows the view of the existence of the notion of a "scientific proof." D-ID 3 argues that evolution, even without "empirical evidence" prevails as a scientific explanation in academia.

The view that *science makes use of the scientific method* appears most clearly in D-ID 3 and 13. The first literally cites the steps that make up the scientific method. It is mentioned that intelligent design:

> [...] follows the parameters of the scientific method. The scientific method is generally described as a four-step process involving observations, hypotheses, experiments, and conclusion. (D-ID 3)

D-ID 13 quotes what would be a characteristic of the scientific method: mathematization. The absence of this feature in evolutionary biology, in turn, is used as an argument to question its scientific character. A similar argument is also used to criticize evolution: it is understood, in a way, that *science produces laws* (another subcategory) and, therefore, if a scientific model does not produce laws, it is wrong. This argument is used to deny the scientific character of evolutionary biology, as, according to D-ID 11, it produces virtually no laws.

The understanding that *naturalism should not be an assumption of science* arises in D-ID 5 and 18. It is alleged that naturalism (philosophical or methodological) would make science blind (D-ID 5) and would prevent it from reaching the truth (D-ID 18). The critique of naturalism (philosophical or methodological) appears in the documents, and also in other categories of analysis, for example, in the category *criticism of evolution*, which is not addressed in this work.

The understanding that *scientific explanations/rules of science may change* appears in a contradictory manner in documents 3, 6, and 18. In D-ID 3, both "scientific truth" and "changes in explanations" are mentioned. Reinforcing this second understanding and thus clashing with the first, D-ID 6 (authored by the same person as the D-ID 3, the President Emeritus of the TDI-Brasil), says: "In science there is no *theoria perennis.*" D-ID 18, in turn, argues that the rules of science may change, but not the facts/evidence.

From D-ID 4 and 10 emerged a subcategory that refers to an idea not mentioned in other documents. First, *science is secular* and this understanding or its defense, in a way, would be constitutionally guaranteed. Thus, on the PCBDI's website it is clarified:

> Committee members understand that, above all, they provide good service to the nation and to the Brazilian people – the financier of Brazilian Science – for compliance with its greatest constitutional obligation to defend and promote a fully secular Science and that, being

> secular, should encourage the free debate of ideas, not assuming a priori a preconception about how the Universe and the Life are made or would necessarily have to be formed. (D-ID 10)

In a way, this passage also implicitly introduces a critique of naturalism as an assumption of science, which is understood by ID proponents as a type of religious belief. Finally, from D-ID 10 two other ideas emerged: *refutability is a feature of science*, which again refers to a Popperian perspective of science; and *science involves freedom of thought*, which is understood in the document as a constitutional right related to freedom of choice of the problem to be investigated and how to conduct research.

From the category *ideas about science*, as in D-EVO, some interesting insights emerge. We understand that there is a contradiction in defending the notion that science can change (subcategory *scientific explanations/rules of science may change*) and that, therefore, there would be no *theoria perennis* (D-ID 3), and at the same time affirming that science proves or that it is a way to reach the truth (D-ID 3). In D-ID, an empirical-inductive, nontheoretical, and rigid view of science prevails. Also in D-ID, the predominance of Popper's view of science can be noted. Here, however, misunderstandings regarding the thought of Karl Popper are evident. This is very clear in D-ID 12. It is argued, using Popper, that evolution is not a theory or a fact and a theory itself is not science. A theory, when proven, becomes a fact, and is then incorporated into science. This latter understanding that "a theory when proven becomes a fact" is considered an alternative conception of NOS (Smith 2010). In the D-ID, it is further argued that biology, and evolution in particular, do not produce laws or they are not mathematicized, and, in this sense, would not be science. The documents often discuss science from the perspective of the physical and chemical sciences, showing ignorance of biology specificities.

Summarizing, we could say that, in general, it is argued in the D-IDs that methodological and philosophical naturalism should not be considered an assumption of science; that science makes use of the scientific method composed of pre-defined stages, thus instigating a "more rigid" view of it; that science proves, being a way of reaching the explanatory truth about a certain phenomenon through the production of laws, for example. On the other hand, it is argue in the D-IDs that science is characterized by the refutability and by the possibility of changing the explanations/rules it produces; that science must be secular, understanding that this would imply the free discussion of ideas and that, therefore, science would involve freedom of thought. Tables 9.4 and 9.5 below highlight which subcategories of the category *ideas about science* have emerged in each of the D-ID documents.

9.4 Concluding Remarks

The analysis of the D-EVO and D-ID documents, with regard to the category *ideas about science* – and its subcategories – highlights a number of misunderstandings about the nature of science. Some of them appear in both groups, such as the view

Table 9.4 Subcategories that emerged from D-ID 1 to 10

	D-ID									
Subcategories	1	2	3	4	5	6	7	8	9	10
Science is a way of reaching the truth			X	X	X					
Science proves			X						X	
Science makes use of the scientific method			X							
Science produces laws										
Naturalism should not be an assumption of science					X					X
Scientific explanations/rules of science may change			X			X				
Science is secular				X					X	
Refutability is a feature of science										X
Science involves freedom of thought										X

Table 9.5 Subcategories that emerged from D-ID 11 to 21

	D-ID										
Subcategories	11	12	13	14	15	16	17	18	19	20	21
Science is a way of reaching the truth								X			
Science proves		X									X
Science makes use of the scientific method	X		X								
Science produces laws	X										
Naturalism should not be an assumption of science								X			
Scientific explanations/rules of science may change								X			
Science is secular											
Refutability is a feature of science											
Science involves freedom of thought											

that science proves, verifies, and that it is a way of reaching the truth. The idea of "scientific truth," however, arises more explicitly in the D-IDs, although it is often implicit in the D-EVOs.

The view of the existence of a single scientific method, rigorous, comprising fixed and rigid steps, without the guidance of a priori theories and hypotheses, that is, an empirical–inductive, nontheoretical and rigid view, also arises in both groups. This view, which has been classified as a distorted understanding of science (e.g., by Gil-Pérez et al. 2001), it is more prevalent in G-ID, because we can detect nuances in the view of the method in G-EVO, as occurs, for example, in the documents produced by the field of paleontology, as we have seen. These documents mention the existence of experimental sciences and historical sciences and cite the hypothetical–deductive method. Overall, we believe that the G-EVO presents a "broader" view of science in relation to the existence of the scientific method, whereas the G-ID brings a view that we consider "narrower".

The idea of temporariness, in turn, is poorly understood for both groups. Among the G-EVOs, the prevailing idea is that temporariness is linked to an extension of knowledge and not to its change. On the other hand, in the G-ID, the temporariness is understood predominantly as change, not extension, which means that the ID members do not understand, for example, the latest research on biological evolution, arguing that this area is in crisis. However, in both groups, the ideas and arguments in this regard are sometimes contradictory, clashing with the notions of "fact," proof, and scientific truth.

The ID documents also present distorted views about the role of laws and mathematics in science, tending therefore to question the scientific character of biological evolution and, sometimes, of biology as a whole. In general, we can say, though, that in the G-ID there seems to be difficulty in understanding certain characteristics of contemporary science.

Considering our results and the fact that the debate concerning the biological evolution and the various creationism facets –ID being one of them – must also be present in the classroom, we recommend that questions about NOS should be addressed more strongly in basic education and also in undergraduate courses in the biological sciences, considering the number of misunderstandings about science manifested in the D-EVOs. There is a need to understand science as a form of explanation of natural phenomena; that it is a human enterprise and that it is linked to a social, cultural, and historical context; that it is provisional and therefore subject to change (although not every change represents a "revolution" in the Kuhnian sense). Therefore, there is no valid scientific truth forever, and, in this sense, the idea of the existence of scientific proof is misplaced too. On the other hand, it is also important to emphasize that science has characteristics of its own; its explanations are based on substantiated evidence obtained through its own methodologies linked to the particularities of the various areas that comprise scientific knowledge – in this sense, there is no single way of doing science, or a single scientific method. Regarding undergraduate courses in the biological sciences, we recommend a stronger presence of epistemological discussions, whether in the form of a specific discipline of the philosophy of science, or throughout the whole teacher training process. We need to discuss, more emphatically, what science is, how it is produced, how it develops/developed historically and relate it to other socially constructed knowledge, and to have a deeper understanding, for example, about what theories, hypotheses, models, and laws are, considering, of course, the specificities of biology.

Our work highlights how the opposing sides of the EVO versus ID debate bring simplified and limited elements with regard to the construction of more satisfactory views of science. On the other hand, we agree that there is no consensus on what science is and how it develops. Despite all this, it is extremely important to deal with this discussion in science education, either at a basic or at a higher level of education. For this purpose, we should consider issues and questions to be explored by the teacher (Martins 2015), whose training for this task needs to contemplate specific knowledge in the field of the philosophy of science.

References

Almeida, D. d. F. (2012). Concepções de alunos do ensino médio sobre a origem das espécies. *Ciência & Educação, 18*, 143–154.

Bizzo, N., & El-Hani, C. (2009). Darwin and Mendel: Evolution and genetics. *Journal of Biological Education, 43*, 108–114.

Blancke, S., de Smedt, J., de Cruz, H., Boudry, M., & Johan Braeckman, J. (2012). Implications of the cognitive sciences for the relation between religion and science education: The case of evolutionary theory. *Science & Education, 21*, 1167–1184.

Costa, L. d. O., da Cunha Melo, P. L., & Teixeira, F. M. (2011). Reflexões acerca das diferentes visões de alunos do ensino médio sobre a origem da diversidade biológica. *Ciência & Educação, 17*, 115–128.

de Souza, S. (2009). *A goleada de Darwin: sobre o debate Criacionismo/Darwinismo*. Rio de Janeiro: Record.

D-EVO 1. (2014). *Manifesto da Sociedade Brasileira de Genética sobre ciência e criacionismo*. http://media.wix.com/ugd/b703be_94c3e31c579b49e881d8c593e3fe6c60.pdf. Accessed 10 Jan 2016.

D-EVO 10. (2008). *Educação e Discurso Científico*. http://www1.folha.uol.com.br/fsp/opiniao/fz0612200808.htm. Accessed 10 Jan 2016.

D-EVO 11. (2015). *Transformando ignorância em sabedoria*. http://observatoriodaimprensa.com.br/jornal-de-debates/_ed832_transformando_ignorancia_em_sabedoria/. Accessed 10 Jan 2016.

D-EVO 2. (2014). *Manifesto da Sociedade Brasileira de Paleontologia sobre a validade da Evolução Biológica e seu ensino nas escolas do país*. https://www.dropbox.com/s/ybfcoy-s6exg0wk2/Manifesto%20final.pdf?dl=0. Accessed 10 Jan 2016.

D-EVO 3. (2014). *Notícia sobre a Nota de Repúdio do Conselho Federal de Biologia ao Projeto de Lei 8099/2014*. http://www.cfbio.gov.br/artigos/CFBio-repudia-projeto-que-torna-obrigatorio-o-ensino-do-Criacionismo-nas-escolas. Accessed 10 Jan 2016.

D-EVO 4. 2014. Carta da Sociedade Brasileira para o Progresso da Ciência aos deputados sobre o Projeto de Lei 8099/2014. http://www.sbpcnet.com.br/site/arquivos/arquivo_402.pdf. Accessed 10 Jan 2016.

D-EVO 5. (2014). *Carta aberta da Associação Brasileira de Pesquisa em Educação em Ciências e da Associação Brasileira de Ensino de Biologia sobre o Projeto de Lei 8099/2014*. http://www.sbenbio.org.br/wordpress/wp-content/uploads/2014/11/Carta-Aberta-SBEnBio-ABRAPEC.pdf. Accessed 10 Jan 2016.

D-EVO 6. (2014). *Réplica de professores e alunos (Grupo Balduíno Rambo de Estudos em Filosofia, Ciência e Sociedade) da Universidade Federal do Rio Grande do Sul sobre o Manifesto da Sociedade Brasileira do Design Inteligente*. http://adagadeoccam.blogspot.com.br/2014/11/resposta-ao-manifesto-da-sociedade.html. Accessed 10 Jan 2016.

D-EVO 7. (2012). *Blogue do Mestre Chassot - Design Inteligente: evolucionismo ou criacionismo?* http://mestrechassot.blogspot.com.br/2012/04/02-design-inteligente-evolucionismo-ou.html (1); http://mestrechassot.blogspot.com.br/2012/04/03-design-inteligente-2-parte.html (2). Accessed 10 Jan 2016.

D-EVO 8. (2009). *A árvore da vida não está morta*. http://ciencia.estadao.com.br/noticias/geral,a-arvore-da-vida-nao-esta-morta-afirma-biologo,360509. Accessed 10 Jan 2016.

D-EVO 9. (2009). *Evolução e religião*. http://www.cienciahoje.org.br/noticia/v/ler/id/4330/n/evolucao_e_religiao. Accessed 10 Jan 2016.

D-ID 1. (2014). *Manifesto público da Sociedade Brasileira do Design Inteligente (TDI-Brasil) sobre o ensino da Teoria da Evolução e da Teoria do Design Inteligente nas escolas e universidades públicas e privadas*. http://darwinedeus.blogfolha.uol.com.br/2014/11/18/sai-o-manifesto-do-design-inteligente/. Accessed 10 Jan 2016.

D-ID 10. (2014). *Lecture in PCBDI: Liberdade científica: o direito constitucional de discutir e investigar o design inteligente no meio acadêmico*. https://www.youtube.com/watch?v=OF4yQ4oOb-c. Accessed 10 Jan 2016.

D-ID 11. (2014). *Lecture in PCBDI: Evidências do ajuste ultrafino no universo*. https://www.youtube.com/watch?v=KmcDypKXUrY. Accessed 10 Jan 2016.

D-ID 12. (2014). *Lecture in PCBDI: Preconceitos e fraudes no ensino de evolução e design inteligente na sala de aula: visão de um educador*. https://www.youtube.com/watch?v=7Sl446P7S7A. Accessed 10 Jan 2016.

D-ID 13. (2014). *Lecture in PCBDI: A ideia de design inteligente: dos filósofos gregos antigos aos teóricos atuais*. https://www.youtube.com/watch?v=_bSC6XcNMdE. Accessed 10 Jan 2016.

D-ID 14. (2014). *Lecture in PCBDI: Design Inteligente: um pressuposto fundamental e primordial da ciência*. https://www.youtube.com/watch?v=EuorC6aAIuc. Accessed 10 Jan 2016.

D-ID 15. 2014. Lecture in PCBDI: Darwin e o falso dilema. https://www.youtube.com/watch?v=RVSk-xi9xc0. Accessed 10 Jan 2016.

D-ID 16. (2014). *Lecture in PCBDI: Um mundo imperfeito com Bad Designers: evidência a favor da evolução ou do design inteligente*. https://www.youtube.com/watch?v=pmGnYXJC3YI. Accessed 10 Jan 2016.

D-ID 17. (2014). *Lecture in PCBDI: O projeto fantástico com complexidade mega irredutivelmente complexa dos cefalópodes: evolução ou DI?* https://www.youtube.com/watch?v=COWjN4Emcig. Accessed 10 Jan 2016.

D-ID 18. (2014). *Lecture in PCBDI: Understanding inteligente design: The myths and the reality*. https://www.youtube.com/watch?v=aufHIvHioDw. Accessed 10 Jan 2016.

D-ID 19. (2014). *Lecture in PCBDI: A TDI e a química: as bases termodinâmicas da impossibilidade de existência de sistemas vivos ordenados sem design inteligente*. https://www.youtube.com/watch?v=43jry0ocy6U. Accessed 10 Jan 2016.

D-ID 2. (2014). Tréplica da TDI BRASIL à réplica de professores e alunos da Universidade Federal do Rio Grande do Sul sobre o Manifesto da Sociedade Brasileira do Design Inteligente. http://darwinedeus.blogfolha.uol.com.br/2014/12/18/treplica-dos-defensores-do-design-inteligente/. Accessed 10 Jan 2016.

D-ID 20. (2014). *Lecture in PCBDI: Evidências do Design Inteligente na química da vida: processos bioquímicos ao nível molecular?* https://www.youtube.com/watch?v=hDFSidBAnI8. Accessed 10 Jan 2016.

D-ID 21. (2014). *Lecture in PCBDI: Evidências de Design inteligente na biologia: mimetismo e camuflagem*. https://www.youtube.com/watch?v=yAyR2Dk_rLU. Accessed 10 Jan 2016.

D-ID 3. (2015). *O ouro de tolo*. http://observatoriodaimprensa.com.br/feitos-desfeitas/_ed835_o_ouro_de_tolo/. Accessed 10 Jan 2016.

D-ID 4. (2014). *Primeiro Congresso Brasileiro do Design Inteligente*. http://www.designinteligentebrasil.com.br/. Accessed 10 Jan 2016.

D-ID 5. (2014). *Entrevista com o Presidente Executivo da TDI-Brasil*. http://www.anajure.org.br/entrevista-prof-marcos-eberlin-fala-sobre-a-teoria-do-design-inteligente-1-inscricoes-abertas-para-o-1o-congresso-brasileiro-do-di/. Accessed 10 Jan 2016.

D-ID 6. (2009). *O debate científico que ainda não ocorreu*. http://observatoriodaimprensa.com.br/feitos-desfeitas/o-debate-cientifico-que-ainda-nao-ocorreu/. Accessed 10 Jan 2016.

D-ID 7. (2009). *Origem de Deus é questão absurda*. http://vida-estilo.estadao.com.br/noticias/geral,origem-de-deus-e-questao-absurda,360645. Accessed 10 Jan 2016.

D-ID 8. (2014). *Lecture in PCBDI: A química da vida e suas evidências ao nível molecular: origem espontânea da vida ou Design Inteligente?* https://www.youtube.com/watch?v=cGTV7BWlBb8. Accessed 10 Jan 2016.

D-ID 9. (2014). *Lecture in PCBDI: O terceiro elemento da vida: prova irrefutável por leis naturais de design inteligente*. https://www.youtube.com/watch?v=XaAVbVNN7jg. Accessed 10 Jan 2016.

Fleck, L. (2010). *Gênese e desenvolvimento de um fato científico*. Belo Horizonte: Fabrefactum.

Gil-Pérez, D., Montoro, I. F., Alís, J. C., Cachapuz, A., & Praia, J. (2001). Para uma imagem não deformada do trabalho científico. *Ciência & Educação, 7*, 125–153.

Groto, S. R. (2016). *O debate evolução versus design inteligente e o ensino da evolução biológica: contribuições da epistemologia de Ludwik Fleck*. PhD Dissertation in Education. Universidade Federal do Rio Grande do Norte, Natal, Rio Grande do Norte, Brazil.

Holsti, O. R. (1969). *Content analysis for the social sciences and humanities*. Reading: Addison-Wesley.

Jablonka, E., & Lamb, M. (2010). *Evolução em quarto dimensões: DNA, comportamento e a história da vida*. São Paulo: Companhia das Letras.

Lüdke, M., & André, M. (1986). *A Pesquisa em educação: abordagens qualitativas*. São Paulo: EPU.

Martins, A. F. P. (2015). Natureza da Ciência no ensino de ciências: uma proposta baseada em temas e questões. *Caderno Brasileiro de Ensino de Física (Online), 32*, 703–737.

Mazzur, A. (2005). Believers and disbelievers in evolution. *Politics and the Life Sciences, 23*, 55–61.

Meyer, D., & El-Hani, C. (2005). *Evolução: o sentido da biologia*. São Paulo: Editora da UNESP.

Miller, J. D., Scott, E. C., & Okamoto, S. (2006). Science communication: Public acceptance of evolution. *Science, 313*, 765–766.

Moraes, R., & do Carmo Galiazzi, M. (2011). *Análise textual discursiva*. Ijuí: Ed. Unijuí.

Numbers, R. L. (2006). *The creationist: From scientific creationism to intelligent design* (Expanded ed.). Cambridge, MA: Harvard University Press.

Oliveira, G. d. S., & Bizzo, N. (2011). Aceitação da evolução biológica: atitudes de estudantes do ensino médio de duas regiões brasileiras. *Revista Brasileira de Pesquisa em Educação em Ciências, 11*, 57–79.

Scott, E. C. (2009). *Evolution vs. creationism: An introduction* (2nd ed.). Westport: Greenwood Press.

Silva, M. G. B. da. (2011). *Um estudo sobre a evolução biológica como eixo norteador do processo de formação de professor de biologia*. Master Thesis in Teaching, Philosophy and History of Sciences. Universidade Federal da Bahia e Universidade Estadual de Feira de Santana, Brazil.

Smith, M. U. (2010). Current status of research in teaching and learning evolution: II. Pedagogical issues. *Science & Education, 19*, 539–571.

Tidon, R., & Vieira, E. (2009). *O ensino da evolução biológica: um desafio para o século XXI*. Campinas: Comciência. http://www.comciencia.br/comciencia/handler.php?section=8&edicao=45&id=535. Accessed 15 Oct 2012.

Chapter 10
Multiple Narratives As Cognitive and Political Bridges to Understanding the Nature of Scientific Knowledge

Alcira Rivarosa and Carola Astudillo

10.1 Introduction

Research into science education in recent decades has highlighted the need to complement and integrate the training of future teachers and scientists in specific disciplinary fields with content of metascientific reflection, such as history, philosophy, STS (science,–technology–society). The aim is to contribute to building a more humane, relative, and contextualized image of science, beyond the classic standards and dogmatic views that contradict the emancipatory goals of science education.

Considering these premises, this chapter presents the arguments, the theoretical references, and the characteristics of certain activities and didactic materials designed to promote reflection on the nature of scientific knowledge in the training of researchers and teachers of biological sciences.[1] Moreover, the analysis of the results of their implementation with groups of university students are provided, highlighting the main learning achieved.

In particular, these activities and teaching resources have been selected and designed with the intention of activating a critical reflection of the *cuisine* of scientific research. We have been especially concerned with the processes of reflection, reasoning, and conceptual arguments that real scientists conduct during practices in their contexts. On another level, it has interested us to reflect on these contexts from their historical and non-neutral nature, regaining their rich connections to other

[1]The *epistemology and history of science course* is in the first year of study for teachers and Bachelor of Biological Sciences (1999 -Resol. 196/97/2013) Faculty of Exact, Physical-Chemical and Natural Sciences of the University of Río Cuarto (UNRC), Cordoba, Argentina.

A. Rivarosa (✉) · C. Astudillo
Department of Natural Sciences, Faculty of Physical-Chemical and Natural Sciences, National University of Río Cuarto, PICTOR Project – Science and Technology, Cordoba, Argentina
e-mail: arivarosa@exa.unrc.edu.ar; castudillo@rec.unrc.edu.ar

M. E. B. Prestes, C. C. Silva (eds.), *Teaching Science with Context*, Science: Philosophy, History and Education,
https://doi.org/10.1007/978-3-319-74036-2_10

fields, such as politics, economics, religion, art, and technology. Finally, with some specific activities, we have focused on understanding the processes of construction and development of explanatory models in biological sciences.

These didactic designs have one element in common: they appeal in all cases to (verbal, graphic, and audiovisual) texts with a narrative nature. We have found in the stories of various kinds a set of possibilities for epistemological reflection that are of particular interest in our teaching goals (Ellis and Bochner 2000). Among them, we can mention the following: promoting divergent thinking, contributing to the interpretation of reality, retrieving the contextual and cultural nature of human practices, facilitating the construction of mental images of unfamiliar experiences, accessing the description of the mental world of the characters, promoting emotional connections with the reader. As we have developed, these potentials have been studied and validated by multiple educational research projects exploring the complex relationships between thought and language (Vigotsky 1995; Scolari 2007, 2013).

10.2 Epistemological Reflection on the Training of Future Scientists and Educators in Biological Sciences: Obstacles and Possibilities

Today there is a broad epistemological consensus on a dynamic, contextual and relative conception of scientific knowledge and its construction. In this regard, experts refer to it as an institutionalized and collective social practice that, through the construction of theories and explanatory models in constant evolution and review, realizes those aspects of reality that it seeks to address (Palma and Wolovelsky 2001; Matthews 1991).

The international educational movement (Jenkins and Pell 2006) reflected in the proliferation of magazines devoted to science education for the past 40 years highlight the significant ideological shift: goals that traditionally focused on the development of theories and concepts of each disciplinary domain were gradually modified with the inclusion of new objectives and content. These include: the "doing" of the scientist, the question of method, the semantic structure of the disciplines, the development of science, the ideological assumptions underlying scientific practices, the economic and ethical dimensions involved, the complex and demanding relationship science–technology–consumption, among many others (Latour and Woolgar 1995; Acevedo et al. 2005).

First, educational research has been pointing out that the integration of this type of metascientific content promotes greater conceptual understanding of disciplinary issues, especially when students are coming to understand the origin and nature of knowledge (Adúriz-Bravo et al. 2002; Lemke 2006; Carrascosa et al. 2006).

Second, the relationship between the philosophical and historical contributions helps to understand the scientific development as a complex process, nonlinear, creative, and contextualized in the structure and social dynamics. In this way, it helps to demystify the vision of progress in science as always being successful and upward.

Third, this previous epistemological reflection can help us to differentiate the theoretical, semantic, and axiological aspects of scientific arguments and the representational systems (graphics, symbols, images) that accompany the various conceptual developments.

Fourth, the problematical and contextualized view of the nature of knowledge leads to addressing the ethical and political dilemmas that modern research practices go through, the complexity of the techno-scientific system, and the public nature of science, among other relevant topics (Datri 2006; Matthews 2009).

Faced with such potentials, some experts recognize that there are still few and limited proposals or formative experiences in this line (Gallego Torres and Badillo 2007). As referred to by Stephen Jay Gould (in Sacks et al. 1996), we know very little about what the conceptual, cognitive, and practical courses are. These are articulated in the work of the scientist to build explanatory and challenging assumptions about a particular fact or phenomenon, such as a genetic model, evolutionary theory, ecological balance (Wolovelsky 2008).

Furthermore, as we have been stating as trainers in the field of university science courses, students have serious difficulties in addressing humanistic content and philosophical texts (Rivarosa et al. 2014). Among the most significant obstacles are:

1. Although they are able to define epistemological concepts, they fail to make a reflection of second order that allows them to integrate and transpose these concepts when analyzing specific situations
2. They fail to identify and argue positions on real processes of the construction of scientific knowledge
3. They still hold the linear and routine idea of scientific research practices.

We believe that, assuming possibilities and obstacles, the most powerful way is to propose a reflection that restores the profoundly human nature of the activity and scientific knowledge. In this regard, addressing Bruner (1991), we recognize the undeniable role that the narrative provides in the history of humanity and of our own existence as human beings. Specifically, the documentary and fictional narrative about situations, cases, and historical episodes act as a mediator resource between the theoretical content of the epistemological field and the actual practices of building knowledge in the field of biology, and natural Sciences in general (Toulmin 1997; Adúriz-Bravo 2009). They are significant and powerful as they:

1. Provide a deeply human sense of reflection on science
2. Clarify and challenge notions about the nature of knowledge providing alternative perspectives
3. Demonstrate a contextualized image of the scientific enterprise from a primarily socio-historical perspective
4. Allow us to connect cognition and emotion.

In this regard, Kornhaber et al. (2004) suggest the need to provide students with materials that offer these multiple dialogic interactions to challenge, in our case, the epistemic reflection on scientific practices. Similarly, Furinghetti and Morselli (2009) emphasize the need to integrate affective and cognitive factors in the proposed scientific training, motivating the problematization of attitudes and ethical commitments sedimented from the historical and cultural "doing" of scientists.

Retrieving these principles, we share some of the didactic designs that we have developed and implemented in the field of epistemology and the history of biology for teacher training students and bachelor's degree students in biological sciences.

10.3 How Do We Build Powerful Epistemological Meanings from Narratives?

> The narrative is more or less important depending on its capacity to cause readers to broaden their horizons, reflect critically on their own experience, immerse themselves empathetically in experiences different to their own, and actively participate in the dialogue in as for [...] different perspectives and viewpoints. Invited to pick up a story and use it, readers become co-actors, auto-examined through the evocative power of the narrative text. (Ellis and Bochner 2000, p. 748)

10.3.1 The "Cuisine" of Scientific Research in Biographical and Autobiographical Narratives of Real Scientists

We start by assuming that the evocation of a story of scientific research offers the opportunity to integrate events, knowledge, and mental states in a joint representation (Aguirre 2012). In this sense, the narrative unifies the personal and social awareness of doing, the technical language, the contexts of practice, and the explanatory models that are constructed. Thus, it enables students to think about how their own practice will be like that of future professional scientists, recognizing its *values* and *complexities* (Wertsch 1999).

In line with the challenges already expressed, we have proposed to incorporate the critical reading of biographical and autobiographical recounts of real scientists from different narrative formats: written texts and audiovisual resources. Both are seen as powerful teaching scenarios for understanding the nature of knowledge and particularly the reconstruction of the processes involved in scientific research practices (Astudillo et al. 2011).

Specifically, we have selected the autobiographical narratives of two Argentine researchers (Gorla and Rivarosa 2013; Polop 2015). Both evoke, in the first person, two separate paths of research (40 and 15 years respectively) in evolutionary ecology and environmental health in the Argentine and Latin American context. In one case, the research focuses on the problem of Chagas disease, whereas the other case is concerned with the environmental conflict with population zoonoses (Hantavirus).

Narratives offer a rigorous documentation and experimental study designs, with their argumentative writing processes. The voice of the protagonists approaches a more human and real perspective of scientific practices (dedication, passion, frustration, risk, uncertainty, error, surprise, humor). Therefore, we believe that this facilitates the understanding of the complex conceptual frames of the issues raised and the progressive construction of explanatory models about it. It thus helps to demystify a long tradition that conceives the scientific practice as a linear, predetermined, successful and formalized process (Echeverría 2003).The stories (Collins 1999) presented in a colloquial language, with illustrations and photographs, become powerful scaffoldings for imagination and for the reconstruction of assumptions and models about the facts. The procedural skills of scientists are then made explicit, as well as their creative, interpretative and reasoned nature.

Specifically, we propose the student to read the selected narratives problematizing them and considering activities such as:

1. Reading a fragment of a narrative in small groups and identifying the problem that guides and motivates the research
2. Outlining the interpretation model that defines the starting point of each research
3. Outlining the identified narrated methodological processes:
   (a) Decisions and arguments
   (b) Design reformulations
   (c) Obstacles or problems and ways of addressing them
   (d) Emerging questions during the process
   (e) Resources used
4. Retrieving the initial interpretation model and reformulating it incorporating new variables, assumptions, and understandings that can be identified in the complete reading of the narrative
5. Identifying fragments of the text read that account for or deal with some of the following ideas: creativity, error, random, affectivity, culture, doubts, collaboration, and uncertainty.

Such approaches help students to gradually build a shared epistemological language (Vigotsky 1995; González García 2013). Precisely, it is the specific disciplinary contents (the kissing bug, mice, predators, population dynamics, zoonoses, evolution, reproduction, and birth, etc.) that are plotted in each described event and acquire meaning in the context of genuine problems of investigation.

Moreover, both stories hint at the contextual and ideological background of scientific work, the nuclei of meaning and the tensions that define the complexity of the researcher's task: a task that is both singular and collective, free and conditioned, methodical and anarchic, creative and routine.

Finally, the narratives offer a socio-political and territorial contextualization of the research problem, revealing edges of health and environmental vulnerability. This helps to demystify the apparent apathy and lack of ethical and political commitment of the scientific practice, allowing students to think about the meaning of their future practice as teachers or researchers in science.

In the same line of problematization, we have appealed to a second type of narrative device: *the biographical film*. We know that film productions have been of particular interest to science education specialists concerned about the influence that the meanings circulating in this type of media can have on the understanding of science. The overlaps between reality and fiction are used in didactic research to identify the possible positions and perceptions, the coherence and techno-scientific plausibility, the interpretation of the historical context that places the genesis of the story, the ethical problems emerging, among many others (Barnett et al. 2006; Sierra 2007; Piassi and Pietrocola 2001; García 2008).

In this sense, it is recognized that film production has a great potential for integrating multiple dimensions (temporal, economic, cultural, personal, geographical, ideological) and different languages that give strength and dynamism to the story. The emotional response that scripts cause also gain pedagogical relevance, leading to multiple sensitizations and identifications linked to values and attitudes.

Bearing in mind these possibilities, we have designed didactic scripts for viewing film productions of drama, which recreate *real cases and stories* of scientific research (e.g., "And the band played on," "Creation," "Casas de Fuego," "Agora"). Students are invited to enjoy viewing and then prepare a comment that integrates a wide range of categories:

1. The socioeconomic context
2. The cultural context
3. The role of the institutions involved
4. The personality and social life of the protagonists
5. The scientific methodology
6. The resources for research
7. The scientific community.

10.3.2 Classical Literature and Science Images

The didactic design that we share in this section is aimed at challenging the dichotomy *discovery* vs *scientific innovation*. We know that the imaginary representations about science often permeate the thinking of common sense and can be recurrent in the popular circulation of the figure of scientists and their practices. We are used to creations of science fiction and even to some dissemination products (such as documentaries or articles of mass circulation) that recreate successful, simplified, linear, and fast visions of practical science research.

Behind these popular recreations, a very strong idea is nestled: knowledge is *discovered*. Considering this, science means in terms of sudden realization, often the result of trial and error practices. It seems then that science refers to the task of *uncovering* or revealing facts that, remaining encrypted or hidden, *would be there* waiting to be found. It is the image that sneaks behind the representations of the crazy/genius scientist and the inventor scientist.

This type of representation also ends by communing with a linear interpretation of the progress of scientific knowledge: a progress that would be the mere accumulation of successful discoveries.

Recognizing this framework of meanings, we intend to explore and make explicit the intuitive ideas of students about the process of building developments in science from analyzing an episode taken from classical literature. We have chosen an excerpt from the novel *The Strange Case of Dr. Jekyll and Mr. Hyde*, written by Robert Louis Stevenson and first published in English in 1886.[2] Specifically, we have considered the first part of the last chapter that corresponds to the confession of the main character on the procedure by which he arrived at his great and dangerous discovery. We appeal to science fiction literature recognizing the power of narrative to mobilize the reader's imagination and enable the interpretation of a world that, perceived as unpublished, may be subject to multiple interpretations. We also recognize the nonlinear structure of the fictional narrative that favors the construction and deconstruction of educationally powerful analogies (Aguirre 2012).

After presenting the content and context of the work, students are invited to read the fragment chosen and answer the following questions: In what aspects is the story similar to the real science, and in what aspects is it different from it? Below there is a list of issues that students can identify while reading and categorize, with some differences, as *real* or *fictitious*.

1. Attributes of the scientist: wise, hard-working, ambitious, honest, caring, wealthy, virtuous
2. An intimate insight is the starting point of scientific research
3. Research process as approaching the truth
4. Science as the realization of individual desires and a miracle solutions provider
5. Scientific activity as the manipulation of nature for their own benefit
6. The scientific activity as solitary work that requires isolation
7. Trials and subsequent laboratory tests sharpen a perception that leads to the discovery of the desired formula
8. There is a component of incompleteness and fallibility in the research results
9. The scientific activity implies a level of transgression and danger
10. The results of the scientific application are unpredictable; some consequences go beyond the control of the investigator
11. The scientific research is conducted outside institutions and groups.

The discussion about it promotes the exchange of views on the limits and possibilities of the scientific enterprise, the non-neutral character of the researcher's task,

[2] Summary taken from Villacañas (2001): The play tells the story of Dr Jekyll, an elderly, wise, and respectable man, though with likely not very edifying pleasures, who after taking his own potion, undergoes physiochemical changes that will turn him into another being opposite to him in temperament and appearance. His two personalities, through their scientific methods, are split, resulting in two different creatures, the usual Dr Henry Jekyll and the sinister Edward Hyde, small, agile, young, repulsive, and frightening.

its collective and institutional nature, the role of rigor, error, and creativity, among others. Finally, and to move toward the central target, a second question is proposed to students:

> It is clear that the story we have shared illustrates a process that culminates with the definition of a product or new achievement: How would you outline this process? We also know that throughout history, scientists have participated in the construction of innovations or new products of various kinds. Retrieving the preceding analysis: What characteristics and moments would you modify or incorporate to arrive at a more accurate version of the real processes of scientific innovation?

This second activity allows, in a general way, the complex and recursive relationship between the design and the experimental evidence and the theoretical reference model to be introduced. Similarly, this attempts to reflect on the notion of the scientific problem as a starting point in the research process. Finally, we return to the reflection on the classic naive realist conception that denies or minimizes the theoretical load that scientists use to represent the world to model it (Adúriz-Bravo 2014).

10.3.3 The Historical Account As a Source of Analogies

We now present a new didactic design thought to reflect with students on how the explanatory models in science change and how much they are transformed throughout history. In this regard, we intend to problematize the classical linear, cumulative, and individual vision of scientific progress. Therefore, we propose the following stages:

1st Stage During the class, students are contextualized in the second half of the twentieth century, collecting some of historical events that led to the critical review of the relationships among society, culture, science, and technology. Thus, we contextualize the movement that has been called the new philosophy of science and we focus on new discussions of the classic question: how does science progress?

From here, the intention is to introduce the proposal of Thomas Kuhn and the concept of scientific revolution as one of the most influential perspectives of the moment. As a teaching strategy, we have chosen to bring a caricature of Charles Darwin that we believe may serve as a trigger resource analogy between scientific revolution and political revolution; an analogy that the same Thomas Kuhn offered as scaffolding in the presentation of its central thesis. The proposed activity begins as follows:

Then, the students are proposed some questions as: What ideas does this cartoon suggest to you? And the expression that goes with it? Why do you think that Charles

Undoubtedly, and even without knowing his work in depth, we recognize in Charles Darwin a prominent figure in the history of the natural sciences. Immediately, some images come to our minds of this character among which surely the white-bearded old man is the first. This seems to be the most popular image movement. Anyway, if we are to dive among the many ways of representing Darwin through graffiti, cartoons, advertisements, etc., we can discover some surprises as in Fig. 10.1.

Fig. 10.1 "Che-Darwin" caricature. (*Source*: Illustration by Sergio Villar)

Darwin is associated with the idea of revolution? What other figure or historical situation is associated with Charles Darwin in this representation? What do they have in common?

2nd Stage After collecting and systematizing the preliminary ideas of the students, we intend to define the concept of scientific revolution that Thomas Kuhn proposes from building the analogy with the notion of political revolution. To this end, students are given a *simple story* about the case of apartheid in South Africa as an example of political revolution. Based on this example, students develop *a scheme integrating the components and moments involved in the story (PR) and the relationships that can be established between them.*

It is important to note that the account given was specially developed for this didactic design. The case was selected for its relevance to the construction of the analogy proposal and required successive consultations with a specialist in political philosophy. Figure 10.2 is a possible resolution scheme of the proposed activity.

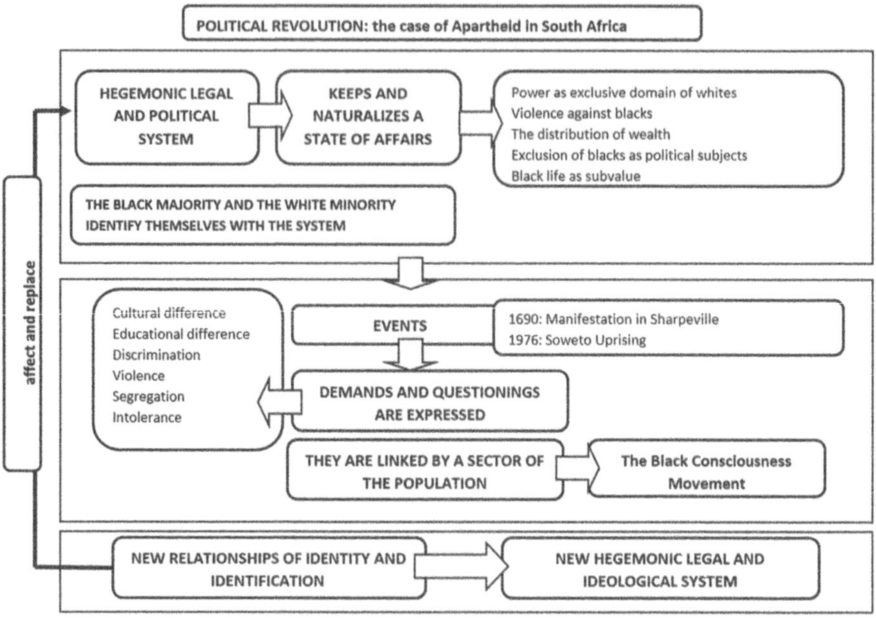

Fig. 10.2 Scheme of the apartheid case

3rd Stage Keeping this scheme in mind, students are invited to imagine how a process of scientific revolution could be represented reflecting on questions such as: What changes in a process of scientific revolution? Why would these changes happen? How does this process occur?

Then, a short text (UBA 1997) is given to the students that defines in a simple way the notion of scientific revolution (SR) from notions of physics, proposed by Thomas Kuhn, in the 1960s. Students, in light of the text, are asked to revise their initial ideas and try to build an analogy between the two notions (PR and SR), considering its central components. In the case of CR, students should consider at least the following elements: dominant paradigm, normal science, anomalies, emerging paradigm.

4th Stage Finally, the cartoon presented at the beginning is brought back together with the initial question: Why is Charles Darwin associated with the idea of revolution? With some specific readings, the latter activity enables the revolutionary nature of the contribution of Charles Darwin to the model of evolution to be analyzed. This analysis leads to rebuild briefly:

1. The general and main components of the fixist paradigm – creationist
2. Some of the most important anomalies and, in response
3. The main contributions of the evolutionary model.

10.3.4 Science in the News

This latter didactic design is aimed at problematizing the idea that science is the source of absolute truth about the universe and therefore scientists are the owners of that truth. We are concerned about the persistence of such representations that carry a set of serious omissions:

1. The unfinished nature of scientific knowledge
2. The component of uncertainty of the scientific activity
3. The limits and scope of science as a human construction, among others.

Moreover, this vision of science truth is consistent with the widespread image of the scientist as neutral, objective, capable of the abstracting of passions to become a mirror of reality (authors). In other words, people see the scientist as the owner of truth and knowledge, abstracted of interest or affection, and as the one who has clear criteria in decision-making processes about the world that has been revealed to himself.

On these ideas and images, we want to reflect with students from reading a piece of local news: "The UTN gives its approval to the new landfill" (available in *www.lavoz.com.ar/content/la-utn-le-da-el-visto-bueno-al-nuevo-relleno*). The story recounts the situation experienced in 2012 about the discussion of the installation of a new sanitary landfill in the southern sector/area of the city of Cordoba, Argentina. In particular, the news presents the position of a group of academics expressed in a technical report on the subject.

The activity is organized into two parts:

1st Part After reading the story, students reflect on the following questions:

1. Who gives the "thumbs up" to the new sanitary landfill in the southern city of Cordoba? What would be the criteria underlying that statement?
2. How do you interpret the following expressions? What questions could be asked about them? "[...] (the landfill site) meets certain minimum requirements from a technical point of view" "[...] downplayed the relationship between the burial and the harmful effects it may have on nearby communities" "I cannot find a real impact, scientific, provable of the burial in the population"
3. How are the voices of other actors involved in the problem considered? What other impact prospects could be considered?

From these questions, the discussion progresses in consideration of possible simplifications of the problem of reductionism that could be behind the report, from the conception of knowledge that seems to be behind some expressions. The discussion always arises in hypothetical terms recognizing that this is a deliberate cut of reality and the processes referred to. The main intention is that the news is a trigger to emerge and problematize common sense ideas about the activity of scientists, their speech, and their civic responsibility (Palma and Wolovelsky 2001; Echeverría 2003).

2nd Part After this first discussion, students use a brief narrative that reconstructs the history of the conflict spanning the last 40 years. The narrative introduces the different dimensions of the problem, its articulation with other problems of the sector, its historical evolution, and the participation of the different sectors involved: municipalities, judiciary, universities, government agencies, companies' waste collection, neighbors and neighborhood organizations (Astudillo et al. 2013). The story culminates with the satellite image of the conflict zone indicating the locations of old and new landfills.

After this reading and with the intention of concluding new contrasts and dimensions of analysis, students are posed a final series of questions: Can you identify other voices there, perspectives and dimensions of the problem? What aspects seem to leave out the proposition of specialists in the news analyzed? How does the story seem to be continued? What other ideas are suggested by the conflict map?

10.4 As a Final Reflection and to Summarize

In his article "Scientific Training and Philosophical Reflection" (Cupani 2001 p. 150), Alberto Cupani interrogates himself about what must be the objective, strategy, and the benefit of including epistemological reflection on the training of future scientists: "I think the goal can be characterized, synthetically, saying that it consists of doing research more lucidly and responsibly". Thinking also about future science teachers, we believe that the answer may be the same: the aim will be to teach more lucidly and responsibly.

The didactic alternatives presented are based in this idea, using the philosophical ideas as a tool for analysis and meta-reflection on scientific practice in the context of technological, political and socio-cultural breakthroughs. The alternatives described try to recover – within the framework of instances of historical re-contextualization – the controversial and complex nature of research on the biological sciences; not only because of the status of development of their theories, but by the broad technological impact that today involves its transposition into society and culture (Technoscience).

In this regard, our own investigations (Astudillo et al. 2012a) into the implementation of these types of didactic scenarios show how students gradually approach and overcome cognitive and epistemological obstacles regarding the work of research as institutional business, in dialogue with society, history, and culture. These new meanings are those that lead to some mobilization and changes in their ideas, such as:

1. Denaturalization of absolutist conceptions of scientific knowledge that enables the argumentation of *relativistic* positions
2. Understanding of the progress of knowledge in terms of breaks, subjectivity, contrasting ideas, and progressive integration

3. Recognition of the necessary interplay of method and theory and the collaborative and collegial nature of science research
4. Adoption of a *law* perspective with regard to scientific work, assuming the public nature of science and the net power relations of various kinds
5. Recognition of the principle of methodological pluralism in terms of triangulation of procedures, creative imagination, with ethical and political implications.

Our concern for educational processes in science (Astudillo et al. 2012b, 2013) demand a critical and profound reflection, not only on the science we do, but on the processes of training. In this regard, we believe that this approach can make a contribution to the construction of less dogmatic and radical positions and bring a more humane and humanizing perspective to scientific activity (Latour and Woolgar 1995; Palma and Wolovelsky 2001).

Our challenges continue along the lines of designing and promoting new forms of activities capable of enabling:

1. An immersion in scientific culture as dilemmatic, creative, and political work
2. Anchoring in their own disciplinary knowledge
3. Contextualization in the specific problems posed by the work of the researcher and the teaching
4. Problematization from a scientific perspective and civic literacy
5. Concern about the major social–scientific dilemmas of our time.

References

Acevedo, J. A., Vasquez, A., & Paixão, M. F. (2005). Mitos da didática das ciências acerca dos motivos para incluir a natureza da ciência no ensino das ciências. *Ciencia y Educacao, 11*(1), 1–16.

Adúriz-Bravo, A. (2009). La naturaleza de la ciencia “ambientada” en la historia de la ciencia. *Enseñanza de las Ciencias*, N° extra VIII Congreso Internacional sobre Investigación en Didáctica de las Ciencias, Barcelona, 1178–1181.

Adúriz-Bravo, A. (2014). La historia de la ciencia en la enseñanza de la naturaleza de la ciencia: Maria Sklodowska-Curie y la radioactividad. *Educació Química Eduq, 16*, 10–16.

Adúriz-Bravo, A., Perafán, G., & Badillo, E. (2002). Una propuesta para estructurar la enseñanza de la filosofía de la ciencia para el profesorado de ciencias en formación. *Enseñanza de las ciencias, 20*(3), 465–476.

Aguirre, R. (2012). Pensamiento narrativo y educación. *Educere, 16*(53), 83–92.

Astudillo, C., Rivarosa, A., & Ortiz, F. (2011). Naturaleza de la ciencia y enseñanza: Un aporte para la formación del profesorado. *Revista de Educación en Biología, 14*(2), 13–23.

Astudillo, C., Rivarosa, A., & Ortiz, F. (2012a). La reflexión metacientífica a través del cine: un estudio sobre los saberes docentes. *Revista Eureka Sobre Enseñanza y Divulgación de las Ciencias, 9*(3), 361–391.

Astudillo, C., Rivarosa, A., & Ortiz, F. (2012b). *Progresión del conocimiento profesional de docentes de ciencias naturales: Un estudio en trayectos de formación inicial y continua*. Doctoral dissertation, Universidad Nacional de Córdoba, Córdoba, Argentina.

Astudillo, C., Rivarosa, A., & Ortiz, F. (2013). Historia y actividad experimental: una discusión relevante que interpela la enseñanza. In C. C. Silva & M. E. B. Prestes (Eds.), *Aprendendo ciência e sobre sua natureza: abordagens históricas e filosóficas* (pp. 363–376). São Carlos: Tipographia.

Barnett, M., Paul, W., Gatling, A., Anderson, J., Houle, M., & Kafka, A. (2006). The impact of science fiction film on student understanding of science. *Journal of Science Education and Technology, 15*(2), 179–191.

Bruner, J. (1991). *Actos de significado: Más allá de la revolución cognitiva*. Madrid: Alianza.

Carrascosa, J., Perez, D. G., Vilches, A., & Valdés, P. (2006). Papel de la actividad experimental en la educación científica. *Cuaderno Brasileiro de Ensino de Física, 23*(2), 157–181.

Collins, F. (1999). The use of traditional storytelling in education to the learning of literacy skills. *Early Child Development and Care, 152*, 77–108.

Cupani, A. (2001). Formación científica y reflexión filosófica (acerca de la utilidad de la filosofía de la ciencia en la formación del científico). *Context, 4*(5), 148–153.

Datri, E. (2006). *Una interpelación desde el enfoque CTS a la privatización del conocimiento: Política, ideología y tecnociencia*. Río Cuarto: Universidad Nacional de Río Cuarto.

Echeverría, J. (2003). *La revolución tecnocientifica*. Madrid: Fondo de Cultura Económica.

Ellis, C., & Bochner, A. (2000). Autoethnography, personal narrative, reflexivity: Researcher as subject. In N. Denzin & Y. Lincoln (Eds.), *Handbook of qualitative research* (2nd ed., pp. 733–768). Thousand Oaks: Sage.

Furinghetti, F., & Morselli, F. (2009). Every unsuccessful problem solver in unsuccessful in his or her own way: Affective and cognitive factors in proving. *Educational Studies in Mathematics, 70*, 71–90.

Gallego Torres, A., & Badillo, R. G. (2007). Historia, epistemología y didáctica de las ciencias: unas relaciones necesarias. *Ciência & Educação, 13*(1), 85–98.

García, F. (2008). Bienvenido míster cine a la enseñanza de las ciencias. *Revista Eureka sobre Enseñanza y Divulgación Científica, 6*(1), 79–91.

González García, J. (2013). Imagination: A creative tool to achieve meaningful understanding of new information. *Global Journal of Human Social Science Linguistics & Education, 13*(8), 70–79.

Gorla, D., & Rivarosa, A. (2013). *Aventuras de un biólogo: Desandando los caminos de la investigación científica*. Anillaco: CRILAR.

Jenkins, E. W., & Pell, G. (2006). *The relevance of science education project (ROSE) in England: A summary of findings*. Leeds: University of Leeds.

Kornhaber, M., Fierros, A., & Veenema, S. (2004). *Multiple intelligences: Best ideas from research and practice*. Boston: Allyn & Bacon.

Latour, B., & Woolgar, S. (1995). *La vida en el laboratorio: la construcción de los hechos científicos*. Madrid: Alianza Editorial.

Lemke, J. (2006). Investigar para el futuro de la educación científica: nuevas formas de aprender, nuevas formas de vivir. *Enseñanza de las Ciencias, 24*(1), 5–12.

Matthews, M. (1991). Un lugar para la historia y la filosofía en la enseñanza de las ciencias. *Comunicación, Lenguaje y Educación, 11*(12), 141–145.

Matthews, M. (2009). *Science, worldviews and education: Reprinted from the journal science & education*. Dordrecht: Springer.

Palma, H., & Wolovelsky, E. (2001). *Imágenes de la racionalidad científica*. Buenos Aires: Eudeba.

Piassi, P., & Pietrocola, M. (2001). De olho no futuro: ficção científica para debater questões sociopolíticas de ciência e tecnologia em sala de aula. *Ciência & Ensino* 1. Número especial.

Polop, F. (2015). *¿Cómo se aprende a ser investigador?: Entre pasiones y opciones*. Río Cuarto: UniRío.

Rivarosa, A., Polop, J., Astudillo, C., Valdano, S., & Trefs, F. (2014). Las narrativas sobre prácticas científicas, como puente para la reflexión epistemológica y metodológica de futuros científicos. Ponencia presentada en la *III Conferencia Latinoamericana del Grupo Internacional de Historia, Filosofía y Enseñanza de las Ciencias*. Santiago de Chile, 17 al 19 de noviembre 2014.

Sacks, O., Kevles, D., Lewontin, R. C., Gould, S., & Miller, J. (1996). *Historias de la Ciencias y del Olvido*. Madrid: Edición Siruela.
Sierra, C. (2007). Fortalezas epistemológicas y axiológicas de la ciencia ficción: un Potosí pedagógico mal aprovechado en la enseñanza y divulgación de las Ciencias. *Revista Eureka sobre Enseñanza y Divulgación de las Ciencias, 4*(1), 87–105.
Scolari, C. (2013). *Narrativas transmedia: Cuando todos los medios cuentan*. Bilbao: Deusto.
Toulmin, S. (1997). *La comprensión humana: el uso colectivo y la evolución de los conceptos*. Madrid: Alianza.
UBA XXI. (1997). *Introduction to scientific thinking: The progress of science. Module 3*. Buenos Aires: Eudeba.
Vigotsky, L. (1995). *Pensamiento y lenguaje*. Barcelona: Paidos.
Villacañas, B. (2001). De doctores y monstruos: La ciencia como transgresión en Dr. Faustus, Frankestein y Dr. Jekyll and Mr. Hyde. *Asclepsio, LIII, 1*, 197–211.
Wertsch, J. (1999). *La mente en acción*. Buenos Aires: Aique.
Wolovelsky, E. (2008). *El siglo ausente: manifiesto sobre la enseñanza de la ciencia*. Buenos Aires: Libros del Zorzal.

Part III
History of Science

Chapter 11
Abraham Trembley (1710–1784) and the Creature That Defies Classification: Nature of Science and Inquiry Through a Historical Narrative

Filipe Faria Berçot, Eduardo Cortez, and Maria Elice de Brzezinski Prestes

11.1 Introduction

This chapter offers a historical narrative for biology teacher training courses, and for science or biology classrooms in secondary school. The theme of the narrative is the study of the Genevan naturalist Abraham Trembley (1710–1784) about the nature of "freshwater polyps" (*Hydra*), in the first half of the eighteenth century.

After studying theology, philosophy, and mathematics in Geneva, Trembley moved to the Netherlands in 1732, where he worked as a private teacher to the sons of the English count William Bentinck, resident in The Hague. It was at the earl's mansion, in Sorgvliet, southern Netherlands, while teaching the young Antoine and Jean Bentinck, that Trembley studied the hydras, in the summer of 1740.

As a better way to teach the boys about natural history, he conducted several series of observations and experiments with these organisms. From these studies, Trembley went on to describe new findings on the movements and nutrition of hydras, and, what caused the greatest impact at the time on its forms of reproduction, by regeneration and by budding.

F. F. Berçot (✉)
Biological Sciences Graduate Program (Genetics) of the Institute of Biosciences of University of São Paulo, São Paulo, SP, Brazil
e-mail: bercot@ib.usp.br

E. Cortez
Science Teaching Inter-Unit Graduate Program of the University of São Paulo, São Paulo, SP, Brazil

M. E. B. Prestes
Department of Genetics and Evolutionary Biology, Institute of Biosciences, University of São Paulo, São Paulo, SP, Brazil
e-mail: eprestes@ib.usp.br

M. E. B. Prestes, C. C. Silva (eds.), *Teaching Science with Context*, Science: Philosophy, History and Education,
https://doi.org/10.1007/978-3-319-74036-2_11

The regeneration of parts of an organism, such as the legs or claws of a lobster, had been known since Antiquity, but the formation of two new individuals from one was a novelty not imagined before. Likewise, the budding was known as a phenomenon for the multiplication of plants only, not for animals. Thus, the description of reproduction by regeneration and sprout of the hydras, represented much more than two specific empirical discoveries: it ended up as the reconceptualization of animal reproduction. In the midst of this, Trembley also contributed to the discussion of the criteria that separate animals from plants, overcoming the preliminary challenge that was to identify the nature of that organism so little known at the time: was the hydra in fact an animal or was it a vegetable?

Through historical studies on reproduction such as those performed by Trembley, students may be able to learn how to ask fundamental questions about reproductive processes of living beings, and become more motivated to understand all varieties of reproduction known today. Historical narratives can pave the way for engaging students in the inquiry process, formulating science-guided questions, using evidence to answer questions, formulating evidence-based responses, assessing explanations in the light of alternatives and justifying and communicating proposed explanations (NRC 2000).

Historical episodes, if well-chosen and well-prepared, may be useful for providing problematic situations, and thereby introducing epistemic questions, in particular, about how scientific knowledge is developed and comes to be accepted as reliable (Allchin 2013, p. 20). In other words, by accompanying and discussing the historical research questions, through a process of investigation, such as posed in the narrative presented here, students can deepen their understanding of the nature of science (NOS), for instance, of the scientific claims and its construction through tests and the trial and error process (Allchin et al. 2014, p. 474).

In short, episodes of history of biology such as Trembley's research may represent good opportunities to foster the so-called "contextual teaching" of science with the more general objective of preparing students to become citizens able to make well-informed decisions on the scientific subjects that pervade their everyday life.

Yet, applying episodes of the history of science (HOS) in the classroom is not a very simple task to accomplish. Among the most important conditions, it requires the training of teachers in themes of history, philosophy, and social studies of science – subjects that are still not common in undergraduate courses of science teacher training in Brazil. The history of biology is also an emergent field of research in the country, with few materials in Portuguese that can be used on teacher training courses and in secondary schools.

The elaboration of such historical instructional materials, in turn, also requires hard work. Following the methodology of HOS, they must be based on primary sources, i.e., original texts of the researchers of the past, in dialog with the secondary sources, i.e., interpretations of the original works already undertaken by experts. Such studies require the time and investment of researchers, who bring together knowledge from the related fields of history, philosophy, and sociology of science.

Simultaneously, for application in secondary school, such historical resources must also be appropriate to the school culture and the specific group with which these teachers work.

Before presenting the historical narrative on Trembley's research itself, some theoretical perspectives adopted for the construction of such instructional resources are discussed in the following four sections.

11.2 History, Story, and Inquiry

One of the first particular challenges that were faced in the present work is related to the promotion of a desirable didactic manipulation of past time without losing historical acuity. In other words, defining how to adjust *history* into a *story* (Klassen 2007, p. 337). To ensure historical acuity and to avoid pseudo-histories, it is necessary, first of all, for the story to be placed in its original context, doing "justice to original sources and sound historical interpretation"; besides this "conceptual point," history is intended to introduce the "humanistic element into the process of learning science" (ibid.). Bringing the student closer to past researchers as real human beings, not geniuses, portrayed not only by his achievements but also by his limitations and failures, produces a desirable increase in interest in sciences. To achieve this purpose, some level of flexibility or poetic license is allowed. A crucial condition is that if one wants to create a history-based story, those imaginary details "must be consistent with the historical record" (ibid.).

Accordingly, whether the objective is to motivate students, history may be a backbone on which to frame problems and to illustrate scientists at work – a real science-in-the-making work, not the stereotypical scientific work shown in cartoons or in biased historical anecdotes. It is, however, important to avoid using history merely as a illustrative context or a list of ordered chronological events about a unified subject, like a "chronicle" (Norris et al. 2005, p. 538).

To accomplish that and to make this approach meaningful to learning, many science educators and researchers have sought to combine history with an inquiry process (Monk and Osborne 1997; Rudge and Howe 2009; Allchin 2013; Fouad et al. 2015). In fact, history can provide an investigative pathway that ultimately reaches a known solution and a stable closure to the inquiry, conferring security for instructors (Allchin 2011, p. 2).

To ensure that the historical episode would be structured by an inquiry environment was the second challenge faced in this work. The chosen alternative for accomplishing this purpose was the so-called "interrupted narrative" (Allchin 2011, p. 12), inspired in the notion of "punctuate" narrative (Roach and Wandersee 1993). Namely, the story happens around the successive inquiry activities – the core moments for active learning. The narrative is alternately formed by starts and ends, carefully prepared to frame the students' own thinking and then develop it further. Here, besides nourishing an "appetite for the narrative" or for knowing more story (Norris et al. 2005, p. 535), the students may become interested in trying to achieve the outcome themselves, or in participating in creative problem-solving (ibid).

The moments of interruption designed for inquiry, expressed in the narrative on Trembley's studies on hydra, were formulated as "Think questions." It follows the example provided in the historical narrative about Christiaan Eijkman and the cause of beriberi (Allchin 2013, Chap. 10) and worked along the discipline "Teaching History and Nature of Science" taught by Douglas Allchin, at the Institute of Biosciences of the University of São Paulo, in 2015.

These "Think questions" are punctuated breaks interrupting the narrative to produce opportunities for students to reflect actively on their own learning while exploring selected features regarding aspects of NOS. While answering these questions, students are invited to step into the shoes of the scientist and try to find possible solutions within the original historical context in which the story is happening. It means that, to achieve answers, students must engage in explicit reflections within their group, considering epistemic scientific–procedural problems, social and funding biases, challenges regarding communication of ideas and discoveries, and other NOS-related features (Allchin 2013, pp. 165–166).

In these moments, the narrative works the tension between inquiry teaching, characterized by the openness that allows the students to imagine different possibilities of investigation, and the documented historical routes, with their own paths that have been effectively traced (Allchin 2013). This means that any kind of anticipated answer, or any biased clues for the ultimate response should not be prematurely given. It is important to create an "open-ended plot," enabling students to experience the aspects of NOS present on lessons, following the "blind way" of the "science-in-the-making" (Allchin 2014b, p. 11).

For each "Think question" present in the narrative, there is also a corresponding "Teaching note." These notes are intended to work as pedagogical support, guiding teachers among the possibly diverse responses. "Teaching notes" do not indicate exclusively correct answers; they are more like tools for the teacher "to fish for answers" and, whenever possible, encourage and stimulate dialogue from various perspectives among students to shed light on the pertinent dimensions of reliability in scientific practice, or whole science (Allchin 2013, p. 177). History itself may be a benchmark and it is suggested that the whole narrative be shared with students at the end of the work.

11.3 Explicit and Reflective Approach

It is noteworthy that even a proper interaction between history and inquiry may not be effective for learning, if proposed with an implicit approach. Recent studies have presented evidence that only "explicit" and "reflective" approaches improve students' conceptions of aspects of NOS.[1]

As highlighted by Khishfe and Abd-El-Khalick (2002), one important feature to consider what the "explicit" label refers to is that the stated, express discussion of

[1] For explicit and reflexive approaches see: Abd-El-Khalick and Lederman (2000), Adúriz-Bravo and Izquierdo-Aymerich (2009), Rudge and Howe (2009), Rudge et al. (2013), Duschl and Grandy (2012), Vesterinen and Aksela (2013).

aspects of NOS is not something casual that emerges spontaneously in the lessons. On the contrary, it must be "intentionally targeted" (ibid.) as a designed instructional objective accompanied by specific activities, along with the instructor's orientation, to ensure that student attention is drawn to them (Khishfe and Lederman 2007). In turn, the label "reflective" must be linked to the instructional elements. It means that educational activities must provide students with real opportunities to analyze historical scientific problems from the different perspectives of NOS. Students can be stimulated to identify relations between their own educational activities and the practices undertaken by scientists, for example, and, in the end, to be able "to draw generalizations about a domain of knowledge", for example, epistemology of science (Khishfe and Abd-EL-Khalick 2002, p. 555).

Thus, a narrative that uses a historical background, when explicitly interrupted by inquiry questions enabling students to reflect on their learning, may help these same students to acquire a better understanding of scientific knowledge and about *how we know*, and *how science works* (Allchin 2014a; Dagher and Erduran 2014, p. 114).

11.4 Key Points for Developing a Historical Narrative

As Metz and collaborators recall, based on other authors, in a general sense, a narrative might be defined by "anything that tells a story, in whatever genre" or "telling someone else that something happened" (Jahn 2001 and Herrenstein-Smith 1981 apud Metz et al. 2007, p. 315 respectively). From that general conception, Norris and collaborators have proposed some key points to organize and develop a narrative. As Metz and collaborators pointed out, a first group of these keys is formed by the "event-tokens," the "narrator," the "past time," and the "reader" (or listener). These elements are defined as follows:

> Event-tokens: are the particular occurrences covering particular actors at a particular place and time; they are chronologically related, involving unified and interconnected themes.
>
> Narrator: the agent relating a narrative and is who determines the point and purpose of the story; selects the sequence of events they will be told so that they adjust in a significant whole.
>
> Past time: the time of a narrative concern to past, however, it is allowed to manipulation of time, if needed.
>
> Reader: is the one that interpret the text as a narrative in order to approach it with appropriate expectations and anticipations. (Norris et al. 2005, p. 545)

Moreover, combined with these previous factors, there are others that can be of great help in improving the narrative, namely, the "narrative appetite," the "structure," the "agency," and the "purpose," which are defined as follows:

> Narrative appetite: the desire created in the readers or listeners to want to know what will happen; it might be based on several possibilities allowing them to make anticipations and generating suspense.

> Structure: usually, narratives start with imbalances and are introduced complications; at the final, there must be success or failure; it is structured around two independent time moments – sequence of plot events and sequence that they are relating.
>
> Agency: narratives involve human beings, such as actors, which cause and experience events and moments; they are responsible for their own actions;
>
> Purpose: aims to help us better understand the natural world and humans' place in it. Thereby, as claimed by Witherall "stories giving us possibilities to imagine and feel the experience of others". (Norris et al. 2005, p. 543)

These key elements should not be taken as a *required guideline* to develop a narrative, but as a *guidance helping to structure* it to achieve better efficacy. Although not discussed in detail in this chapter, these features, along with validations accomplished with different groups of pre-service biology teachers, served as reference points for calibrating the writing in its different phases, from the initial to the current version of the narrative.

Moreover, the Trembley narrative that follows took into consideration Michael Clough's arguments (2006, p. 474) that effective teaching of aspects of NOS through history requires inclusion of, among other things, connections between historical and contemporary conjuncture to prevent students from perceiving past events as "wrong-thinking" that has now been corrected. Clough also suggests, whenever possible, the use of the "voice" of scientists to emphasize and to provide authenticity to NOS aspects.

Yet, as a material for the classroom, and as Allchin pointed out in his course at the University of São Paulo, the narrative presented here moves away from a standard academic text in some respects. The narrative contains neither footnotes nor quoted sources – although the main ones are provided at the end, in a list of bibliographical references consulted. Also, unlike the usual HOS, the narrative is written in the present tense, i.e., in the historical time shared by the community of readers, to establish a sense of communion between the reader and the characters. This simple strategy facilitates promotion of a timelessness that means the connection between "being in time" and "telling about it" (Colby 2008, p. 64).

The narrative combines historical report with pedagogical issues, following the sense proposed by Metz and collaborators:

> [...] by broadening perceptions of the manner and contexts [...] so as to include imaginative and manipulative elements that provide interactive experiences for students that are more conducive to implementation by science teachers. (Metz et al. 2007, p. 313)

Thus, the narrative takes the liberty of introducing some related fictional elements such as the creation of dialogues or the expression of plausible sensations experienced by the characters – avoiding, of course, distortions that compromise the image of how science works.

Finally, like any instructional material, this resource is configured as a prototype, an example that allows adaptations, alterations, rearrangements, to fit the teaching–learning purposes of different school cultures and pedagogical objectives. It is up to the teacher, in dialogue with the pedagogic project of the school where he/she works, to decide on the opening of this space in the curricular planning.

The narrative can be associated with other resources and different activities that lead to a productive application of the HOS in teaching. As part of the doctoral research in progress of the first author of this chapter, the narrative is one of the products of a teaching learning sequence (TLS), in the sense proposed by Méheut and Psillos (2004). Having detailed the theoretical assumptions for development of the historical narrative, in the next section follows a discussion of the educational guidelines of the TLS' objectives through the adoption of design-based research (DBR), as proposed by The Design-Based Research Collective (2003).

11.5 The Narrative Development Through a Series of DBR Iterations and Its Design Principles

Design-based research has been accounted in the recent decades as a new research methodology aimed at improving educational interventions through iterative cycles of analysis, design, development, and implementation, based on collaboration among researchers and practitioners in real-world settings, leading to contextually sensitive design principles (Wang and Hannafin 2005). Such interventions can be exemplified by an instructional resource, or an assessment, or a technological tool, namely, testing the effectiveness of a particular learning environment or instrument (Anderson et al. 2014).

In addition, DBR clarifies how an intervention might work in the real teaching context by providing specific characteristics, the "design principles," that increase, or somehow, ensure that the educational objectives can be reached (Design-Based Research Collective 2003). In short, it can be expressed as follows:

> If you want to design an intervention X for the purpose/function Y in context Z, then it is advisable that the intervention has the characteristics A, B, and C, because of arguments P, Q, and R. (van den Akker 1999)

Design principles are the outputs, the heuristic statements (characteristics A, B, and C) concerning knowledge produced about whether and why an intervention might work in a specific context. Regarding the credibility of this methodology, it can be remarked that the iteration process engenders too much feedback data, and the classroom tests can be considered a kind of validation (constructing the arguments P, Q, and R). Therefore, it is important that data are collected from the design group, i.e., the teacher' planning and students' activities (intervention X for the purpose/function Y in context Z).

In other words, through these research techniques, it is possible to obtain credible information concerning the attendance of educational objectives in particular situations and contexts. Thus, DBR assumes that useful feedback can be gathered through assessment of the quality, novelty, and usefulness of artefacts such as the narrative presented here (Juuti and Lavonen 2006).

Guided by DBR methodology, the historical narrative presented in this work is the result of two iterations, i.e., educational interventions with similar groups of pre-service biology teacher students of public universities in the city of São Paulo, in the southeast region of Brazil.

Exposing some details about the iterations is relevant to discuss the changes between the initial and current versions of the narrative that led to the acquirement of the design principles described below.

At first, a prototype was developed, consisting of a short text exploring just part of the story, structured with the guidance of the key points discussed by Norris et al. (2005), and followed by a few questions concerning aspects of NOS. The prototype was applied to pre-service teachers enrolled in a São Paulo public university discipline, in 2015 (University A), devoted to the development of strategies for teaching biology in secondary school. The undergraduate students read the narrative in small groups and wrote group responses to posed questions. The analysis of data collected in this first iteration showed that, as result of the type of questions posed, the answers were simple interpretations of the text, without any deep reflection on the aspects of NOS.

Thereafter, facing these first collected data, the first author of this chapter proceeded to make a review and a new design for a second prototype of the narrative. Changes were performed in the narrative to privilege the inquiry learning and the parameters discussed in the previous section, notably the ones proposed by Allchin (2013, 2015) and Metz and collaborators (2007). The second iteration was carried out with a similar group of pre-service biology teachers, from another public university in the city of São Paulo (University B). New incomers from this group were considered in the elaboration of the third prototype of the Trembley narrative, the one that is presented in this chapter.

The current version of the narrative contemplates some of the design principles developed over the two iterations. For instance, to create an environment of more engagement and immersion for students, the whole story was completely reviewed. New elements of the historical episode were added to the narrative, in particular, those conveying a feeling of dead ends. More than just added, dead ends were even occasionally emphasized to put forward challenges and create opportunities of inquiry and reflection among students, as illustrated by Think questions 1 and 4. This feature, *the presence of dead ends*, is one design principle for the present historical narrative.

A second design principle derived from the two iterations resulted in detachment from the strict chronology and historical detailing to explore fictional dialogues and arguments among characters, resulting in the *creation of imaginary but plausible scenarios* to provide both fluidity to the story and "appetite" to readers.

However, the most significant refinement and redesign concerned the number and the way of posing inquisitive questions about aspects of NOS. The first two prototypes presented only a few questions about NOS, and they were posed at the end of the narrative. In this third version, by means of the above-mentioned "Think questions," *inquiry questions were placed interrupting and interweaving the story*, at different moments of the text, constituting a third design principle.

Regarding the didactic strategy, the pre-service teachers to whom the historical narrative was applied were organized into small groups to perform a collective reading. The instructor (the first author) led the classroom dynamics, instructing participants to interrupt the reading at each "Think question" and they were challenged to discuss the answer in their respective groups. The instructor's role was to guide the groups to avoid large deviations from the main objectives – both inquiry and aspects of NOS. In addition, during the activity, the instructor acted to facilitate the group discussions and mainly to guide students to explicitly and reflectively consider how the examined historical case-study, in the context of their own work, could help their understanding of NOS. As suggested by Rudge and Howe (2009, p. 569), at no time during the group work did the instructor give "correct answers" or "valid" interpretations regarding the aspects of NOS. Instead, the instructor sought to show them the strengths or limitations in their responses and how tenuous the interpretations could be.

The active engagement experienced by pre-service teachers during the activity is noteworthy. They took the group discussions seriously and the motivational factor provided by the story plot can be clearly realized. The data that demonstrate these conclusions are discussed in the doctoral thesis by the first author of this chapter.

In the next section, the historical narrative is presented, interrupted by "Think questions" and their respective "Teaching notes."

11.6 The Historical Narrative: "Abraham Trembley and the Creature That Defies Classification"

This story takes place in the first half of the eighteenth century, in the Netherlands, one of the European countries that had a tradition of natural history research. At Sorgvliet, a district near The Hague, a young Genevan Abraham Trembley (1710–1784) is serving as tutor to the sons of the British Count Willem Bentick: Antoine, 6 years old and Jean, 3 years old. Trembley has studied theology, philosophy, and mathematics, but his passions are the subjects of natural history (minerals, plants, and animals) and the very art of teaching (the theme of one of his books). Capturing insects and observing the germination of seeds or activities in a beehive are some of the strategies he uses to hold the attention and incite curiosity of the children. Thus, at the same time, he teaches literacy, morality, religion, math, and, of course, natural history.

It is a summer morning in 1740. As usual, they are in the backyard of the Bentincks' property, a lush place, composed of a series of well-designed gardens. Also, there are many trees trimmed in conical shapes, hedge-covered tunnels, a symmetrical system of interconnected paths from one part of the gardens to another and an elegant ornamental system of ponds containing several species of exotic fish.

There, Trembley and his pupils are looking for aquatic insects in the ponds of the property (Fig. 11.1). Fitted with a net trap, they collect everything that seems puzzling so that they might observe it closely later.

Fig. 11.1 Trembley and his pupils collecting aquatic insects on a pond. (Source: Trembley 1744)

Fig. 11.2 Trembley and the boys at his *cabinet*. (Source: Trembley 1744)

Inside the mansion, there is a space dedicated to the workplace and, all over the windowsill, there are many glass containers where Trembley keeps insects and aquatic plants. The boys are observing what they had taken from the pond (Fig. 11.2). With hand magnifiers, they notice some tiny green shapes attached to the aquatic plants they have just collected.

It is possible to imagine the boys pointing out their fingers impatiently and showing Trembley: "Look at that! It seems that there is a tiny leaf of grass over a small plant!"

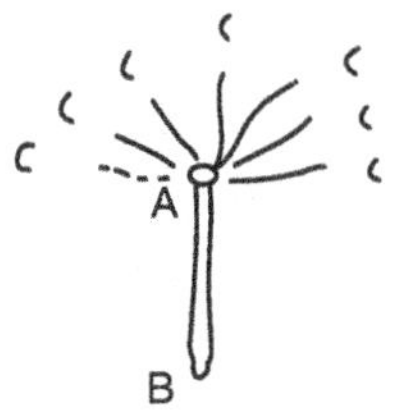

Fig. 11.3 Trembley's draft of the organism collected. (Source: Trembley 1740)

Readily, Trembley places a magnifying glass over the "things" and what he sees looks like some kind of parasitic green plant. Also, he realizes that they were cylindrical bodies measuring less than 2 mm, with arms or threads at the top of the body (Fig. 11.3). Although like quite a tiny green plant, as its shape also resembled a sea octopus, they referred to them as "polyps."

So, he shakes the container gently and places it facing the sunlight, to see whether something happens. Thereafter, he realizes that agitation of the water decreases the threads, and when movement ceases, they slowly extend to their original size. This situation intrigues him very much and Trembley draws the children's attention: "Look at this, boys! These 'things' apparently have some kind of voluntary movement." Thereupon, Trembley writes:

> Thinking that the polyps were plants, I could scarcely imagine that this movement was their own. Yet, those slender threads projecting from one of their extremities did appear to move by themselves and not in response to the agitation of the water. The more I followed of these arms, however, the more it seemed that it had to come from an internal cause and not from an impetus external to the polyps. [Besides,] this contraction and all the movements I saw the polyps make as they extended once again roused sharply in my mind the image of an animal. I likened them at first to snails and other creatures that contract and extend. [However] I was still influenced, I admit, by their shape and their green color. I thought it not impossible that they might be sensitive plants [...]. (Trembley 1744 apud Lenhoff and Lenhoff 1986, pp. 5–6)

Apparently, Trembley is facing a dilemma: the creature's body has a shape that resembles some kind of a plant and an animal; it has a green color, like a plant; yet, it seems to have voluntary mobility, like an animal!

What does this dilemma mean? To understand it, we need to imagine the context at the time. Let us remember that we are in Western Europe, in the mid-eighteenth century. Several people are interested in the so-called "cabinet of curiosities," making observations of plants and animals, drawing them, and discussing their findings in the receptions usually carried out at the wealthy nobilities' salons. Some of them are amateur scholars; others are natural philosophers who devote themselves to systematic studies on natural history. Some seek the pleasure of knowledge about the world; others are motivated by the purposes of admiring and knowing God's creations. A shared thought among many of them is the belief in a natural and immutable order imposed by the Creator. Thus, there is a consensus that each living being is a live representative of the first organisms created by God at the beginning of the world.

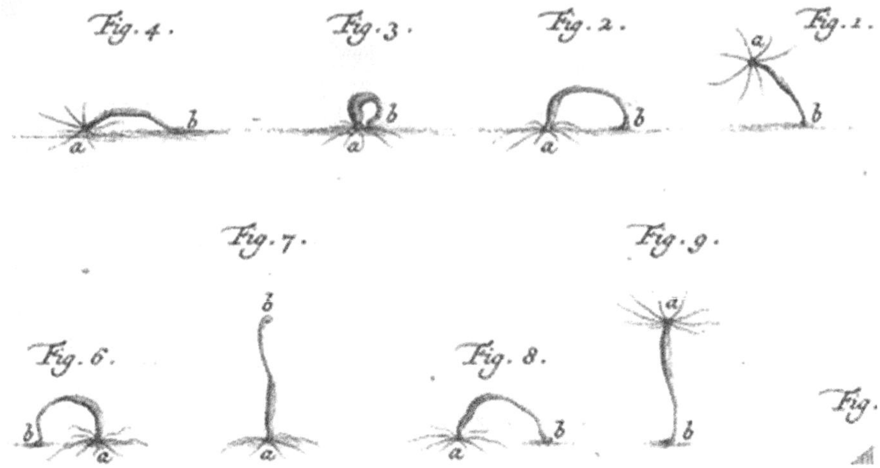

Fig. 11.4 Modes of locomotion of the polyps: worm-like movement represented in the top line, from the right to the left (Figs. 1–4), and by somersaults, in the bottom line, from the left to the right (Figs. 6–9). (Source: Trembley 1744)

In addition, they share a very old idea that the very nature of all creatures on earth, from minerals to men, rests on a scale of continuity and gradation. This idea was known as the "chain of beings." Largely, the work of the naturalists aims to explore this natural order and express it by means of classification systems. Among many different proposals of the time, one of these systems was acquiring significant adepts throughout Europe: that developed by the Swedish physician and naturalist Carl von Linné (1707–1778).

His proposal for classification was presented in a small book, more like a leaflet, containing only 11 pages in its first edition, in 1735: *Systema Naturae*. Linné would publish many enlarged editions of this work, incorporating descriptions of plants newly discovered and collected around the world. In 1770, he publishes the 13th edition, containing about 3,000 pages! In addition, Linné seeks to understand behavior and to know how divine creations were able to perpetuate from the beginning. Some perplexing questions that naturalists had to deal with arise from such studies, such as: "If the chain of beings is continuous and gradual where are the connection points among them? There would be 'missing links' in the chain of beings? Would 'zoophytes' (a name for organisms that share characteristics of both plants and animals) be those missing links?"

Proceeding with their daily observations, Trembley and his students realize that, beside the ability to contract or extend the threads or arms, the creatures also change places! Indeed, they noticed that the creatures even have *two particular modes of locomotion*. Trembley records both kinds of locomotion in a wonderful drawing (Fig. 11.4). The organism exhibits a movement similar to a worm, after the anterior end *a* is fastened securely, the posterior end*b* detaches and draws close to anterior end *a* (Figs. 1–4 in Fig. 11.4). Other times, the creature seems to prefer to move by somersaults (Figs. 6–9 in Fig. 11.4). Those are definitely animal features! But they are green and have a plant shape as well!

Look at the drawings from right to left, from left to right... Nowadays, science books adopt a single pattern from left to right. In addition, it is intriguing that the illustration published by Trembley in 1744 is very much like those that appear in current zoology text books?

Think (1) Facing this deadlock, plant or animal, Trembley must take a stand. At this point and based on these data, what could you say about the nature of these organized bodies? What would justify the choice in favor of one trait or another? What might "carry more weight" or motivate such an option?

Teaching Note We expect the students to perceive that from Trembley's perspective, there is an equivalent amount of arguments in favor of the animal or plant nature of the polyp. In this case, any decision must consider factors other than the evidence. Here, we have a place for students to discuss the role of personal decisions in science-making and how such decisions are taken. They are expected to discuss such aspects of NOS, while they themselves must make a decision.

At this point, according to what he tells us in his book, Trembley decides to consider them animals because of their ability of locomotion, i.e., he places more value on this trait than on its green color and its form of an aquatic plant, or the capacity of contraction as sensitive plants. However, still challenged by this apparent dilemma, he decides to continue studying these unusual creatures. And then a new problem emerges.

From the initial observations, Trembley noticed that the "arms" could appear in unequal numbers. Thus, it was almost impossible not to compare those arms with the branches and roots of plants, which do not maintain an equal pattern either. Again, the analogy with plants! Then, if the "arms" were truly similar to plants' branches or roots, they should be able to grow again after having been cut off.

Reflecting upon this, he decides to carry out an experiment. Rather than cutting off just one "arm" to watch how it grows again, he divided the organism into two pieces – maybe because, it is worth remembering, he was handling on an organism of less than 2 mm in length.

> I conjectured that if a polyp were cut in two and if each of the severed parts lived and became a complete polyp, it would be clear that these organisms were plants. Since I was much more inclined to think of them as animals, however, I did not set much store by this experiment; I expected to see these cleaved polyps die. (Trembley 1744 apud Lenhoff and Lenhoff 1986, pp. 6–7)

Think (2) Through his previous observations, Trembley believes that the creatures are animals. Yet he performed the experiment of cutting the specimen. How could the results obtained throughout the experiment help to guide him to take a decision?

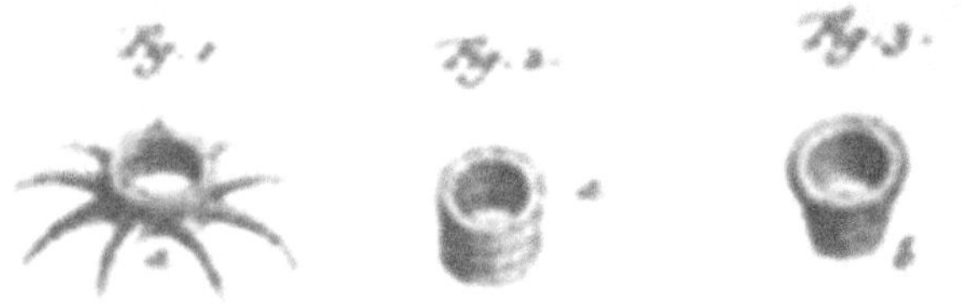

Fig. 11.5 Drawing of an experimental procedure in which Trembley divides a polyp transversally into three parts: Figs. 1, 2, and 3. (Source: Trembley 1744)

Teaching Note While addressing this question, the students explore the balance (or imbalance) between the scientist's personal ideas and the evidence obtained by experiments. To achieve an answer, they must think about how much evidence must exist or how strong evidence must be for a scientist to discard a previous concept or not.

After having carefully cut an individual polyp into three pieces (Fig. 11.5), he placed each piece into a small glass dish containing water. At first, it looked like nothing but three little green dots. But at the end of the day, the results of the experiment were not exactly as expected. This baffled him! The pieces had extended completely. Moreover, 9 days after, neither Trembley nor the boys were able to differ one from another. The pieces became three whole and identical organisms! This kind of multiplication, in which a portion or a bud is detached from the original organism and then grows into a new individual is a well-known feature in plants!

Think (3) Previously, voluntary movement and locomotion seemed to be the most important criteria in considering the organism as an animal. Thus, what is to be done in the face of this new result? How should this new evidence be interpreted in the light of the modes of reproduction?

Teaching Note This question may be used to evaluate consistency between the previous response and this one. The students are experiencing the process lived by Trembley: it is one thing to make predictions about a test constituted in the light of potentially reinforcing one's own ideas – remember that Trembley thought he really would kill the organism with the cutting process – and it is another thing to make an intellectual decision after seeing something unexpected happen. Should the scientist develop another test or should he/she give up his/her previous ideas after one experiment that could prove him/her wrong? This is the dilemma faced by the students after these three "Think questions."

Trembley knew that parts of some animals were able to regenerate, such as lizards' tails or lobster's paws. Easily observed in nature, the phenomenon of regeneration has been known since Antiquity. But a complete regeneration of an animal? No! Animals do not "recreate" themselves. As is known in that period, such an event is commonly referred to only in plants; nowadays, the phenomenon is known as "cutting" (stem, leaf, or root cutting), usual in a manioc (or cassava) or a rose bush, for example.

In Trembley's time, there was a consensus among most naturalists that for the generation of a new animal life, there must be mating between a male and a female, which is sexual reproduction. Also, it had been largely accepted since Antiquity, that some animals (and plants) could appear by spontaneous generation, as seems to happen with the "animalcules" seen through magnifying glasses and microscopes in infusions, or with intestinal worms and worms inside fruits, or with eels. Spontaneous generation seemed to be the best explanation available at the time for such cases. Amid so many new elements, Trembley finds himself disoriented: "Plants do not 'walk'! Animals do not fully regenerate! But these organisms are able of both exploits!"

Think (4) We are facing a dilemma similar to the initial (Think 1), but now it seems worse! Would it be a dead end? It is necessary to adjust observations (more experiments or more information)? Or should you review your criteria of classification of organisms inside the groups of animals and plants? Feel free to put down your thoughts about it!

Teaching Note In contrast to the previous couple of questions, this one does not ask about considerations regarding a scientist's personal ideas, but the ideas shared by the community of scientists – such as criteria for classifications. What happens when these criteria no longer suffice? Is this an example that could lead to what the philosopher of science Thomas Kuhn called a scientific revolution? How long should a scientist insist on a current paradigm? These are questions that rely on the process of developing a response to this "Think" exercise, and, as educators, we expect to assess how our students perceive the role of persistence, personal motivations, theory-laden thinking, and paradigm shifts in science.

Confused, but even more motivated to solve this dilemma, Trembley decides to write to a notorious naturalist of the time. Director of the Academy of Sciences of Paris, the French naturalist René Antoine Ferchault de Réaumur (1683–1757) has developed studies in several areas, in physics (developing a thermometer) and industrial processes (research and production of steel, glass, and ceramics). Moreover, Reaumur has developed many works on natural history, such as on the formation of pearls in bivalves, the anatomy and physiology of plants and animals, the locomotion of snails and sea stars, the regeneration of parts of crustaceans, the electrical discharge in specific fish, spider behavior, and silk production, etc. In addition, Réaumur devoted much attention to insects, whose observations and experiences are detailed in his work *Memoire pour servir a l'histoire des insects* (Memory to serve the history of insects), divided into six volumes and published between 1737 and 1742. The first three volumes of this work have been avidly read by Trembley.

In a letter sent to Réaumur, Trembley detailed his observations and conducted experiments. But Réaumur, as an experienced naturalist, although betting on the animal nature of the creature, made no rash decision until he had personally seen them. Thus, he answers Trembley: "If you, Mr., happen to have many of these organized bodies, perhaps you can send me some of them".

Immediately, Trembley takes 15 specimens, places them within a tightly closed glass bottles, and sends them to Réaumur in Paris. Nevertheless, after nearly a week of travel, the organisms arrived dead.

Think (5) Remembering you are handling small aquatic creatures; why do you think they did not survive the trip? What procedures and care do you think are necessary to avoid the death of these organisms during a trip, by horse or carriage, from The Hague to Paris?

Teaching Note This is an epistemic problem, i.e., students need to create a hypothesis and find a way of testing it. While doing so, they must consider both animal and plant necessities, as it is still not clear what this creature is. Should they prime for common necessities between these two realms of life? Or should they try to guess, once again, whether those creatures are animals or plants before taking a decision? Although reflecting on these details, students are also expected to reflect on what is behind the processes of coming up with a hypothesis and a test.

After the failed attempt, Réaumur writes to Trembley telling him that owing to excessive zeal when closing the bottles with wax, the organisms may have died because of a lack of air. Réaumur advised closing the bottles with cork rather than wax, allowing air diffusion into the container.

Following Réaumur's advice, Trembley captured some new organisms and stored them in three bottles. However, before sending them to Paris, he took the bottles with him on a horseback trip of about 15 km. At the end of the route, he realized that the creatures were still alive. Now he had found a safe way to send the tiny organisms to Paris.

In fact, this time Réaumur receives the creatures alive and immediately starts his studies. He has no more doubts: those small green creatures caught from Dutch ponds were truly animals! And it was Réaumur that suggested naming them "polyps" because of their resemblance to sea octopuses.

Meanwhile, in the Netherlands, during his continuous observations, Trembley notices something like a small excrescence, a dark green dot stuck on the side of a polyp's body. Puzzled by this novelty, and following a procedure taught by Réaumur, he isolates the individual to keep observing it attentively. At that moment, he may have imagined seeing something similar to the growth of a branch. Indeed, the dark green dot grows like a branch. Surprisingly, after some time, the "branch" comes loose from the body where it was being held. Then, the "branch" becomes an isolated body and similar to the original form from which it detached! A whole new individual!

One more time, Trembley uses the procedure to isolate the organism under observation. He separates the detached creature and puts it in another container. Would this just-seen phenomenon be a new form of reproduction? Once again, the new individual performs the same phenomenon under Trembley's own eyes. The "branch-like" structure arises, grows and, as before, detaches, forming a whole new living being (Fig. 11.6).

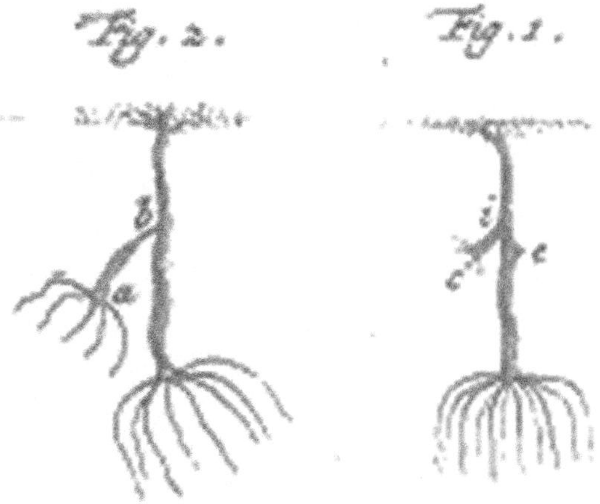

Fig. 11.6 Polyp with an excrescence (Fig. 1) that grows and, after some time, originates another organism, similar to the first one (Fig. 2). Trembley has adopted an ingenious procedure to better observe it: he passed a string over the water surface that served as a substrate so that the polyp could lie fully extended upside down. (Source: Trembley 1744)

What a fascinating creature! At this point, both Trembley and Réaumur are entirely convinced that they have enough evidence to concede that the creature is a real animal. But not without departing from the old conceptions of what plants and animals are, as reproduction by budding is another typical feature of plants!

Think (6) One more time, it seems that we have reached a "non-expected" outcome. How should it be interpreted? How could this observation help you to classify the creature? Would it be time to change your position (developing, for instance, an animal–plant concept)? Explain it.

Teaching Note Once again, we need the students to reflect on the nature of the experiment and on the meaningfulness of a result to the process of taking decisions in science. Throughout this historical narrative, this may be the aspect of NOS that is more frequently addressed by the students. Thus, an instructor may take this chance to assess consistency in the students' thinking process and/or in how much depth the students discuss this point. A question that the instructor might try to answer is "Can my students get deeper and deeper into the discussion on the nature of decisions in science? Or would they stick to their original perspective?" If your students do indeed stick to their first perspective on the animal/plant nature of the polyps, you can provoke them to think about why they are insisting on their first idea, regardless of the evidence against it.

Again, Trembley addresses a letter to Réaumur reporting his observations about this "strange" new form of reproduction observed in the polyps. But Réaumur is not very convinced and responds by telling him that there should not be any budding: "Maybe you are not observing properly. It is possible that there may be some egg hidden somewhere under the body of the mother polyp, as usually occurs in lobsters, for instance" (Fig. 11.7).

Think (7) What would be the reasons "behind" Reaumur's skepticism about generation by budding on polyps? What would have motivated him to not accept that an

Fig. 11.7 Female lobster with its eggs "hidden" at the bottom of the abdomen. (Source: Wikimedia commons)

animal can be generated by budding? What would you do to convince him about Trembley's new observations and the reliability of the data?

Teaching Note Here, the students must address the nature of criticism and skepticism in science. It is a question designed to make them think about the role of these features in the process of making science within a community of researchers, which may also lead them to think about science as a collective enterprise. If students choose to think in a more epistemic way, they are expected to address the question of science-making being theory-laden, i.e., posing questions, developing tests, and interpreting results being steps of the scientific process that can be better understood in the light of a scientist's ideas and under the current theories and paradigms. If a single group of students is able to address both dimensions of NOS – science as human, collective enterprise, and science as a theory-laden process – we can tell that those students are starting to perceive and understand the concept of whole science.

On reading Réaumur's answer, Trembley focuses one more time on these isolated specimens to repeat observations on the unusual form of animal reproduction. To verify such an occurrence, he isolates a few generations of individuals who were born by budding, and always observes the same results: all of them had generated "branches" and subsequently became new organisms. In addition, Trembley draws up an experiment to investigate if there are any eggs "hidden" somewhere in the mother polyp.

To perform the experiment, he takes an individual whose excrescence is still in its early moments and carefully extracts the dot from the body of the mother polyp

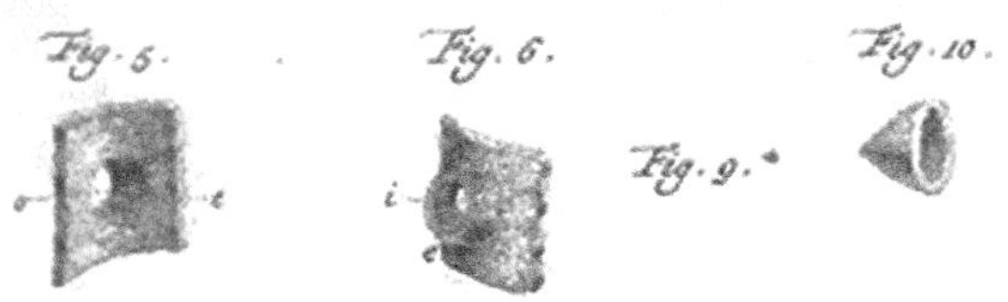

Fig. 11.8 Section of a polyp's body after the appearance of a bud seen in Fig. 11.10. (Source: Trembley 1744)

(Fig. 11.8). Analyzing the two structures under the microscope, he realizes that the excrescence is nothing more than a continuity of the body of the mother polyp and there is absolutely no trace of something that could be considered an egg.

After painstaking observations, would Trembley be able to eliminate the ideas on reproduction by eggs on polyps, as cogitated by Réaumur?

Réaumur's skepticism regarding the ability of the creatures to reproduce by budding lies in the fact that he (like most naturalists at the time), as mentioned before, ignores other forms of animal generation that do not contemplate participation of both sexes, i.e., sexual reproduction, or spontaneous generation.

Moreover, Réaumur is an advocate of the theory of the preexistence of beings, which was also shared by many naturalists of the time. According to this theory, the offspring of each living being has already preexisted in "seeds" (egg or sperm) within the bodies of the parents, since the Creation. Thus, if polyps are animals, there must be eggs somewhere inside their bodies.

Réaumur and Bernard de Jussieu (1699–1777), his friend and another French naturalist, are performing an experiment with a species of tufted polyp (*Lophopus*) (Fig. 11.9). After months of several and continuous observations, they have identified, inside the tufted polyps, something similar to dark granules, which they suspect might be eggs.

Now it is Trembley who asks to have these organisms, which seem so different than those he had been dealing with. As soon as he receives them, he starts his observations and confirms what Réaumur and Jussieu have been able to identify: tufted polyps have eggs! However, the polyps studied by Réaumur have a darkened color; those observed by Trembley were green. During observations, Réaumur had described the dark polyps as having a calcareous part, which he considered a kind of refuge for the animals.

Trying to understand more about these dark polyps, Trembley spent a lot of time studying the creatures. Finally, he was able to claim that they were two distinct species. Despite the efforts, Trembley knew that it would not be enough to convince Réaumur about the absence of eggs on polyps. In other words, it seems another dead end. Will it be?

Both Réaumur and Trembley agreed that polyps are animals because of their movements, which cause them to change their position. Nevertheless, as is well known, animals neither fully regenerate themselves nor are they born by budding. Are there other criteria that must be considered?

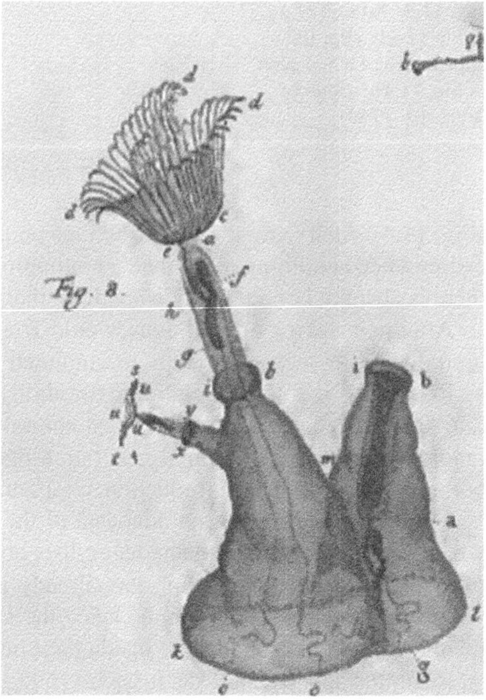

Fig. 11.9 A specimen of a "polyp" that Réaumur and Jussieu are studying and in which they believe to have found eggs. Currently, this organism is considered a different kind animal, a bryozoan, not cnidarian, like polyps. (Source: Trembley 1744)

11.6.1 Trembley Seeks Clues Regarding the Polyps' Nourishment

So far, Trembley had never seen polyps feeding. Therefore, he did not know how (or if) polyps were able to absorb nutrients. In a usual morning of observations, Trembley notices a polyp apparently using its tentacles to capture a "worm" (an insect larva) and making a movement to bring it toward its… "Wait a moment!! It would be a mouth?!?" Trembley perhaps might have thought that way. A few minutes later, the worm captured was within (?!) this polyp! Paying as much attention as possible to this new event, Trembley notices the worm now placed inside a "tube," which he deemed relevant to appoint as a stomach. These tiny creatures have a gap, a mouth, between their tentacles. What a curious organism!

It is worth recalling what has been done up to now. First of all, Trembley had thought that the polyps might be plants because of their green color, their structure, and their apparently stationary way of life. Then, he passed on to considering them as animals owing to their voluntary locomotion. However, continuing with his observations, he realized that those organisms could reproduce in a similar manner to plants. Now, Trembley is aware that these organisms have the ability to take in food, as animals do!

Think (8) Do you think that just the action of "trapping a supposed prey" can be sufficient to make someone consider an organism as an animal? If so, in such cases we can surely say that a carnivorous plant (such as *Dionaea muscipula*) is an animal. Right? No? Maybe? Why? Is this situation another dead end or do you need to obtain more knowledge? What else could you require to take a decision?

Teaching Note This question reflects the matter of establishing criteria and acceptable exceptions for those criteria in science, notably in biological sciences. While considering Trembley's case about feeding in polyps and the fact that he is using this case to reaffirm his animal–nature idea, once more the students must address science as being a theory-laden construct, and reflect on the role of a scientist's point-of-view when he/she is developing his/her studies. Also, when the students are asked about whether they could think about further studies, we expect them to address the role of chemical nutrition in feeding, anticipating the next stage of the story and feeling that they are in Trembley's shoes.

Trembley is aware that the capture is only half a response; once a similar situation is observed in carnivorous plants as well. Yet, he realizes that the polyp had not merely captured the worm; it seemed that it had been ingested and placed inside the polyp's "tube." Thus, Trembley thinks that this case must be studied in the light of nourishment. Of course, both animals and plants absorb nutrients, but in different ways.

To accomplish the task, Trembley uses a theoretical framework published in a work entitled *Elementa Chemie* (Elements of chemistry) published by the Dutch physicist and botanist Herman Boerhaave (1668–1738), an authority on studies of natural history, like Réaumur. In his book, Boerhaave offers a differentiation between plant and animals through aspects of nourishment.

Basically, his ideas were: nourishment in plants occurs because of the absorption of nutrients extracted directly from the soil, through roots; in animals, the action of absorption happens because of the inner vessels. To him, animals were similar to hydraulic machines, whose lacteal vessels were spreading internally through their bodies and were able to transport liquids (nutrients) to all parts. Following Boerhaave's ideas, Trembley performs experiments of dissection to find the lacteal vessels that are so elementary in animals; thus, he may be able to elucidate the form of nourishment of polyps.

The first experiment is to cut a polyp lengthwise, splitting it along the stomach ("tube"; Fig. 11.10). After that, Trembley put the halves under a microscope and, instead of finding the "lacteal vessels," the only thing he could observe were identical small green granules spread in both internal and external sides of the stomach. In the same way, in an experiment with a transversal cut, he places each half in separate and identified glass dishes. Yet, this time, the cut polyp needed only 1 day to regenerate completely rather than 9 days, as happened in the previous experiment.

The second experiment is inspired by ideas proposed in the book *Histoire physique de la mer* (Physical history of the sea), published in 1725 by the Italian naturalist Luigi Marsigli (1658–1730). Marsigli claimed that some marine plants had

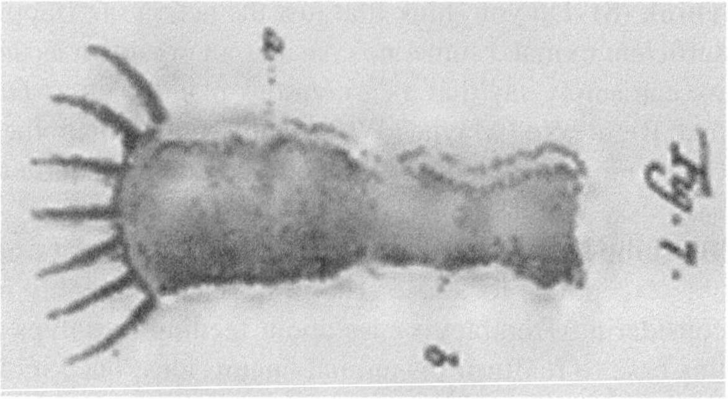

Fig. 11.10 Longitudinal section of a polyp: "stomach". (Source: Trembley 1744)

glands or vesicles able to extract nutrients from sea water that would also work when the organism was immersed in some sort of "nutritive juices." Trembley thought about putting some of the polyps in "nutritious juice" (which may be a kind of solution formed by water and nutrients) to verify if polyps have the ability to absorb nutrients through all the parts of their body. However, he was not able to find substances that could serve as nutritive juices.

Think (9) As seen, Trembley found support in the different knowledge available at the time about the nourishment of animals and plants. Yet, apparently, none of these contributions seem to shed light on the polyp nutrition system. What would have been, then, the role of this knowledge in his research?

Teaching Note With this question, we intend to see how students address confirmation and denial of hypotheses in science and, through their responses, to check once again how well they understand the epistemic NOS.

Nevertheless, Trembley does not give up trying to understand polyps' nourishment. Returning to his observations, he pays attention to the characteristics he had noticed in the previous cutting experiment: the granules spread along the polyp's stomach.

To analyze these granules once more, Trembley induces an isolated polyp to take in a "worm." After some time, he takes this polyp and, observing it under the microscope, he realizes that the granules look swollen and filled with dissolved particles! The observation of these filled granules led Trembley to reason his more industrious experiment: turning the polyp inside out! (Similar to a finger of a glove)! Can you imagine turning inside out a creature less than half a centimeter long by 2 or 3 mm thick?

> It occurred to me that if the granules which were on the external surface of the skin were closer to these nutritive juices, they would be the first to became filled with it, and perhaps the polyp would be nourished as thoroughly as when the juices pass first into the vesicles in the lining the walls of the stomach. (…) [So] I thought of inverting them so that the external surface of their skin would form the walls of their stomach. I had very little confidence that I would see this experiment succeed, but I did not believe [it was a] proper [reason] not to try it. (Trembley 1744 apud Lenhoff and Lenhoff 1986, p. 155)

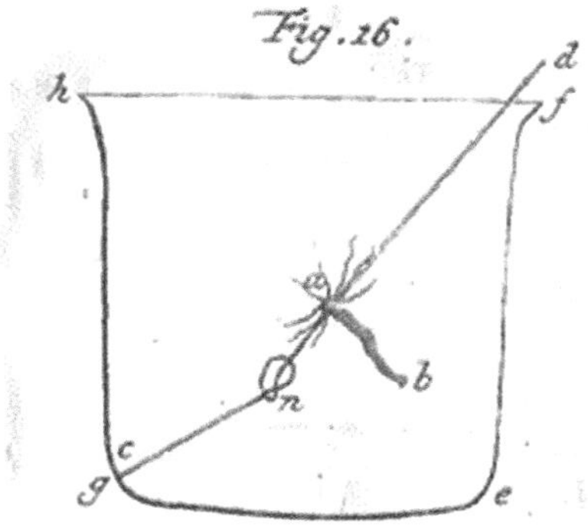

Fig. 11.11 An "inverted" polyp inside a glass container filled with water. The letters (a), (n), (c) and (d) represent the *line* that ties the polyp. (Source: Trembley 1744)

To perform the intricate task, Trembley proceeded through a series of careful steps until the polyp was inverted. In the end, he tied the inverted polyp, situating it on the middle of a glass recipient, in such manner that it was kept away from touching the sides or the bottom of the jar, thus preventing itself from turning back (Fig. 11.11).

Trembley has thoroughly studied his inverted polyp and, one more time, he saw that the creature not only survives, but also keeps reproducing and capturing and ingesting food.

Obviously, this experimental uniqueness would be the subject of doubt and skepticism. Aware of this, Trembley invites other naturalists to testify to his experiment. Among the guests were Pierre Lyonnet (1708–1789), his friend and Dutch naturalist; Jean-Nicolas-Sébastien Allamand (1713–1787), another Dutch naturalist; and Bernhard Siegfried Albinus (1697–1770), a German anatomist who lived in Netherlands and would later become Boerhaave's successor. All of them could witness the experiment of the inversion of the polyp carried out by Trembley.

The analysis of results obtained through extensive experiments and observations in the light of the selected theoretical framework, in particular the studies of Boerhaave, allowed Trembley to explain that polyps did not contain any structures similar to those found in an animal body. Indeed, Trembley found that polyps have neither "lacteal vessels" nor something that looks like roots, such as was claimed by Boerhaave. Yet, the fact that the polyps (even inverted) perform digestion, internally, was in agreement with the definition of internal digestion as known in more complex animals.

Facing all these outcomes, Trembley reaches the main conclusion that the polyp could be a simple animal, formed by a single "nutritive vessel," i.e., a "tube" open at one of its ends. The polyp itself was the nourishing vessel, unlike other animals whose vessels are located inside their bodies, such as worms or caterpillars.

Since the beginning, Trembley was more inclined to believe that polyps were animals, especially because of their movements... But the number of "intermediate" plant characteristics he had found caused mistrust. Thus, he always sought more and more evidence until he could defend his final position. He wrote:

> In order to be able to decide that a particular organism is neither plant nor animal but belongs to some intermediate class, it would be necessary to know precisely all the attributes of plants and animals. As we seen above, we are still very far from such knowledge. Only when we succeed in acquiring it, can [we] create other classifications of organisms. In the meantime, it is much more natural to consider the polyps and various other organisms which have been given the name zoophytes as animals which show more noteworthy similarities to plants than other animals. (Trembley 1744 apud Lenhoff and Lenhoff 1986, p. 185)

11.6.2 Trembley's Polyps: From Ponds to Fame

The unusual property of the regeneration of polyps that leads to the formation of two new individuals aroused the curiosity of many people. Réaumur, for instance, was judged worthy of presenting the phenomenon to the Academy of Sciences, in Paris, with Trembley's permission.

Over a period of days, Réaumur was able to demonstrate the regeneration of polyps to both scholars of the Academy and the people of the court, always leaving the audience stunned with the "marvelous" ability of a small organism to "rebuild" itself.

Yet, the presentation was not restricted only to the French Academy. After some time, many scholars in several places in Europe had become aware of the "incredible" property of regeneration of the "pond's phoenix." This was made possible because of Trembley's "strategy of generosity," namely, he distributed live specimens throughout Europe. For this purpose, he had developed another technical achievement. To ensure that the specimens would arrive alive in the most distant places, Trembley sent the polyps not alone, but inside glass containers with aquatic plants, developing an aquatic system suitably conditioned to withstand long-distance travel. Moreover, this was accompanied by detailed instructions, both for the conservation of the organisms, and for the replication of the experiments. Thus, he has contributed to the consolidation of a procedure that was already shared among the experimental naturalists of the time: the replication of experiments.

However, even after many public presentations, people were still not fully convinced that an animal, after having been cut two or three times, might be able to reproduce. There is heavy distrust among scholars, especially in England. After the subject appeared in the prestigious journal *Philosophical Transactions*, the discovery was a matter of criticism, irony, and skepticism for many people.

Think (10) Why, even after several public presentations witnessed by many people, was there still resistance about accepting the complete regenerative ability of polyps? What might be the reasons for such skepticism? What could be done to respond to unbelievers?

Teaching Note This is another question for students to reflect on the role of skepticism and criticism in science; however, this time, students must also consider a new dimension of NOS: the communication of performed works and trust in the communicated data. The students are not expected to achieve a "right answer," but to consider a whole range of possibilities behind distrust in science, such as personal beliefs, previous personal work that made one famous, the role of nationalism in accepting or not accepting research, among other factors.

In London, fame, skepticism, and disbelief about the unusual features of polyps spread quickly in the halls of the Royal Society. Sir Martin Folkes (1690–1754), President of the Society at the time, trying to put an end to the matter, writes to Count Willem Bentinck, Trembley's boss, asking him to confirm the reliability of Trembley's experiments. Bentinck readily asks Trembley to prepare and to send a consignment of polyps to the President of the Royal Society.

As soon as Folkes receives the organisms, he follows Trembley's instructions and performs the regeneration experiments at his own house, under the scrutiny of 20 Fellows of the Royal Society. Shortly afterward, Folkes repeats the experiment at a meeting of the Society and over 1 week, more than 150 people are able to witness it. Through these public presentations and following others, Folkes managed to minimize the criticism and irony of those "unbelievers," at least, in London.

Now Trembley, a simple tutor of the children of a prominent politician, is seen as a "celebrity," not only in one but two of Europe's greatest intellectual centers: Paris and London. The experimental skills that he has developed, such as the isolation of individuals, the elaboration of experimental series, the replication and variation of experiments on polyps, the sharing of results among the community of scholars, lead him to fame among naturalists and the European scholars of the eighteenth century.

All these elements were worthy of merit. In 1743, 3 years after the beginning of his studies on polyps, Martin Folkes honors Abraham Trembley with the Copley Medal, one of the most important English scientific acknowledgments. He wrote:

> We are not less sensible of your great candor, and the Readiness you have shown not only to transmit to us faithful abstracts of your own experiments, but also to send us the Insects themselves, whereby we have been enabled to examine by ourselves, and see with our own Eyes the Truth of the astonishing Facts you had before made us acquainted with. (Folkes 1743 apud Ratcliff 2009, p. 110)

Nevertheless, Trembley's strategy of generosity and his fame as the naturalist of the discovery of the "phoenix," elicited not only interest, but also jealousy and envy.

Henry Baker is an English naturalist who studies chemistry and has earned a medal from the Royal Society for his skills in observations on crystals. He is very interested in Trembley's works on polyps. As a Fellow of the Royal Society, he has access to the letters exchanged among naturalists, including those interchanged between Trembley and Folkes. Seeing the opportunity to improve his reputation as a naturalist, Baker quickly replicates some of the experiments that Trembley had described in the letters, but had not formally published. Upon obtaining the results, Baker publishes, in 1743, a work entitled "Attempt towards a natural history of the polyp." This work reaches the public shortly before the publication of Trembley's book.

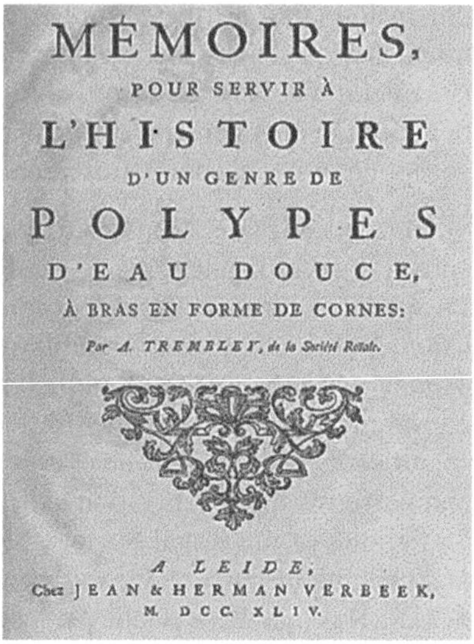
MÉMOIRES,
POUR SERVIR À
L'HISTOIRE
D'UN GENRE DE
POLYPES
D'EAU DOUCE,
À BRAS EN FORME DE CORNES:
Par A. TREMBLEY, de la Société Royale.

A LEIDE,
Chez JEAN & HERMAN VERBEEK,
M. DCC. XLIV.

Fig. 11.12 Frontispiece of the work published by Trembley in 1744

Think (11) What is to be done regarding this real situation of anticipating results obtained by others? Is it plagiarism? Do you think the situation could influence Trembley's "strategy of generosity"? How? What could have been the consequences for Baker?

Teaching Note In this question, we expect students not just to point to plagiarism as a wrong and unethical procedure, but also to present a solution to it, now putting themselves in the shoes of a representative of the Royal Society, for instance. For them to be able to do so, they must reflect on the implications of plagiarism in science and on how ethics assumes an important role in NOS too.

Finally, subsidized by a well-established culture of patronage (a kind of incentive or financial support offered by wealthy naturalists to fund "research" and disclosure expenses), 4 years after having found those tiny green creatures, Trembley publishes, in 1744 in Leiden, his book entitled *Mémoires pour servir à l'histoire d'un genre de polypes d'eau douce, à bras en forme de cornes* (Memoirs concerning the natural history of a type of a freshwater polyp with arms shaped like horns; Fig. 11.12).

In his book, Trembley accurately describes the experiments and observations, illustrated with drawings, as seen in this narrative. In addition, he included tables testifying to having produced systematic records on the day-to-day studies carried out. Within a few months, his work was also published, and widely disclosed, in Paris. His experimental work began to be cited, along with that of Réaumur and others, as a model to be followed by anyone who wants to reveal the functioning of organisms.

11.6.3 Epilogue

In the eighteenth century, when the Enlightenment spread throughout many parts of Europe, when rational thought "fought" against religious dogmatism, the idea of a natural (or divine) order was of paramount importance. Therefore, organizing and classifying the natural universe of living beings was a goal for many naturalists. Nevertheless, some organisms – so-called zoophytes – presented such different characteristics, which caused problems in classifying them.

Guided by such "problems" and underpinned by careful experimental abilities, Trembley conducted many procedures and observations, besides creating and devising new techniques and instruments to try to understand in depth the nature of his polyps.

> I kept, at the same time, several glass jars which contained polyps operated; So I was able to distinguish each jar through a number or letter and I used these same marks in my notebook to discern the polyps. No one but myself handled these jars and I was always careful to avoid making any mistakes when I had to change water. I have always taken precautions with all the polyps used in the experiments recorded in this memory. (Trembley 1744 apud Lenhoff and Lenhoff 1986, p. 156)

However, Trembley did not bestow visibility to the small freshwater creatures only to those persons dedicated to the study of living beings and linked to the scholarly community. Until the end of the eighteenth century, many scholars and well-informed amateurs from various parts of Europe wished to see for themselves what the most interesting feature about the polyps was: *their reproductive modes!* "How is it possible for an animal to regenerate itself completely? Or, even stranger, how is it able to reproduce by budding, like a plant?" Because of these odd features, the polyps – later named "hydra" (*Hydra viridissima)*, by Carl von Linné – became the main subject animating conversations in the erudite salons of the European nobility.

Throughout his journey, Trembley was facing the "zoophyte's dilemma" – "would these tiny organisms be animals or plants?" At some point, Trembley asks himself: "Would it be possible to draw a clear line separating the kingdoms of plants and animals? Have I discovered a new class of organized bodies between the two kingdoms?"

A "planimal"? Nowadays, it would certainly sound like a joke! Would it not?

11.7 Concluding Remarks

As mentioned earlier, the historical narrative presented in this work was developed as part of the instructional resources designed to be used in a TLS intended for pre-service teachers to help to build up informed views about aspects of NOS, providing historical knowledge about discoveries of particular modes of animal reproduction, and creating an environment of inquiry, aiming for meaningful learning through a student-centered approach. Also, it provides an opportunity to present an innovative teaching strategy that, if desired, might be restructured in a collaborative way, with other teachers, and adapted to be aimed at secondary education.

In this study, we developed a combination of a particular episode of the history of biology and inquiry teaching, aiming to help pre-service science and biology teachers, and their future students in secondary school, to deepen their understanding of how scientific knowledge is produced and established, and how the reliability of science is historical and socially constructed.

The experience earned through iterative educational interventions with pre-service teachers and the design principles that oriented the redesign and refining of the historical narrative allows us to establish that it is a valuable means of creating an environment of engagement and motivation among participants toward inquiry, and reflectively and explicitly discuss different aspects of the nature of the whole science.

This is not the place to present a systematic analysis of the results, but a few examples may illustrate what is meant here: during the intervention, the instructor (first author) made an audio recording in each of the participant groups. After preliminary analysis of the transcription, followed by the stage of content analysis (Bardin 1977), it was possible to identify in some excerpts from the records expressions of motivation and interest, together with a feeling of desire to know more (narrative appetite). Here are some examples: "If all stories were told in this way, I'd have much more interest in the history of science" (Student K, University A); "I was a little angry with this Baker guy (the naturalist who had plagiarized Trembley's work) (laugh)! I wonder what happened to him!" (Student B, University B); "I'd love to use this narrative with my elementary school students! I think they would like it very much" (Student F, University B).

We can infer that these results may be linked to the preliminary design principles obtained from iterations, such as: the new elements of the historical episode, in particular, those conveying a feeling of dead ends, which were even occasionally emphasized to put forward challenges and create opportunities for inquiry and reflection among students. Another point was the detachment from the strict chronology and historicity, exploring fictional dialogues and arguments among characters, and to create imaginary, but plausible scenarios to provide both fluidity to the story and "appetite" to readers. However, we believe that most significant refinement and redesign, constituting a third design principle, concerned the number and the way of posing inquisitive questions about aspects of the NOS.

Yet, we are aware of its limitation, once only a few iterations had occurred, contemplating a small and restricted audience. Therefore, the authors, guided by design-based research principles, suggest that new cycles should be carried out in similar contexts and, whenever possible, increasing the number of participants, to obtain more knowledge on the design process, thus consolidating the design principles of the historical narrative.

Acknowledgement The first author thanks The Brazilian National Council for Scientific and Technological Development (CNPq, in the acronym in Portuguese), for research support.

References

Abd-El-Khalick, F., & Lederman, N. G. (2000). The influence of history of science courses on students' views of nature of science. *Journal of Research in Science Teaching, 37*(10), 1057–1095.

Adúriz-Bravo, A., & Izquierdo-Aymerich, M. (2009). A research-informed instructional unit to teach the nature of science to pre-service science teachers. *Science & Education, 18*, 1177–1192.

Allchin, D. (2011). Evaluating knowledge of the nature of (whole) science. *Science Education, 95*(3), 518–542.

Allchin, D. (2013). *Teaching the nature of science: Perspectives and resources*. St. Paul: SHiPS Education Press.

Allchin, D. (2014a). From science studies to scientific literacy: A view from the classroom. *Science & Education, 23*(9), 1911–1932.

Allchin, D. (2014b). The episodic historical narrative as a structure to guide inquiry in science and nature of science education. In *10th International Conference on History of Science & Science Education. Annals...* Minneapolis: University of Minnesota.

Allchin, D., Andersen, H. M., & Nielsen, K. (2014). Complementary approaches to teaching nature of science: Integrating student inquiry, historical cases, and contemporary cases in classroom practice. *Science Education, 98*(3), 461–486.

Anderson, T., Shattuck, J., & Brown, A. (2014). Design-based research: A decade of progress in education research? *Educational Research, 41*(1), 16–25.

Bardin, L. (1977). *Análise de conteúdo*. Lisboa: Edições 70.

Clough, M. (2006). Learners' responses to the demands of conceptual change: Considerations for effective nature of science instruction. *Science & Education, 15*(5), 463–494.

Colby, S. R. (2008). Energizing the history classroom: Historical narrative inquiry and historical empathy. *Social Studies Research and Practice, 3*(3), 60–79.

Dagher, Z. R., & Erduran, S. (2014). *Reconceptualizing the nature of science for science education: Scientific knowledge, practices and other family categories*. Dordrecht: Springer.

Duschl, R., & Grandy, R. (2012). Two views about explicitly teaching nature of science. *Science & Education, 22*(9), 2109–2139.

Fouad, K., Masters, H., & Akerson, V. (2015). Using history of science to teach nature of science to elementary students. *Science & Education, 18*(5), 561–580.

Juuti, K., & Lavonen, J. (2006). Design-based research in science education: One step towards methodology. *Nordina, 4*, 54–68.

Khishfe, R., & Abd-El-Khalick, F. (2002). Influence of explicit and reflective versus implicit-oriented instruction on sixth graders' views of nature of science. *Journal of Research in Science Teaching, 39*(7), 551–578.

Khishfe, R., & Lederman, N. G. (2007). Relation between instructional context and views of nature of science. *International Journal of Science Education, 29*(8), 939–961.

Klassen, S. (2007). The application of historical narrative in science learning: The Atlantic cable story. *Science & Education, 16*(3–5), 335–352.

Méheut, M., & Psillos, D. (2004). Teaching-Learning sequences: Aims and tools for science education research. *International Journal of Science Education, 26*(5), 515–535.

Metz, D., Klassen, S., McMillan, B., Clough, M., & Olson, J. (2007). Building a foundation for the use of historical narratives. *Science & Education, 16*(3), 313–334.

Monk, M., & Osborne, J. (1997). Placing the history and philosophy of science on the curriculum: A model for the development of pedagogy. *Science Education, 81*(4), 405–424.

Norris, S., Guilbert, S., Smith, M., Hakimelahi, S., & Philips, L. (2005). A theoretical framework for narrative explanation in science. *Science & Education, 89*(4), 535–563.

NRC, National Research Council. (2000). *Inquiry and the National Science Education Standards: A guide for teaching and learning*. Washington, DC: National Academy Press.

Roach, L. E., & Wandersee, J. H. (1993). Short story science. *The Science Teacher, 60*(6), 18–21.

Rudge, D. W., & Howe, E. M. (2009). An explicit and reflective approach to the use of history to promote understanding of the nature of science. *Science & Education, 18*(5), 561–580.

Rudge, D. W., Cassidy, D. P., Fulford, J. M., & Howe, E. M. (2013). Changes observed in views of the nature of science during a historically based unit. *Science & Education, 23*(9), 1879–1909.
The Design-Based Research Collective. (2003). Design-based research: An emerging paradigm for educational inquiry. *Educational Researcher, 32*(1), 5–8.
Van den Akker, J. (1999). Principles and methods of development research. In J. Van den Akker, R. M. Branch, K. Gustafson, N. Nieveen, & T. Plomp (Eds.), *Design approaches and tools in education and training* (pp. 1–14). Boston: Kluwer Academic.
Vesterinen, V. M., & Aksela, M. (2013). Design of chemistry teacher education course on nature of science. *Science & Education, 22*(9), 2193–2225.
Wang, F., & Hannafin, M. (2005). Design-based research and technology-enhanced learning environments. *Educational Technology Research and Development, 53*(4), 5–23.

References of the Historical Narrative

Dawson, V. P. (1987). *Nature's enigma: The problem of the polyp in the letters of Bonnet, Trembley and Réaumur, Memoire 174 of the American Philosophical Society*. Philadelphia: American Philosophical Society.
Lenhoff, S. G., & Lenhoff, H. M. (1986). *Hydra and the birth of experimental biology, 1744: Abraham Trembley's Mémoires concerning the polyps*. Pacific Grove: Boxwood press.
Ratcliff, M. J. (2009). *The quest for the invisible: Microscopy in the enlightenment*. Farnham: Ashgate Publishing.
Trembley, A. (1744). *Mémoires pour servir à l'histoire d'un genre de polypes d'eau douce, à bras en forme de cornes*. Leide: Chez Jean & Herman Verbeek.

Chapter 12
Historical Reconstruction of Membrane Theoretical Models: An Educative Curriculum Material

Anna Cassia Sarmento, Cláudia Sepúlveda, and Charbel N. El-Hani

12.1 Introduction

Curriculum reforms in several countries have advocated the inclusion of history and philosophy of science (HPS) in science education (Matthews 2014). In Brazil, this proposal is found, although in a sparse and fragmented way, in the national curriculum document *National Curriculum Guidelines* (Parâmetros Curriculares Nacionais, PCN, Brasil 1999). For instance, this document argues for overcoming ahistorical perspectives on the biological knowledge present in textbooks by considering the historical and philosophical dimensions of scientific work (Brasil 1999, part III, p. 2). It also considers that science learning should allow a discussion of the features of science such as:

1. The limits and possibilities of different explanatory systems
2. The dynamics of scientific knowledge building, including its openness to questioning, tentativeness, and mutability

A. C. Sarmento (✉)
Bahia Military Police School, Salvador, BA, Brazil

National Institute of Science & Technology in Interdisciplinary and Transdisciplinary Studies of Ecology and Evolution (INCT IN-TREE), Salvador, Brazil

C. Sepúlveda
Department of Education, State University of Feira de Santana, Feira de Santana, BA, Brazil

National Institute of Science & Technology in Interdisciplinary and Transdisciplinary Studies of Ecology and Evolution (INCT IN-TREE), Salvador, Brazil

C. N. El-Hani
Institute of Biology, Federal University of Bahia, Salvador, BA, Brazil

National Institute of Science & Technology in Interdisciplinary and Transdisciplinary Studies of Ecology and Evolution (INCT IN-TREE), Salvador, Brazil

M. E. B. Prestes, C. C. Silva (eds.), *Teaching Science with Context*, Science: Philosophy, History and Education,
https://doi.org/10.1007/978-3-319-74036-2_12

3. The nature of scientific models as human constructions that allow us to explain both what we can observe and what we can infer (Brasil 1999, part III, p. 219).

Besides the ahistorical approach, stressed in the curriculum document mentioned above, biological education is often characterized by a "rhetoric of conclusions" (Schwab 1964, 1969), that is, a presentation of outcomes of scientific research without a clue about what problems and inquiries led to those outcomes. Moreover, it is also common for biological topics to be addressed without connection with other pertinent scientific fields and/or with students' daily experiences. As a consequence, rote learning tends to predominate, with students' efforts being erroneously directed to memorizing structures, functions, compounds, etc. (Bastos 1992; Chiatppetta and Fillman 1998; Tsai 2004; Fogaça 2006; Lima and Teixeira 2011). All these aspects lead to students' lack of motivation and learning difficulties in several areas of biology, including cell biology (Dias et al. 2010). This biological field, in particular, brings difficulties to students' understanding because of the abstract and complex nature of the phenomena at stake, which demand an integration of several organization levels for their modeling and understanding, and the use of a plethora of scientific terms (Lazarowitz and Penso 1992; Kurt et al. 2013; Silveira and de Araújo 2014).

These difficulties deserve attention because cell biology deals with key concepts for understanding living systems (Palmero and Moreira 1999). After all, the cell is either the living entity per se, in unicellular organisms, or the basic structural and functional unit in multicellular systems.

Within cell biology, the understanding of the plasma membrane is of central importance. This membrane is the boundary between the cellular system and the external environment. More than that, it is a communication system between intra- and extracellular media, with many receptors involved in signal transduction. Finally, it shows the capacity to selectively allow compounds to enter or leave the cell. Besides the plasma membrane, understanding cellular endomembranes is also important for learning about the compartmentalization and organization of cellular processes.

The present work was motivated, on the one hand, by the importance of student understanding of cellular structures and processes and the difficulties faced by students learning cell biology. On the other hand, it was stimulated by the urge to overcome decontextualized approaches to cell biology education, both in terms of students' experiences and the historical and philosophical dimensions of the scientific endeavor. Given their central role in cellular functioning and the difficulties faced by students when trying to understand the relationship between their structure and functions, we chose cellular membranes as the topic for developing an educational innovation[1] for cell biology education. Moreover, when teaching about

[1] In this work, we use the concept of innovation developed by the group Collaboration in Research and Practice in Science Education (CoPPEC), which gathers together university educational researchers and basic education teacher–researchers, of which the authors are members. According to this concept, educational innovations are intentional, systematic, and participatory processes of producing, recontextualizing, and disseminating novel approaches to teaching that show the capac-

membranes, an interesting opportunity appears to address the construction of scientific knowledge and the relationship among theory, model, and experience.

The educational innovation discussed in this paper is framed as an educative curriculum material for teaching about the plasma membrane in the context of the first year of High School in the Brazilian educational system, attended by students 15–16 years of age. This educative curriculum material, composed of a teaching sequence[2] (TS) with an associated text, addresses the historical development of membrane theoretical models, with the goal of promoting learning about the structure and function of membranes and a more informed view of science. This educative curriculum material is part of a larger study, conducted according to the theoretical and methodological underpinnings of educational design research (The Design-Based Research Collective 2003; van den Akker et al. 2006; Plomp and Nieveen 2009). This study included both the planning and construction of the educative curriculum material and the investigation of the outcomes of its use in the science classroom (Sarmento 2016).

The goal of this chapter is, thus, to present:

1. A brief discussion of educative curriculum materials (ECMs)
2. A set of theoretical guidelines for developing a teaching approach informed by HPS, based on a dialogue between educational knowledge available in the literature and teacher knowledge
3. An ECM offering a historical reconstruction of membrane theoretical models.

This ECM has the form of a description of a TS and an associated text, tested in Brazilian high school classrooms. The analysis of classroom data obtained in this test made it possible to refine this ECM (Sarmento 2016), which also incorporates educative elements that potentially promote teacher learning.

12.2 Educative Curriculum Materials

Educative curriculum materials (ECMs) are planned for promoting both teacher and student learning. Teacher learning involves building an integrated knowledge base about content, teaching, and learning, becoming able to apply that knowledge in real-time instructional decisions, participating in the discourse of teaching, and becoming enculturated into (and engaging in) a range of teaching practices (Davis and Krajcik 2005). According to Davis and Krajcik, teacher learning is situated in teachers' practice, including classroom instruction and planning, lesson changes, assessment, peer collaboration, and communication with parents.

ity to promote professional development of actors in the school community in personal, social, and intellectual terms (Sousa et al. 2016).

[2] A teaching sequence is a set of ordered, structured, and articulated activities, aimed at reaching certain educational goals, with a beginning and end known by both teachers and students (Zabala 1998).

A teacher's knowledge base includes subject matter knowledge, pedagogical content knowledge, and pedagogical content knowledge for disciplinary practices (Davis et al. 2014). Educative curriculum materials show the potentiality to help teachers to develop new ideas in these different knowledge domains. Finally, to the extent that ECMs contribute to increasing teachers' knowledge, informing their practices, they can also help by promoting students' learning (Davis et al. 2014).

The ECMs should include theoretical guidelines for the teachers, which systematically disclose the rational reasons for the teaching proposals included in them. They should also provide orientations for the teachers concerning how to introduce those proposals into their classrooms (Davis et al. 2014). Therefore, ECMs should create for the teachers opportunities to both increase their knowledge of how to make teaching decisions in particular situations and developing a more generalizable knowledge, which can be applied to new situations (Davis and Krajcik 2005).

The construction of ECMs is supported by both educational literature, which provides theoretical and empirical underpinnings for teaching proposals and learning goals, and teachers' experiential knowledge, provided by previous teaching experiences.

We now turn to a presentation of theoretical guidelines of the ECMs discussed in this chapter.

12.3 Theoretical Guidelines for Teachers

12.3.1 Guidelines for Building HPS-Informed Educational Innovations

Building an HPS-informed approach to teaching science content is not a trivial matter. Care must be exercised with regard to several aspects if we expect the approach to reach the planned goals. When dealing with a concept, model or theory constructed throughout many years of investigation, care should be taken to avoid errors such as anachronistic interpretation or the creation of pseudohistory. As Allchin (2004) argues, pseudohistory is an important pitfall because it casts a deeply misleading light over historical events and can contribute to stereotypes and false ideas regarding how science works. Andrade and Caldeira (2009) highlight the challenge of working in the classroom in an HPS-informed manner with scientific ideas that were developed through the participation of many scientists, knowledge fields, diverging theoretical traditions, and large amounts of information. In this case, it is very important to establish the key points to be discussed with the students to make it possible for articulated and meaningful science learning and a more informed and sophisticated view of science to follow from the understanding of both school science and the historical construction of scientific ideas. All these caveats are germane when intending to teach about cell membranes in an HPS-informed manner.

Sequeira and Leite (1988) point out that simplification is an unavoidable aspect of introducing history of science (HOS) into science teaching, as it is impossible to teach all the scientific content included in the curriculum and the full history of their development at the same time. Thus, we can say that, in a sense, when using HPS in science teaching we face a problem found when any scientific idea is moved into school science: didactic transposition entails taking a selective view on what is to be taught, as it is not possible to include all the facets of academic knowledge in school science. This demands, in turn, epistemological vigilance with regard to the distance between academic scientific knowledge and school science knowledge included in the curriculum and taught in the classroom (Chevallard 1989, 1991). We consider that this applies to historical and philosophical aspects of scientific work included in science teaching as much as to the scientific content per se.

Some errors in addressing HPS content in science education are nonetheless quite evident. For instance, it is unacceptable to reduce the historical approach to mere curiosities or biographies of scientists, or to keep HPS ideas secluded from the main science content explained for students (for example, in separate boxes as we commonly see in textbooks), or to focus on asking students to memorize names and dates. Martins (2006) discusses, for example, that to reduce history to scientists' biographies fosters the false idea that science advances thanks to the accomplishments of isolated scientists, great geniuses. In this manner, HOS becomes – misleadingly – no more than a series of isolated episodes of great scientific discoveries. Kampourakis (2013) argues similarly in the specific case of the reduction of the birth of genetics, which resulted from the work of a whole scientific community interested in heredity in the nineteenth century, to the rediscovery of a single scientist's work, namely Mendel (who, incidentally, was not part of that community). Science as a social process is lost from sight in this way and, for each scientific advance, what will look most important to students will be to find out when exactly it was made, as if it were independent of other achievements and could be understood in isolation.

In contrast to this misleading approach to the HOS, it is important to give the students at least a glimpse of how scientific advances are typically built through a long history of collective research (Martins 2006), avoiding romantic and simplistic (not only simplified) views of the process of scientific knowledge construction (Martins and Brito 2006). Even when needing to highlight a scientist because some breakthrough in the history of collective research is linked to his or her name (say, Newton, Darwin, Marie Curie, Einstein, McClintock, and so forth), it is crucial to make it clear how that breakthrough was only possible because he or she was standing upon the shoulders of giants, as Newton himself wrote.[3]

It is also important to address HOS in a manner that makes it clear that the past influences the present (Leite 2002). To put it differently, HOS should show that the

[3] Newton made this famous statement in a letter to Robert Hooke, in 5 February 1676. Newton was alluding to an image by Bernard of Chartres who in the twelfth century argued that "moderns" could only advance further than the "ancients" because they were "dwarfs sitting on the shoulders of giants" (*Nos esse quasi nanos gigantum humeris insidientes*). See Merton (1965).

current construction of scientific knowledge, with theories and models that typically have more explanatory power than past models, is only possible because of the scientific theories and models previously constructed. It is crucial, therefore, to avoid evaluating past knowledge from the perspective of current knowledge, and, also, to ascribe to words used in the past the same meaning they have in the present (Leite 2002). In this manner, a pseudohistory of science is more likely to be avoided (Allchin 2003, 2004), a skewed and anachronistic, whiggish[4] view of the process through which a given scientific content was established.

A quality approach to HOS within science teaching needs to consider the social and intellectual context in which scientific knowledge is produced; to employ adequately metascientific and scientific vocabulary (in the latter case, also avoiding the anachronistic use of current terms as if they were used in the past in the same way); to portray scientific work as a social process and scientific knowledge as a human product and, thus, open to controversy and error.

A quality approach to the philosophy of science, in turn, demands to assume a set of modest goals in science teaching (Matthews 1998) that may allow us to make the complex issues intertwined within the philosophical dimension of the scientific endeavor learnable by students. Otherwise, the treatment of philosophical issues related to scientific knowledge may lead to more confusion than elucidation, even though our goal may sometimes be confusion at a higher level of awareness and complexity (Chalmers 1995). Finally, it is important to avoid, when trying to keep philosophical issues learnable, to reduce them to statements that can lead to misleading views about the nature of science and scientific work, such as, say, that "science is subjective" when attempting to consider that there is an influence from subjective elements in scientific work. Notice that the latter statement is quite different from the former, which loses sight of key aspects such as the intersubjective nature of scientific knowledge, as the product of mutual ongoing criticism of scientific ideas by inquiry communities. Or the trained nature of a scientist's subjectivity, shaped by his or her development and enculturation as a member of a specific scientific community, where a "collective empiricism" (Daston and Galison 2010) develops, that is, a set of epistemic practices and values characteristic of the specific way that community produces knowledge. To become a member of a particular inquiry community, a novice scientist is encultured to be accepted as part of the community, and in this process his or her approach to knowledge construction tends to be marked by intersubjective knowledge, values, and practices.

---

[4]The terms "whig" and "whiggish" were coined by Herbert Butterfield in 1931 to critically describe certain historical reports that treated events and institutions from the perspective of subsequent developments (Prestes 2010), anachronistically evaluating the past through the lens of the present. These terms derive from a British political faction and later party, the Whigs, which advocated a constitutional monarchy as an opposition to monarchical absolutism, defended by another party, the Tories. The tendency of historians to write in favor of the Whigs was called "whiggish," emphasizing a narrative of progress and thus producing a history that ratified the present. Besides anachronism, the adjective "whiggish" also refers to other methodological errors, such as assuming a linear perspective in a historical narrative, in which events in a series are described as simple chains of causes and effects that lead to improved results throughout time (Martins 2010).

12.3.2 Teaching About Cell Membranes Through an HPS-Informed Approach

The cell concept is highly sophisticated and it is also a key structuring concept for understanding biology, alongside other concepts such as evolution, metabolism, and so forth (Palmero and Moreira 2002). As Garcia Barrutia et al. (2002) argue, understanding the structure and function of cell membranes is important for building a consistent knowledge about the cell, as they play a key role in cellular organization and many functions in eukaryotic cells take place in the membranes. The understanding of many processes that connect the extracellular and intracellular media, such as substance transport across the plasma membrane and cell signaling, depends on grasping how membranes operate as dynamic cellular structures (Joglar et al. 2011). Given the central role of these processes in the functioning of organisms, mastering the scientific models about membrane structure and function becomes indispensable for comprehending the living world.

Among students' learning difficulties related to the plasma and other cell membranes, we find: the understanding of the models proposed to explain their molecular structure, the concept of the "unit membrane" that universally assigns to all cell membranes the same basic structure (the lipid bilayer structure), and the diverse functions carried out by membranes (Garcia Barrutia et al. 2002). There is also an important gap in understanding the relationship between membrane structure and function (Garcia Barrutia et al. 2002; Larsson 2008; García Irles et al. 2013).

We noticed in teaching practice the students' difficulty in understanding the relationship between structure and function in cell membranes. We believe that at least part of these problems results from the way in which content related to membranes is addressed in science classrooms and textbooks, in such a manner that the currently accepted plasma membrane model is often presented without being contextualized in relation to its historical process of construction and refinement. Moreover, this model is treated in a manner that gives priority to membrane structure and chemical components, without giving due attention to the relationship between these components and structure and membrane properties and functions in cell processes.

Another experience that underlies the present work concerns the inclusion of a historical approach to the process of building and refining membrane models in the teaching practice. When such an approach is implemented, we perceive that the students are more motivated to participate in the classes, and, given the relationship among motivation, interest, and learning achievements (Pintrich et al. 1993; Wigfield and Cambria 2010), tend to reach a better understanding of the scientific content related to cell membranes. This is in accordance with arguments found in the educational literature (Bastos 1992; Joglar et al. 2011) to the effect that cell biology teaching can be facilitated by an HPS-informed approach, which allows students to know the scientific problems, the models, the evidence for scientific ideas discussed in the classroom, and how they were obtained throughout history.

An HPS-informed approach seems even more relevant when considering epistemological challenges faced by students trying to understand cell structures and functions. They should learn about structures that cannot be observed in their everyday experiences, to which scientific meaning is ascribed through considerably abstract models and concepts. Accordingly, they face difficulties in understanding how the organisms operate at a microscopic level and in building a meaningful cell concept, especially in its dynamic rather than static aspects. It is hard for them to comprehend the living being as a particular kind of (complexly organized) chemical system. Abstract models of rather complex dynamical processes, such as substance transport across the membrane, demand a large amount of cognitive work that the students are not often willing or prepared to do.

Finally, when a teacher uses an HPS-informed approach to address the construction and refinement of membrane models, he/she also gives the students the opportunity to gain a rich experience of learning about how scientific work is done and how scientific knowledge is constructed and justified.

Owing to the scarcity of teaching materials addressing the historical development of cell membrane models that were informed by HPS in a way that is appropriate to our intentions, we decided to build materials to support the development and implementation of a TS. It is important to present the TS and the associated text here in a systematized manner to make it easier for other teachers and researchers to apply them in other contexts. These are the results of the first and second phases of a larger study conducted by Sarmento (2016).

12.4 An HPS-Informed Teaching Sequence on Cell Membranes

For the construction of an educational innovation that was appropriate to the context in which we intended to apply it, we took into consideration three aspects when building the first prototype of the TS:

1. The pertinence of the content to be learnt by the students in the teaching context
2. The students' cognitive maturity as related to their possibilities of learning the proposed content
3. The constraints posed by the functioning of the school where we developed and applied the TS (in terms of its programs, schedules, and, especially, the time devoted to teach on cell membranes in the curriculum).

Accordingly, in the TS, content on cell membranes is addressed during ten classes of 50 min each, using an approach aimed at promoting a continuous effort by the students to reflect upon the historical process of building and refining membrane models. The goal was to foster an understanding of both the membrane structure and function, and a number of features of science.

To analyze the extent to which this goal was reached, we gathered data on the stages of student understanding of the content addressed in the TS by means of three tests. One week before the first class, the students answered a pretest, with 16 closed questions, eight about models and eight about membranes. Two other tests with the same format were applied throughout the TS: in the seventh class (intermediate test), during the TS, and 2 weeks after the end of the TS (post-test). The data of the tests were triangulated with data from the classroom interactions to reach a conclusion about the relations between student's trajectories throughout the three tests and the classroom work (Sarmento 2016). We do not discuss these data here, as our main goal in this chapter is to present the ECM itself, but the version of the ECM presented reflects the conclusions obtained through the analysis of the data, which led to changes in the ECM.

In the sequel, we will suggest a way of implementing the TS using an associated text presented in the following section.

12.4.1 First Class

In the first class of the TS, the students play the game "*Célula adentro*" (Within the cell, available at <http://celulaadentro.ioc.fiocruz.br>). In this game, the students play the role of detectives trying to solve the case "the structure of the plasma membrane," interpreting clues offered by schemes, figures, experiments, and scientific results.

We chose this game to open the TS with the intention of initiating the discussion about the construction and refinement of membrane models in a playful, ludic manner. In particular, we expected that the students could understand how the lipids are organized in cell membranes. The main learning goal was, thus, to understand the lipid bilayer model and scientific findings associated with it.

As a suggestion, the class can be divided into small groups of students, which can play the game in a more organized manner than large groups. After the time given to solve the case, it is interesting that the teacher discusses the outcomes of the game with the students, with the goal of systematizing the knowledge put into play.

12.4.2 Second and Third Classes

The second and third classes are devoted to carrying out a classroom activity about different membrane models, in which the students, divided into groups, can discuss the process of building and refining membrane models along the history of biology.

The learning goals are:

1. To understand science as an activity of building models that are modified throughout time, in an effort to better describe, explain, and predict phenomena

2. To know the chemical constituents of the plasma membrane and understand how they are organized to form its structure, according to each historical model discussed in the activity.

To carry out the activity, it is suggested that the class be divided into four groups. Each group receives a text about a different membrane model, as a support for the students to understand the model, which they will have to defend to the other groups by means of a presentation. Four different texts were derived from the source text presented in the next section, to be distributed to the student groups. Each text addressed a membrane model:

1. Lipid monolayer
2. Lipid bilayer
3. The so-called "sandwich" model
4. The fluid mosaic model (for details, see below).

Although the students are working with the text, it is interesting that the teacher circulates among the groups, solving doubts and guiding the elaboration of their presentations.

When the groups are ready, each one defends its model, and, after the presentations, the teacher conducts a discussion with the whole class about the complexity of the models and the order in which they appeared throughout the HOS.

It is important that the teacher guides the discussion to keep the focus on the chemical constituents of the membrane and how they are organized in each model. From our classroom experience, we noticed how important it is to work with the technical terms employed in the construction of the membrane models in a clear and consistent manner. For instance, it is important to work with the scientific terms "hydrophobic" and "hydrophilic" in an integrated manner with the related terms "non-polar" and "polar," respectively, making all of them available in the social plane of the classroom. Otherwise, the students may face difficulties in using these terms with the meaning ascribed to them in the social language of science, as a way of building a conceptual understanding of the relation between the chemical constituents and the membrane structure, particularly when using texts in which they find only one or other of those related terms.

It is also important to give the students opportunity, during the discussion, to establish by themselves the historical order in which the different membrane models were proposed. Finally, the discussion can be very productive if focused on science as a model-building activity, always addressing the relationship among theory, model, and reality.

12.4.3 Fourth Class

In the fourth class, after a brief synthesis of the previous activities, the teacher makes an exposition about the historical development of membrane models and the chemical composition of the membrane, including moments of dialogical interaction with the students. The nature of the evidence involved in the construction, refinement, and testing of the models is also addressed in this class. The learning goals are the same from the second and third classes.

From our experience, we concluded that it is important for the teacher to build a historical narrative that spans from the preliminary studies that produced evidence used to formulate the first membrane models to Singer and Nicolson's proposal of the fluid mosaic model, in 1972. This means taking into account a number of investigations that contributed to refining the membrane models, through a historical process involving a series of models, four of which seem important to consider in school science:

1. Langmuir's (1917) model of lipid monolayers in the air/water interface, which was not developed through studies on biological membranes, but through research into the behavior of phospholipids in that interface, but had a significant influence on subsequent studies on membrane models
2. The lipid bilayer model, proposed in 1925 by Evert Gorter and F. Grendel
3. The so-called sandwich model, initially put forward by James Danielli and Hugh Davson in 1935, which basically represented the membrane as a protein–lipid–protein sandwich, i.e., as a lipid bilayer, as proposed by Gorter and Grendel, covered in both sides by proteins
4. The mosaic fluid model, proposed by S. Jonathan Singer and Garth Nicolson in 1972, which maintains the lipid bilayer as the basic membrane structure, in which proteins are immersed.

It is also important to discuss that Singer and Nicolson's model brought a number of novel features, such as the asymmetric distribution of proteins in the bilayer and the fluid nature of the membrane, as shown in the name given to the model. This means that the lipid components are in constant motion, while many proteins are also capable of moving within the membrane. If the teacher considers it appropriate, she/he can extend the narrative to consider the development of the concept of the glycolipoprotein microdomains named lipid rafts by Kai Simons and Gerrit van Meer (Simons and van Meer 1988; Simons and Ikonen 1997).

A clear and consistent use of the scientific vocabulary employed when explaining the models is essential, as already discussed above.

12.4.4 *Fifth and Sixth Classes*

The fifth and sixth classes are used to discuss models of substance transport through the membrane, by means of teacher exposition with moments of dialogical interaction.

The fifth class is dedicated to the discussion of passive transport (simple and facilitated diffusion). The learning goals are:

1. To understand the property of selective permeability of the plasma membrane, and how it emerges from part-whole relations in membrane structure

2. To understand that integral proteins are part of the plasma membrane and play a function in substance-facilitated diffusion
3. To understand that the lipid bilayer has the property of being permeable to small molecules soluble in lipids, influencing the phenomenon of simple diffusion
4. To understand the models of mechanisms involved in simple and facilitated diffusion.

We suggest that it might be important to use illustrations presenting didactic models to explain to the students the mechanisms of substance transport through the membrane. The use of examples of simple diffusion, such as the transport of O_2 and CO_2, and facilitated diffusion, such as glucose transport, can also be useful when teaching this topic. Moreover, if the teacher relates these examples to the mechanisms of pulmonary and cellular respiration, she contributes to the students' construction of a more integrated biological knowledge.

In the sixth class, the teacher makes an exposition about models proposed to explain osmosis and active transport, and about the importance of the cholesterol and glycocalyx in animal cells, and the cell wall, in plant cells. The exposition enables moments of dialogical interaction with the students.

This class has the following learning goals:

1. To understand the models proposed to explain osmosis and active transport
2. To understand that glycoproteins and glycolipids are membrane components that constitute the glycocalyx, which plays an important function in cell signaling and adhesion
3. To understand that the cell wall is a structure involving the membrane of the plant cell that has functions related to structural support and protection
4. To understand that cholesterol is a functionally important molecule that is part of the plasma membrane in animal cells.

Using illustrations showing didactic models is also helpful in explaining osmosis and active transport. To explain osmosis, it is useful to appeal to typical school science examples concerning what happens to animal and plant cells in hypotonic, hypertonic, and isotonic solutions. To explain the model of active transport, the sodium–potassium pump offers a good example, already included in school science.

In this class, it is important to discuss with the students the specific functions played by the chemical constituents of the membrane in each kind of transport and how they determine, in an integrated manner, the selective permeability property. This helps the students to build comprehension of the part–whole relations in the membrane.

To work clearly and consistently with scientific terms such as "lipid-soluble" and "water-soluble" is crucial, just as we argued above in relation to the terms "hydrophobic" and "hydrophilic." After all, if the students face difficulties in comprehending these terms, this has a negative impact on their construction of a proper understanding of the processes of substance transport across the membrane.

12.4.5 *Seventh Class*

In the seventh class, the students take the intermediary test, so that the teacher can analyze the development of their conceptual understanding.

12.4.6 *Eighth and Ninth Classes*

The relations between membrane structure and function are discussed in the eighth and ninth classes, through teacher expositions, with space for dialogical interactions with the students.

The learning goals are:

1. To understand the relationship between the structural chemical constituents of the currently accepted membrane model and specific examples of membrane functions (from which the following learning goals emerge):
2. To understand the mechanisms of regulation of blood sugar levels in the human organism
3. To understand the action mechanism of the growth hormone in the human being
4. To know some problems related to membrane malfunctioning, such as cystic fibrosis, diabetes, and breast cancer.

It is important that the teacher makes a proper connection, in these classes, between membrane structure and function, and cell signaling mechanisms involved in cellular processes associated with blood sugar regulation and the action of the growth hormone. This is helpful for students in understanding how membrane structure is related to function and, moreover, how important the membrane is for physiological functioning.

The association between membrane malfunctioning and health problems is also a key topic for students to understand how relevant it is to understand the structure–function relation in the membrane. In the classroom, three diseases – cystic fibrosis, diabetes, and breast cancer – are explained by means of models that highlight how changes at the level of the membrane are associated with those diseases, what causes the membrane changes, and what the consequences are for the human organism.

Another crucial feature to consider in the classroom is the need to explicitly discuss with the students part–whole relations in the membrane, to favor their understanding of how systemic properties, at the level of the membrane as a whole, emerge from the integration of specific functions played by the membrane chemical constituents.

12.4.7 *Tenth Class*

In the tenth class, the students read a popular science text about the plasma membrane, which prompts a discussion about membrane plasticity, a topic not often considered in school science, but of key importance in understanding how membranes function.

The learning goals are:

1. To understand how knowledge about the structure and functions of the plasma and other cell membranes continues to grow
2. To understand membrane plasticity, i.e., that despite all cell membranes being formed by phospholipids and proteins, different cell types have membranes with different characteristics, including chemical constituents, and that the plastic structure is related to plastic membrane function. This leads to the emergence of specific properties in different cell types, for instance, that some cells of the immune system, like macrophages, are capable of phagocytosis, whereas many other cell types are not.

In our experience, we used the text "Fronteiras fluidas: propriedades elásticas da membrana variam segundo o tipo e a função da célula" (*Fluid frontiers: membrane elastic properties vary according to cell type and function*. Zorzetto 2013, available at: http://revistapesquisa.fapesp.br/2013/11/18/fronteiras-fluidas/), which was read and discussed by the students in small groups. Evidently, each teacher has to select a similar popular science text in her/his own language that is also accessible to the students.

After working with the text, each group has to prepare a summary of each paragraph. The teacher then randomly chooses the groups to present the paragraphs in the same order in which they appear in the text. We consider it important that the teacher strives to systematize school science knowledge throughout the students' presentations, and verifies how they comprehend what is discussed in the text, correcting possible mistakes in relation to the perspective of school science.

It is important that the text makes it possible to discuss with the students that even though there is a scientifically accepted model that explains the structure of the cell membranes in all cell types, this model shows limitations when it comes to explaining membrane dynamics and function in particular cell types or cell membranes. To explain how membrane functions dynamically in different cell types, more specific and complex models are needed, taking into account particular features of those cell types. It is also important that the students understand that biological membranes are complex and dynamic, and that models can generalize to different extents, depending on their adequacy for the explanatory tasks at hand. Thus, to account for plastic and dynamic features of membranes in particular cell types, models built – as any model – by means of abstraction and idealization should become more complex. They also become, conversely, less generalizable than a model like the fluid mosaic, which explains the general structural features of all biological membranes, but at the expense of properly explaining their plasticity and dynamics in specific cell types or cell membranes.

12.4.7.1 Source Text for the Teaching Sequence

12.4.7.1.1 Cell Membranes: How Did We Come to Understand Them?

Membranes are dynamic structures present in all living beings, except viruses. They delimit the cells, keeping the intracellular content distinct from the extracellular content. In eukaryotic cells, membranes also delimit organelles, in which cellular processes important to the maintenance of life take place. Without cell membranes, it is likely that the current life forms could not have evolved from earlier living systems.

Today, we have a wealth of scientific knowledge about the structure and function of cell membranes. But this has not always been the case. Many earlier studies were necessary to support current research into the membrane. The construction of scientific models is possible because scientists make use of knowledge previously produced by the scientific community to explain phenomena, raise doubts, identify explanatory limits, and propose new hypotheses.

We can ask: how have the studies about the membrane begun? How has the currently accepted membrane model been constructed?

12.4.7.1.2 Earlier Studies About Membranes

The British naturalist William Hewson (1739–1774)[5] did experiments in 1773 to study the behavior of red blood cells in water, observing the phenomenon currently known as osmosis (for more details about the currently accepted model of osmosis in blood red cells, see Box 12.1).

When Hewson noticed that the blood red cells (erythrocytes) changed their shape, swelling up or shrinking when placed in solutions with different concentrations in relation to their cytoplasm, he was able to infer the existence of a membrane surrounding the cell.

Stronger evidence for the presence of a membrane in blood red cells was obtained by C. H. Schultz (1798–1871) in 1836, when he stained erythrocytes with iodine and was thus capable of visualizing the plasma membrane and, also, estimating its thickness.

Today, we know that biological membranes are too thin and, therefore, impossible to see with the naked eye, even when using an optical microscope. However,

[5] We recommend that teachers, when reproducing or adapting this material for classroom use, include images of all the scientists mentioned, to allow students to consider their identities of gender, race, and social class. Images of the scientists mentioned can be found in Stillwell (2013) and via the links http://en.wikipedia.org/wiki/Kenneth_Stewart_Cole (Kenneth Stewart Cole); https://www.kcl.ac.uk/lsm/about/history/heroes.aspx#JDanielli (James F. Danielli); http://www.physoc.org/sites/default/files/page/Hugh%20Davson.pdf (Hugh Davson)

http://www.heuserlab.wustl.edu/experience/Robertson%20obit.pdf (James David Robertson); http://beyondthebandaid.com.au/dr-garth-nicolson/ (Garth L. Nicolson); https://www.mpi-cbg.de/research-groups/alumni/ (Kai Simons).

Box 12.1: Model of Osmosis in Blood Red Cells, Showing How These Cells Behave in Isotonic, Hypertonic, and Hypotonic Media

Osmosis in blood red cells. Osmosis is a special case of diffusion in which water molecules move across semipermeable membranes. In this case, the water molecules diffuse in such a manner that they dilute the more concentrated medium, that is, water molecules diffusion always takes place from the less concentrated (hypotonic) to the more concentrated medium (hypertonic). This is a spontaneous process that does not use energy. When the internal and external concentrations are equal (A), that is, isotonic, the blood red cells keep their normal shape. When a blood red cell is placed in a more concentrated medium (B), water from the internal medium diffuses to the exterior, making it shrink. Contrariwise, when a blood red cell is placed in a medium that is less concentrated (C), water from the external medium diffuses to the interior of the cell, which swells up.

A – isotonic medium, B – hypertonic medium, C – hypotonic medium

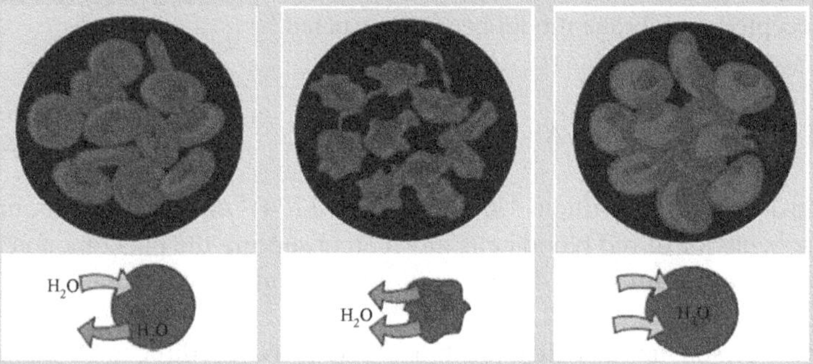

Source: https://upload.wikimedia.org/wikipedia/commons/7/76/Osmotic_pressure_on_blood_cells_diagram.svg (by Mariana Ruiz Villarreal. E-Mail: MRV LadyofHats.com)

when the external surface of the membrane is stained with a diversity of dyes, it is possible to detect its presence.

In 1855, the Swiss botanist Karl von Nägeli (1817–1891) explained the phenomenon of osmosis in plant cells (for more details about the currently accepted model of osmosis in plant cells, see Box 12.2).

As we saw above, the studies by Hewson on erythrocytes and by Nägeli on plant cells provided evidence for the existence of osmosis. In 1877, the German botanist Wilhelm Pfeffer (1845–1920) used artificial semipermeable membranes to measure the osmotic pressure of solutions of sugar cane (sucrose) and other organic molecules. Based on his studies on osmosis, he concluded that the cell membrane was thin and semipermeable.

Box 12.2: Model of Osmosis in Plant Cells, Showing How These Cells Behave in Isotonic, Hypertonic, and Hypotonic Media

Osmosis in the plant cell: osmosis also occurs in the plant cell. When the internal and external concentrations are equal (A), that is, isotonic, the water movement between the intracellular and extracellular media takes place with the same intensity in both directions. When a plant cell is placed in a more concentrated medium (B), water from the internal medium diffuses to the exterior and the cytoplasm retracts, detaching the plasma membrane from the cell wall, which suffers no deformation. Contrariwise, when the plant cell is placed in a medium that is less concentrated (C), water from the external medium diffuses to the interior of the cell. However, the cell wall prevents cell lysis.

A – isotonic medium, B – hypertonic medium, C – hypotonic medium

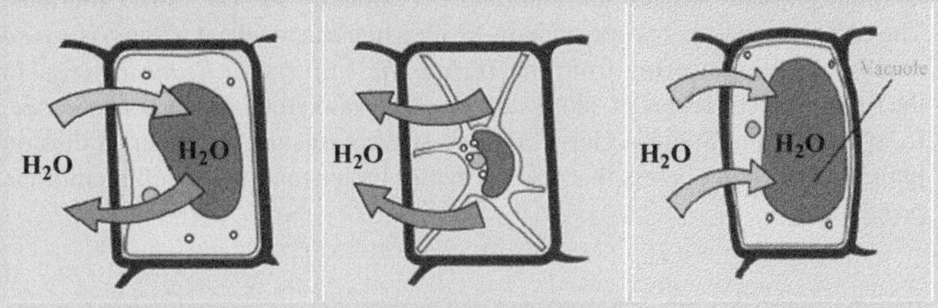

Source: https://commons.wikimedia.org/wiki/File:Plasmolyse_Pflanzenzelle.svg#/media/File:Turgor_pressure_on_plant_cells_diagram.svg (by Mariana Ruiz Villarreal. E-Mail: MRV LadyofHats.com)

Scientific studies on the behavior of oil in water were also important to the development of scientific knowledge on membranes. These studies were initially performed by the North American statesman Benjamin Franklin (1706–1790) in 1772. They became very popular in the period. In the 1880s, the British physicist John William Strutt, 3rd Lord Rayleigh (1842–1919), developed such studies further.

In his experiments, Franklin poured a small amount of oil onto the water surface of a small pond in England and immediately realized that the oil formed a thin film on the surface, making it as smooth as a mirror. Franklin's observations appeared in a letter he wrote to a friend on 7 November 1773. He did not know why the oil soothed the small waves in the water surface. Subsequent studies showed that the oil decreased the surface tension of the water, when it forms a single thin film on the surface.

Lord Rayleigh won the Nobel Prize for Physics in 1904 for his studies on gases and his discovery of argon. He was skillful in mathematics and physics, and this was helpful in his several experiments involving oil and water. He was capable of calculating both the area over which a known volume of oil would spread and the thickness of the film formed by the same volume when in contact with water.

On 10 January 1891, Lord Rayleigh received a letter from a German woman called Agnes Pockels (1862–1935). In this letter, Pockels told Lord Rayleigh that she had built a device capable of measuring the exact area that a certain volume of oil occupies when spreading over the water surface. When he realized the importance of Pockels' work, Lord Rayleigh helped her to publish her first paper in the prestigious journal *Nature*. Lord Rayleigh adopted Pockels' methods in his subsequent research. Her work provided great scientific contributions, and had a lasting impact on science (for more details on Agnes Pockels, see Box 12.3).

Box 12.3: A Bit More About Agnes Pockels
During her life, Agnes Pockels published around 15 papers. She published two more papers in *Nature* (in 1892 and 1894) and continued to publish in German popular science journals and occasionally in specialized journals. The recognition of her work came in 1932, when she received a *honoris causa* doctorate in Engineering from the *Technische Universität Braunschweig*. In the same year she received, alongside Henri Devaux, the Laura Leonard Prize, from the Colloid Society. Her work was of great relevance to later studies on phospholipid monolayers that led to a better understanding of cell membrane properties

Around 1895, the North American scientist Charles Ernest Overton (1865–1933) was doing his doctoral studies in botany at the University of Zurich, developing studies on permeability that were important for understanding membranes. In a period when it was thought that the membrane was permeable only to water, Overton found that it was permeable to other substances too. Moreover, he showed that this permeability was common to diverse cell types. His studies also showed that lipid-like substances passed through the membrane with great velocity. Thus, he considered that the capacity of a substance to cross the membrane was related to its chemical nature. This idea made him elaborate the hypothesis that biological membranes were similar to lipids, say, to olive oil, and, therefore, there would be a lipid barrier between the cell cytoplasm and the extracellular medium. Certain kinds of substances (the lipids) would move across this barrier through dissolution in the lipids constituting it.

12.4.7.1.3 Development of Membrane Models: From the Monolayer to the Fluid Mosaic

The study of monolayers produced by the interaction between oil and water was important to the construction of knowledge of biological membranes. The North American chemist and physicist Irving Langmuir (1881–1957) carried out studies on the behavior of oil in water and the monolayers resulting from oil–water interaction. He refined Pockels' device and was able to accurately measure the surface occupied by known quantities of oil on the water surface. In 1917, Langmuir published a paper describing the behavior of phospholipids in water (for more details on the structure of phospholipids, see Box 12.4).

Box 12.4: Knowing the Phospholipids

Structure of the phospholipids: the phospholipids are the major structural constituents of the cell membranes. These biomolecules are formed by a glyceride (composed of glycerol and fatty acids) combined with an electrically charged phosphate group. Due to their chemical nature, the phospholipids contain a polar (hydrophilic) and a nonpolar region (hydrophobic). The polar or hydrophilic region interacts with water, whereas the nonpolar region does not interact with water

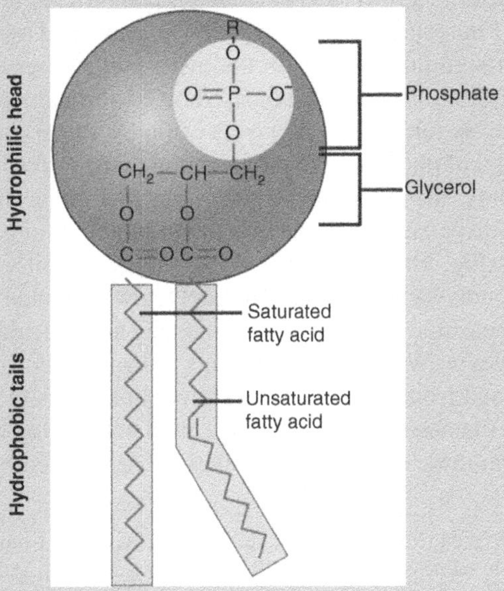

Source: OpenStax College. The Cell Membrane.https://legacy.cnx.org/content/m46021/1.11/

Langmuir observed that when phospholipid molecules enter into contact with water, they organize in a single layer (monolayer), with the polar (hydrophilic) region of the phospholipids interacting with water and the nonpolar region facing out, in contact with air (Fig. 12.1). In 1932, Langmuir won the Nobel Prize for Chemistry for his discoveries and investigations on surface chemistry, including the studies just described.

Langmuir's studies of phospholipid monolayers were key to the understanding of biological membranes. His model of phospholipid monolayers in the air/water

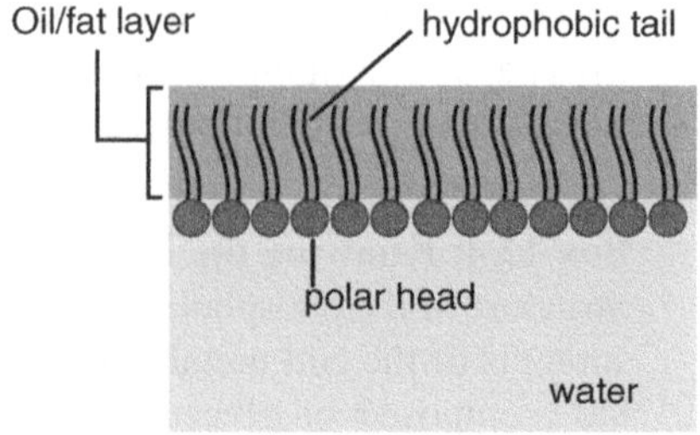

Fig. 12.1 Behavior of phospholipids in water. Figure elaborated by the authors

interface[6] was not built to explain cell membranes, but showed how these molecules form single layers owing to their chemical nature, influencing later studies on membranes.

In 1925, Evert Gorter (1881–1954), a Dutch pediatrician and scientist, and his research assistant, François Grendel (1897–1969), published a small paper regarded as one of the most significant works in the history of studies on biological membranes. In this work, Gorter and Grendel showed through experiments that the plasma membrane was formed of two layers of lipids, rather than a single layer, as in Langmuir's model of phospholipid organization in the air/water interface. In their classical experiments, they extracted erythrocyte membrane lipids using acetone and then carefully dispersed the lipids over water, compressing the lipid layer with a movable barrier (Fig. 12.2). As expected, given Langmuir's previous work, the lipids formed a monolayer. Gorter and Grendel knew how many extracted erythrocytes produced a measured monolayer area and they also calculated the surface area of a typical erythrocyte. With these parameters at hand, they found that the area of a monolayer obtained from extracted erythrocyte lipids was twice the surface area of the erythrocytes themselves. In the passage cited below, Gorter and Grendel show how they draw from these results the hypothesis that the membrane was formed of a lipid bilayer[7]:

> If chromocytes are taken from an artery or vein, and are separated from the plasma by several washings with saline solution, and after that extracted with pure acetone in large amounts, one obtains a quantity of lipoids that is exactly sufficient to cover the total surface of the chromocytes in a layer that is two molecules thick. [...]. We therefore suppose that every chromocyte is surrounded by a layer of lipoids, of which the polar groups are directed to the inside and to the outside. (Gorter and Grendel 1924, p. 439)[8]

Gorter and Grendel's work offered an important basis for the construction of later membrane models by other scientists.

[6] Historical membrane models can be found in Eichman (1999), available in shipseducation.net/9-2/membrane.htm, accessed at Jan 17th 2018. We recommend that teachers include figures showing these models, when reproducing/adapting this material for classroom use.

[7] A representation of this model can be found in Eichman (1999), available at shipseducation.net/9-2/membrane.htm, accessed on 17 January 2018.

[8] "Chromocyte" is a term used as a synonym of "erythrocyte." "Lipoid" is an alternative spelling for "lipid."

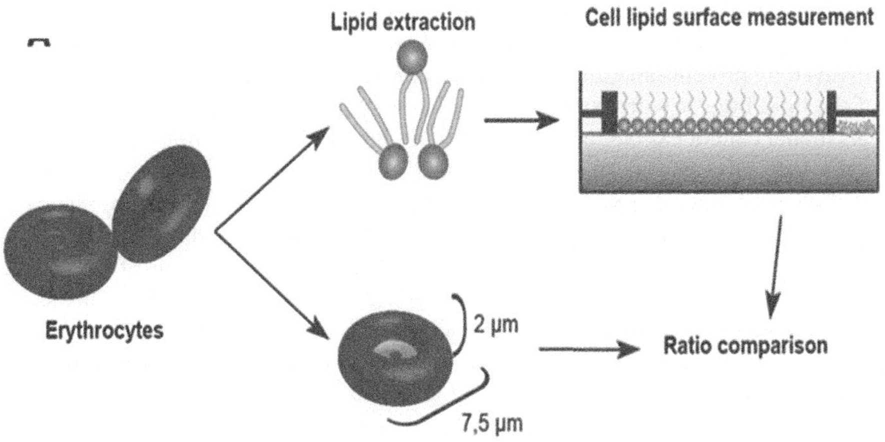

Fig. 12.2 Experimental device used by Gorter and Grendel. (*Source*: Lombard 2014)

Also in 1925, J. B. Leathes and H. S. Raper suggested, in their book *The fats*, that phospholipids could be the key structural elements in cell membranes.

In 1932, the North American biophysicist Kenneth Stewart Cole (1900–1984) concluded from his studies about the membranes of sea urchin eggs that the cell membranes contained other substances besides lipids.

A new membrane model was then proposed by Danielli and Davson in 1935, and was accepted by most of the scientists of the time. James Frederic Danielli (1911–1984) was an English biologist who developed many studies on the plasma membrane, in collaboration with his personal friend and physiologist, Hugh Davson (1909–1996).

For 2 years (between 1933 and 1935), Danielli worked in Princeton University (USA) with E. Newton Harvey (1887–1959), who studied the cell surface. Danielli and Harvey showed that there was another class of molecules in cell membranes, namely, proteins.

In 1935, Danielli returned to England, to work at the University College in London, where he continued to investigate membranes, alongside Davson. Their first membrane model consisted of a sort of "sandwich" in which a lipid bilayer – as proposed by Gorter and Grendel – was covered on both sides by globular proteins.[9] Later, Danielli and Davson modified the model, including proteins that form pores.[10] This modification became necessary when they noticed that polar substances needed by the cell could not pass through the lipid bilayer if membranes had the structure proposed in their first model.

---

[9]A representation of this model can be found in Eichman (1999), available at shipseducation.net/9-2/membrane.htm, accessed 17 January 2018.

[10]A representation of this model can be found in Eichman (1999), available at shipseducation.net/9-2/membrane.htm, accessed 17 January 2018.

This static membrane model was accepted by the scientific community for a long time. However, this model is awkward because the cell membrane is completely involved by proteins and, thus, there is no contact of the polar or hydrophilic region of the lipid bilayer with extracellular or intracellular aqueous solutions. Therefore, this model cannot explain how phospholipids are arranged in a bilayer.

In 1957, the North American biologist James David Robertson (1932–1995) used electron microscopy to obtain the first images of biological membranes (for more details on microscopy, see Box 12.5). These images[11] showed that all cell membranes exhibited a trilaminar structure (Fig. 12.3), consisting of two darker outer sheets and a lighter inner region. Robertson suggested that the two outer sheets corresponded to the protein layers, whereas the inner region amounted to the lipid bilayer. This suggestion provided great support for Danielli and Davson's model.

Robertson's studies established that biological membranes in several cell types, including the plasma membrane, organelle membranes, and the nuclear membrane, all showed the trilaminar structure. He concluded that all biological membranes shared a similar underlying structure, which he called the "unit membrane."

Later studies showed the limitations of Danielli and Davson's "sandwich" model and in 1972, two scientists proposed a new membrane model. They were the North American biochemists Seymour Jonathan Singer (1924–) and Garth L. Nicolson (1943–). Singer and Nicolson called their proposal the fluid mosaic model.[12] This is the currently accepted membrane model. Singer and Nicolson's model maintains the lipid bilayer initially proposed by Gorter and Grendel and subsequently modified by Danielli and Davson, and by Robertson. However, the proteins are immersed

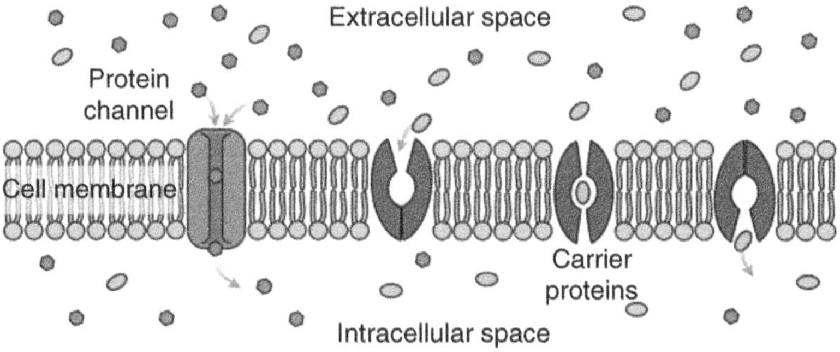

Fig. 12.3 Channel and transport integral proteins. (*Source*: Villarreal, Mariana Ruiz (E-Mail: MRV LadyofHats.com) [Public domain], via Wikimedia Commons. https://commons.wikimedia.org/w/index.php?curid=3981034)

[11] The electromicrography of the plasma membrane can be found in Stillwell (2013).

[12] Graphic representations of this model are commonly reproduced in textbooks. When using this educative curriculum material, the teacher can include one of these representations or use the original figures from Singer and Nicolson (1972).

in the lipid bilayer, instead of externally and internally involving the bilayer, as in the "sandwich" model. As Singer and Nicolson write:

> In this model, the proteins that are integral to the membrane are a heterogeneous set of globular molecules, each arranged in an amphipathic structure, that is, with the ionic and highly polar groups protruding from the membrane into the aqueous phase, and the nonpolar groups largely buried in the hydrophobic interior of the membrane. These globular molecules are partially embedded in a matrix of phospholipid. The bulk of the phospholipid is organized as a discontinuous, fluid bilayer, although a small fraction of the lipid may interact specifically with the membrane proteins. (Singer and Nicolson 1972, p. 730)

According to the fluid mosaic model, biological membranes are basically formed by phospholipids and proteins. The phospholipids have a polar (hydrophilic) and a nonpolar region (hydrophobic). This property of the phospholipids leads to their spontaneous arrangement in the membrane, forming a double layer: the polar (hydrophilic) region of the external layer faces the water present in the extracellular medium, whereas the hydrophilic region of the internal layer faces the water present in the intracellular medium. Nonpolar (hydrophobic) regions of the phospholipids face one another and form a fluid barrier.[13]

Box 12.5: Microscopy

Microscopes are equipment used to visualize structures that are invisible to the naked eye, such as the vast majority of cells and organelles. Basically, there are two kinds of microscopes: optical and electron microscopes. Contemporary optical microscopes work with light beams that cross the material prepared for visualization, providing images magnified 100–1,500 times with good resolution. Electron microscopes work with electron beams that cross the prepared material and provide images magnified 5–100,000 times with good resolution. As we can see, electron microscopes make it possible to obtain images with much higher magnification than optical microscopes, allowing more detailed studies

Two kinds of proteins are included in the fluid mosaic model: integral and peripheral.[14] The two kinds of proteins are unevenly (asymmetrically) distributed across the membrane. Integral proteins are found in larger amounts in biological membranes and are immersed in the phospholipid bilayer, keeping contact with the hydrophobic regions of the phospholipids.

Integral proteins can form channels or pores that allow the passage of small polar molecules such as water and ions through the membrane (Fig. 12.4). In this case, there is no binding between the small molecules being transported and the protein

[13] A representation of this model can be found in Eichman (1999), available in shipseducation.net/9-2/membrane.htm, accessed at Jan 17th 2018.

[14] A representation of this model can be found in Eichman (1999), available at shipseducation.net/9-2/membrane.htm, accessed on 17 January 2018.

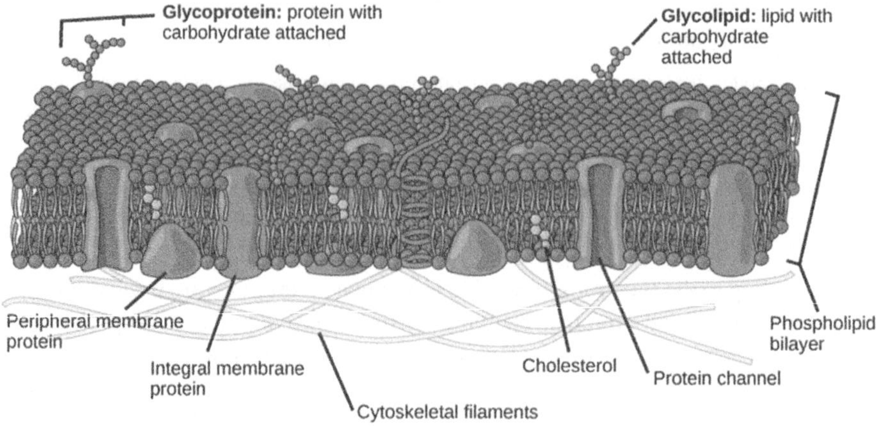

Fig. 12.4 Fluid mosaic model. (*Source*: OpenStax College. The Cell Membrane. https://legacy.cnx.org/content/m46021/1.11/)

channels. Other integral proteins, called transporters, bind to specific substances transported across the membrane, such as glucose, carried to the intracellular medium. Integral proteins cannot be easily separated from the phospholipids.

Peripheral proteins are not immersed in the lipid bilayer, but adhere to integral proteins or polar regions of the phospholipids, with no contact with the hydrophobic regions. They can more easily attach to, or detach from the membrane.

Besides phospholipids and proteins, the fluid mosaic model also predicted the existence of carbohydrates bound to the proteins (forming glycoproteins) and lipids (forming glycolipids). These carbohydrates are found in the external surface of the membrane, constituting the glycocalyx, which is responsible for cell recognition and union, giving origin to areas of large cell adhesion.

Singer and Nicolson's fluid mosaic model is shown in Fig. 12.4. Besides the asymmetric distribution of proteins in the bilayer, a key novelty in this model is the fluidity of the whole membrane. That is, both phospholipid molecules and integral proteins freely move within the bilayer.

The fluid mosaic model has undergone changes since 1972, when it was first proposed, but is still accepted by the scientific community as an explanation of how biological membranes are structured. It also offers a basis for understanding how membranes function.

In 1988, Kai Simons (1938–) and Gerrit van Meer (1953–) proposed the model of lipid microdomains currently known as lipid rafts (Fig. 12.5). Lipid rafts are subdomains of the plasma membrane containing high concentrations of cholesterol and glycosphingolipids, which are more ordered and tightly packed than the surrounding fluid bilayer, and can freely float in the latter. Lipid rafts serve as organizing centers for the assembly of signaling molecules, influencing membrane fluidity and membrane protein trafficking, and regulating key cellular processes, such as cell signaling, neurotransmission, and programmed cell death. They are also involved in bulk transport mechanisms, such as endocytosis and exocytosis.

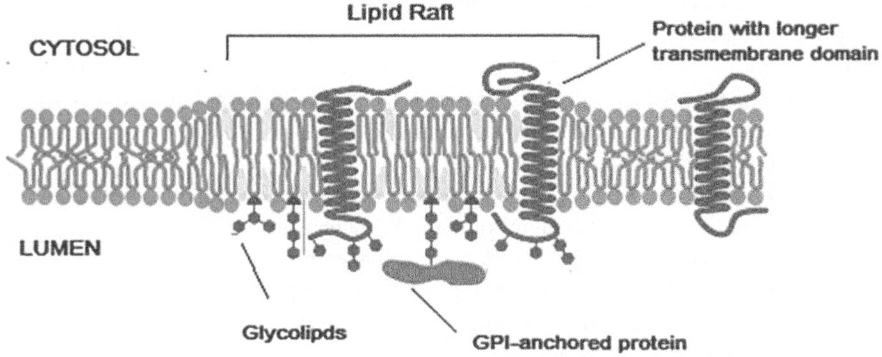

Fig. 12.5 The lipid raft model, a refinement of the mosaic fluid model. Figure elaborated by the authors, based on a figure found in https://en.wikipedia.org/wiki/Lipid_raft

For many researchers, the lipid rafts model is just a refinement of Singer and Nicolson's fluid mosaic model, that is, it does not fundamentally break with the membrane structure proposed by that model.

As we can see, studies on biological membranes began much earlier than the advent of electron microscopy, through investigations into the interactions between oil and water. Historically, the knowledge obtained by those studies led to the development of membrane models, which made it possible to understand the structure, properties, and functions of cell membranes.

12.5 Final Remarks

This work presents an ECM resulting from a larger study aimed at developing an educational innovation for cell biology teaching at the high school level (Sarmento 2016), oriented by the theoretical–methodological guidelines of educational design research. In this study, we strived to solve problems found in the real context of high school biology teaching, in particular, the lack of a historically and philosophically contextualized approach to scientific ideas, and the difficulties faced by students in understanding cell biology content, more specifically, that related to biological membranes. The educational innovation presented in this chapter, including a TS and an associated source text (ST), takes as a starting point a focus on the historical development of membrane models to favor learning about membrane structure and function in the tenth grade.

Through a dialogue between educational literature and teacher knowledge, we established a set of theoretical guidelines to support the design and construction of the educative curriculum material (including the TS and ST). This material was, in turn, refined by means of empirical evidence obtained in the study carried out by Sarmento (2016).

We consider that the educational innovation presented in this work contributes to the contextualization in historical and philosophical terms of teaching and learning about membrane structure and function. The analysis of different membrane models, developed and refined throughout the HOS, shows great potential in providing the students with opportunities to understand the process of knowledge construction on biological membranes.

Another important aspect of this approach is the discussion of the explanatory power and limitations of each membrane model, which can contribute to increasing student understanding about issues such as substance transport between the intracellular and extracellular media, membrane-selective permeability, and cell signaling. This approach also makes it possible for students to develop a more informed and richer view on the role of models in science.

It is also important to highlight that to understand biological membranes and, in particular, their structure and function, is key to student learning about cell biology, as membranes have crucial functions in several vital processes of living beings, such as glucose metabolism, cell proliferation, growth processes, cell signaling. Membranes are also important for understanding a number of diseases related to changes in their chemical constituents, structure, and function, such as diabetes, cancer, dwarfism and gigantism, and cystic fibrosis.

Given the contributions discussed above, we believe that the educative curriculum material presented in this chapter has the potential to be used by other teachers in their teaching contexts, with the necessary adaptations to each school context, multiplying its contribution to cell biology education.

Finally, it is important to stress that although the educational innovation discussed here was developed and implemented for teaching on biological membranes, the historical and philosophical approach used can be transposed to teaching other school science content. For instance, in biological education, this approach can be helpful in addressing other cell biology content or content related to genetics, evolution, ecology, etc., contributing both to learning about them and to the construction of a more informed and reflective view of science.

References

Allchin, D. (2003). Scientific myth-conceptions. *Science & Education, 87*, 329–351.

Allchin, D. (2004). Pseudohistory and pseudoscience. *Science & Education, 3*, 179–195.

Andrade, M. A. B. S., & Caldeira, A. M. A. (2009). O modelo de DNA e a biologia molecular: inserção histórica para o ensino de biologia. *Filosofia e História da Biologia, 4*, 139–165.

Bastos, F. (1992). O conceito de célula viva entre alunos do segundo grau. *Em Aberto, 11*, 63–69.

Brasil. (1999). Ministério da Educação, Secretaria de Educação Média e Tecnológica. In *Parâmetros Curriculares Nacionais: Ensino Médio*. Brasília: Ministério da Educação.

Chalmers, A. F. (1995). *O que é ciência afinal?* São Paulo: Brasiliense.

Chevallard, Y. (1989). On didactic transposition theory: Some introductory notes. In *Proceedings of the International Symposium on Selected Domains of Research and Development in Mathematics Education*, 51–62. http://yves.chevallard.free.fr/spip/spip/article.php3?id_article=122. Accessed 20 July 2016.

Chevallard, Y. (1991). *La transposición didáctica: del saber sabio al saber enseñado*. Buenos Aires: Aique.

Chiatppetta, E. L., & Fillman, D. A. (1998). Clarifying the place of essential topics and unifying principles in high school biology. *School Science and Mathematics, 98*, 12–18.

Danielli, J. F., & Davson, H. (1935). A contribution to the theory of permeability of thin films. *Journal of Cellular Physiology, 5*, 495–508.

Daston, L., & Galison, P. (2010). *Objectivity*. New York: Zone Books.

Davis, E. A., & Krajcik, J. (2005). Designing educative curriculum materials to promote teacher learning. *Educational Researcher, 34*, 3–14.

Davis, E. A., Palincsar, A. S., Arias, A. M., Bismack, A. S., Marulis, L. M., & Iwashyna, S. K. (2014). Designing educative curriculum materials: A theoretically and empirically driven process. *Harvard Educational Review, 84*, 24–52.

Design-Based Research Collective. (2003). Design-based research: An emerging paradigm for educational inquiry. *Educational Researcher, 32*, 5–8.

Dias, M. A. d. S., Núñez, I. B., & Ramos, I. C. d. O. (2010). Dificuldades na aprendizagem dos conteúdos: uma leitura a partir dos resultados das provas de biologia do vestibular da Universidade Federal do Rio Grande do Norte (2001 a 2008). *Revista Educação em Questão, 37*, 219–243.

Fogaça, M. (2006). *Papel da inferência na relação entre os modelos mentais e modelos científicos de célula*. Dissertação de Mestrado em Ensino de Ciências e da Matemática, Faculdade de Educação da Universidade de São Paulo, São Paulo, Brasil.

Garcia Barrutia, M. S., Artacho, C. J., Diaz, J. F., Perez, J. F., & Redondo, B. T. (2002). Evolución de conceptos relacionados con la estructura y función de membranas celulares en alumnos de enseñanza secundaria y universidad. *Anales de Biologia, 24*, 201–207.

García Irles, M., Huertas, Y. S., & Ortells, J. M. S. (2013). Aprendizaje basado en problemas en biología celular: una forma de explorar la ciencia. *Revista de Educación en Biología, 16*, 67–77.

Kampourakis, K. (2013). Mendel and the path to genetics: Portraying science as a social process. *Science & Education, 22*, 293–324.

Kurt, H., Ekici, G., Aktas, M., & Aksu, Ö. (2013). Determining biology student teachers' cognitive structure on the concept of "diffusion" through the free word-association test and the drawing-writing technique. *International Education Studies, 6*, 187–206.

Langmuir, I. (1917). The constitution and fundamental properties of solids and liquids. II. Liquids. *Journal of the American Chemical Society, 39*, 1848–1906.

Larsson, C. (2008). *Critical aspects of understanding of the structure and function of the cell membrane: Students' interpretation of visualizations of transport through the cell membrane*. Thesis, Linköping University, Sweden. http://www.divaportal.org/smash/get/diva2:25027/FULLTEXT01.pdf. Accessed 20 Oct 2016.

Lazarowitz, R., & Penso, S. (1992). High school students' difficulties in learning biology concepts. *Journal of Biological Education, 26*, 215–224.

Leite, L. (2002). History of science in science education: Development and validation of a checklist for analysing the historical content of science textbooks. *Science & Education, 11*, 333–359.

Lima, G. P. S., & Teixeira, P. M. M. (2011). Análise de uma sequência didática de citologia baseada no movimento CTS. In *Atas do VIII Encontro Nacional de Pesquisa em Educação em Ciências*. Campinas: Associação Brasileira de Pesquisa em Educação em Ciências.

Martins, R. A. (2006). Introdução: a história das ciências e seus usos na educação. In C. C. Silva (Ed.), *Estudos de história e filosofia das ciências: subsídios para aplicação no ensino* (pp. xvii–xxxx). São Paulo: Editora Livraria da Física.

Martins, R. A. (2010). Seria possível uma história da ciência totalmente neutra, sem qualquer aspecto whig? Boletim de História e Filosofia da Biologia, 4(3), 4–7.

Martins, L. A. C. P., & Brito, A. P. O. P. M. (2006). A história da ciência e o ensino da genética e evolução no nível médio: um estudo de caso. In C. C. Silva (Ed.), *Estudos de história e filosofia das ciências: subsídios para aplicação no ensino* (pp. 245–264). São Paulo: Editora Livraria da Física.

Matthews, M. R. (1998). In defense of modest goals when teaching about the nature of science. *Journal of Research in Science Teaching, 35*, 161–174.
Matthews, M. R. (2014). *Science teaching: The contribution of history and philosophy of science, 20th Anniversary Revised and Expanded Edition*. New York: Routledge.
Merton, R. K. (1965). *On the shoulders of giants*. Chicago: The University of Chicago Press.
Palmero, M. L. R., & Moreira, M. A. (1999). Modelos mentales de la estructura y el funcionamiento de la célula: dos estudios de caso. *Investigações em Ensino de Ciências, 4*, 121–160.
Palmero, M. L. R., & Moreira, M. A. (2002). Modelos mentales vs esquemas de célula. *Investigações em Ensino de Ciências, 7*, 77–103.
Pintrich, P. R., Marx, R. W., & Boyle, R. A. (1993). Beyond cold conceptual change: The role of motivational beliefs and classroom contextual factors in the process of conceptual change. *Review of Educational Research, 63*, 167–199.
Plomp, T., & Nieveen, N. (Eds.). (2009). *An introduction to educational design research*. Enschede: SLO – Netherlands Institute for Curriculum Development.
Prestes, M. E. B. (2010). Whiggismo proposto por Herbert Butterfield. *Boletim de História e Filosofia da Biologia, 4*(3), 2–4.
Sarmento, A. C. de H. (2016). *Como ensinar citologia e promover uma visão informada da ciência no nível médio de escolaridade*. Dissertação de Mestrado em Ensino, Filosofia e História das Ciências, Universidade Federal da Bahia e Universidade Estadual de Feira de Santana, Salvador, Brasil.
Schwab, J. J. (1964). The structure of the natural sciences. In G. W. Ford & L. Pugno (Eds.), *The structure of knowledge and the curriculum* (pp. 31–49). Chicago: Rand McNally.
Schwab, J. J. (1969). *College curriculum and student protest*. Chicago: University of Chicago Press.
Sequeira, M., & Leite, L. (1988). A história da ciência no ensino: aprendizagem das ciências. *Revista portuguesa de educação, 1*, 29–40.
Silveira, M. L. d., & de Araújo, M. F. F. (2014). Alternativas didáticas para o ensino: uma revisão considerando a citologia. In *Atas do V Encontro Nacional de Ensino de Biologia (V ENEBIO) II e Encontro Regional de Ensino de Biologia, (II EREBIO)* (Vol. 7, pp. 5607–5617). São Paulo: SBEnBio.
Simons, K., & Ikonen, E. (1997). Functional rafts in cell membranes. *Nature, 387*, 569–572.
Simons, K., & Van Meer, G. (1988). Lipid sorting in epithelial cells. *Biochemistry, 27*, 6197–6202.
Sousa, A. E. d. A., Muniz, C., & Sarmento, A. C. d. H. (2016). O processo heurístico da construção do conceito de inovações educacionais por um grupo colaborativo de pesquisa. In C. Sepúlveda & M. Almeida (Eds.), *Pesquisa Colaborativa e Inovações Educacionais no Ensino de Biologia* (pp. 97–125). Feira de Santana: Editora UEFS.
Tsai, C.-C. (2004). Conceptions of learning science among high school students in Taiwan: A phenomenographic analysis. *International Journal of Science Education, 26*, 1733–1750.
Van Den Akker, J., & Gravemeijer, K. (2006). In S. Mckenney & N. Nieveen (Eds.), *Educational design research*. London: Routledge.
Wigfield, A., & Cambria, J. (2010). Students' achievement values, goal orientations, and interest: Definitions, development, and relations to achievement outcomes. *Developmental Review, 30*, 1–35.
Zorzetto, R. (2013). Fronteiras fluidas: Propriedades elásticas da membrana variam segundo tipo e função da célula. *Pesquisa FAPESP, 213*, 50–53.

References Used to Elaborate the Source Text

Amabis, J. M., & Martho, G. R. (2010). *Biologia*. v. 1. 3a ed. São Paulo: Moderna.
Bizzo, N. (2011). *Novas bases da biologia*. v 1. São Paulo: Ática.
Edidin, M. (2003).The state of lipid rafts: from model membranes to cells. *Annual Review of Biophysics and Biomolecular Structure, 32*, 257–283.

Eichman, P. (1999). *From the lipid bilayer to the fluid mosaic: A brief history of membrane models*. http://shipseducation.net/9-2/membrane.htm. Accessed 20 Oct 2016.
Gorter, E., & Grendel, F. (1925). On bimolecular layers of lipids on the chromocytes of the blood. *Journal of Experimental Medicine, 41*, 439–443.
Heimburg, T. (2007). Membranes: An introduction. In T. Heimburg (Ed.), *Thermal biophysics of membranes* (pp. 1–19). Weinheim: Wiley-VCH Verlag.
Joglar, C., Quintanilla, M., Ravanal, E., & Brunstein, J. (2011). El desarrollo histórico del modelo científico de membrana plasmática: Perspectivas didácticas. In *Atas do VIII Encontro Nacional de Pesquisa em Educação em Ciências (ENPEC)*. Campinas: ABRAPEC.
Leathes, J. B., & Raper, H. S. (1925). *Monographs on biochemistry: The fats* (2nd ed.). London: Longmans.
Linhares, S., & Gewandsznajder, F. (2011). *Biologia hoje* (Vol. 1, 1st ed.). São Paulo, Ática.
Lombard, J. (2014). Once upon a time the cell membranes: 175 years of cell boundary research. *Biology Direct, 9*, 32.
Nelson, D., & Cox, M. (2005). *Lehninger principles of biochemistry* (4th ed.). New York: W.H. Freeman.
Singer, S. J., & Nicolson, G. L. (1972). The fluid mosaic model of the structure of cell membranes. *Science, 175*, 720–731.
Stillwell, W. (2013). *An introduction to biological membranes: From bilayers to rafts*. London: Elsevier Science.
Zabala, A. A. (1998). *Prática educativa: como ensinar*. Porto Alegre: Artmed.

Chapter 13
The History of Chemistry in Latin America

José A. Chamizo

13.1 Introduction

> The word (context) originates from the Latin language in the verb "*contexere*," "to weave together." In its related noun "*contextus*," the word expresses "coherence," "connection," and/or "relationship." Thus, the function of "context" is to describe such circumstances that give meaning to words, phrases, and sentences. (Gilbert 2006, p. 959)

There are many difficulties in teaching the history of chemistry. Recently, Höttecke and Silva (2011) recognized four of them: teachers' skills, epistemological and didactical attitudes and beliefs, institutional framework of science teaching, and available textbooks. Generally speaking, scientific content is taught, but Schwab's (1962) interpretation of science teaching as a dogma or as "a rhetoric of conclusions" remains. Therefore, if scientific competence is not worked out, we cannot say that scientists are being trained. There are different positions on this, but, to sum up, it is possible to recognize that scientific teaching requires more "context" (Gilbert 2006) and reflection (Matthews 2014).

The main purpose of this paper is to show how chemistry teachers, using history of chemistry, could teach chemistry in the Latin American context. It means something more than an undifferentiated mass of names and dates, i.e., a difficult balance between over-simplification versus over-elaboration, and also seeks to recognize how chemistry has been developed in Latin America. Hence, this paper follows the initial proposal of Jensen (1998) and reconstructs the history of chemistry in terms of five revolutionary moments (Chamizo 2016, 2017). These moments are considered in terms of the Kuhnian notion of "exemplar" rather than "paradigm." This approach enables the incorporation of instruments (Holmes and Levere 2000) and concepts into the revolutionary process and provides a more adequate representation

J. A. Chamizo (✉)
Facultad de Química, Universidad Nacional Autónoma de México, México, DF, México
e-mail: jchamizo@unam.mx

M. E. B. Prestes, C. C. Silva (eds.), *Teaching Science with Context*,
Science: Philosophy, History and Education,
https://doi.org/10.1007/978-3-319-74036-2_13

of such periods of development and consolidation (Chamizo 2014). Moreover, accepting the premise of the "scientific revolutions," it recognizes better the continuity of scientific endeavor once the transitions are closer and less sharp.

13.2 About Context

> America was [...] the instance that made it possible, within Western culture, to extend its world picture to the whole earth and the concept of universal history to all mankind. (O'Gorman 1958)

The antiquity of human societies in the American continent is subject to great scientific controversy and ranges from about 15,000 to 60,000 years ago. In the same way, there are several proposed routes for the settlement of the continent, although it is accepted that there was a major migration from Asia through the Bering Strait.

For Europeans, the American continent was discovered, although for many it was invented (O'Gorman 1958). Then, we were the colonies, dependent on a few European metropolis, some of which still have a presence in this part of the world. The victorious wars of independence of many Latin American countries against the European powers were followed by defeats against their main representative on the continent, the USA, that since 1823 coined Monroe's doctrine "America for the Americans." Some of them were explicit defeats, such as the annexation of half of the former territory of New Spain in the first half of the nineteenth century, some others, implicit defeats such as the separation of Panama from Colombia in the early twentieth century to build the canal. It was during the second half of the nineteenth century in Napoleon III's France, where he warranted the Austrian Emperor Maximilian in his military expedition in Mexico, that the concept of Latin America was invented. France saw itself as the Latin power in the world. A few years later, Spain coined a term to mark their own leadership role in the region: *Ibero-América.* Today, many of us are Latin-American, Ibero-Americans or even Hispanics (for the USA), or from a developing country, or from a Third World region. With all the precautions that this subject deserves, and making an effort to clarify for a book on science education, it turns out that, behind that pluralistic identity, there are common roots: many of us were born and live south of Rio Bravo, although the emigration outside this first geographical boundary is increasing; we are racial mestizos (of three different races, indigenous, black, and white) or cultural mestizos; our mother tongue is Spanish or Portuguese and English, but also Quechua or Nahuatl or Tupi-Guarani; in our national societies economic and knowledge accumulation prevails, generating extreme inequality. In metaphorical terms, an entire archipelago. In this century, we are tensioned in several directions: one is manifested as an interesting puzzle, the promises of global cosmopolitanism and loss of national projects (García-Canclini 2002); another in defense of secularism and public education, i.e., the separation from the national states of the churches; and finally, for science teachers the "essential tension," which is present in the advancement of our professions

(Kuhn 1977), ranging from the singular self-reflection of commitments, in favor of others, a plurality that will be invented. A huge challenge.

13.3 About Philosophy

> Scientific knowledge (is) primarily… a human product, made with locally situated cultural and material resources, rather than as simply the revelation of a pre-given order of nature. (Golinski 2005, xvii)

The way in which chemistry is taught all around the world implies a unique philosophical position that can be characterized as logical positivism (Van Berkel et al. 2000; Van Aalsvoort 2004). Chemistry education practice has not been driven to any great extent by educational, historical, or philosophical research findings. Besides we learn about the world mainly by learning about how to intervene in it. This means through technoscience (Tala 2011) or technochemistry (Chamizo 2013). The know-how involves learning processes of the practice whose explanation lies not only in the internalization of declarative statements and facts, but also in the "locally situated cultural and material resources" (Golinski 2005, xvii). Therefore, a place for the history and philosophy of sciences (or technosciences) using Latin American experiences is possible.

Unlike the various positivist approaches that dominated philosophy of science until the twentieth century, Thomas Kuhn's ideas about scientific revolutions, introducing history into philosophy of science, have been widely discussed and for many scholars accepted in general terms (Kindi and Arabtzis 2012). For Kuhn, scientific revolutions are "taken to be those non-cumulative developmental episodes in which an older paradigm is replaced in whole or in part by an incompatible new one" (Kuhn 1970, p. 92). This change requires a reconstruction of the historical commitments of a particular scientific community. The commitments shared by groups or communities are characterized by the use of the word "paradigm," which means a "criterion for choosing problems that, while the paradigm is taken for granted, can be assumed to have solutions" (Kuhn 1970, p. 37). Over the same historical period, different scientific communities have shared the same paradigm and research, and teaching based on this paradigm is known as "normal science." When there is a scientific revolution, the community changes its paradigm, thereby changing the activities related to "normal science." Based on their previous achievements normal or paradigmatic science is employed by a specific community in its daily activities, and is taught in textbooks. Scientific progress in normal science is cumulative or gradual. Revolutionary science develops when a crisis occurs in normal science. The result of a revolutionary process in science is the emergence of a new paradigm that displaces the previous one and has been traditionally identified with changing theories. Therefore, the communities that assume different paradigms find significant difficulties in communicating with each other. Competing paradigms lack a common measure because they use different concepts and methods to address different problems – they are, in Kuhn's terminology, incommensurable. However, at the end of his career, Kuhn recognized a different meaning of incommensurability, indicating that the emergence of new sub-disciplines within a discipline was a result of a scientific revolution, and accepted by new textbooks, caused the split of scientific communities (Kuhn 1992). In particular, he indicated that incommensurability is important for the growth of scientific knowledge because it isolates practitioners' communities by creating communication barriers that promote the proliferation of specialties.

The term "exemplar" represents a specific historical community's collection of solved problems and is generally found in its professional literature, and especially in its textbooks. It is narrower than the term "paradigm" and avoids some of the ambiguities that the latter has acquired. Making explicit the role of instruments in normal science reduces the gap between normal and revolutionary science.

Challenging the common idea that science is fundamentally theory, in the last few decades no one has contributed more than the Canadian philosopher Ian Hacking to articulating the variety of roles that experiments play in the production of knowledge. He recognized that much of our empirical knowledge results from interventions with instruments (Hacking 1983). Only by intervening is it possible to discover the material resistances of the world.

Allowing specific historiographical variations among historians, the above considerations suggest that there might be at least five acceptable answers to the question "What is a scientific revolution?" According to what has been said previously, these are:

1. A radical reinterpretation of existing thought recognized by contemporaries themselves
2. The resolution of a long-standing debate, the solution of which revolutionizes the kind of problems scientists are able to successfully attack on a routine basis
3. The use of new instruments changes the way in which its practitioners look at and work in the world
4. The opening of a new level of theoretical understanding that subsumes older theories as special cases
5. The opening of new sub-disciplines that produces separate scientific communities.

There are many ways to approach the past. One is through the exclusive biographies of famous people (which can become hagiography). Another uses scientometry (Kragh 1987; Hernández et al. 2016). A third takes into account different ways of knowing (Pickstone 2000). However, and starting from anthropology and sociology, another way of acknowledging the past incorporates other actors, i.e., appreciating more the social context in which our discipline develops (Golinski 2005; Latour 2012; Shapin and Shaffer 2011). Unfortunately, nowadays, too little can be done to teach the history of Latin America's chemistry from those other approaches; thus, there is no remedy for collecting the few accessible data about some celebrities and sorting them properly, so that they can be taught.

13.4 About the History of Chemistry in Latin America

About 5000 years ago in Mesoamerica, corn, beans, and turkeys had been domesticated. With wild birds and fish, the food base was complete, although animal protein intake was under discussion. There were no large animals neither to be eaten nor to drag wheel carts. The only wheels known in our continent were used in some of the Mexican toys. On the contrary, the llama and alpaca, the largest American mammals, lived in the Andean region, where no wheels have been documented. The potato was domesticated there. The two great cultural areas of our continent, the Andean and the Mesoamerican, developed and practically disappeared without any contact. In other parts of the world, the domestication of plants and animals, the use of wheel carts, metallurgy, written language, and gunpowder were easily communicated from one place to another. In different cultures of the American continent, a wide variety of products were developed, many of them solutions to the demands of daily life at that time. As far as we know, these deeply religious cultures interpreted the world only through their gods. However, as in the case of European Alchemy, empirical knowledge was huge (Bond 1993). Apart from an old tradition about medicinal plants and Andean embalming techniques (del Busto 1970), it is sufficient to mention that before the mid-eighteenth century in pre-colonial and colonial America (Chamizo 2004) the next chemistry-related developments occurred:

1. Besides the use of different metals known in Europe, platinum is documented in today's Colombia (Aristizábal-Fúquene 2015)
2. Cochineal, along with "purple snail" and indigo, were the most important dyes from Mesoamerica. On the other hand, the brasil tree, from which bright red dye was extracted, gave its name to an entire country
3. Latex (the colloidal suspension of rubber particles in water) was extracted mainly from the guayule bush and a tree from the jungles of Mexico, Peru, and Brazil.

13.4.1 First Chemical Revolution (1756–1808)

In Tables 13.1, 13.2, 13.3, 13.4, and 13.5 the main historical and philosophical characteristics of each of the five chemical revolutions are summarized. It also recognizes its main protagonist and some participants from Latin America.

José Estevez was the first Cuban chemist, who graduated originally in medicine in 1795 at the Real Universidad de la Habana (Gonzalez-Mora 1983). From 1802 to 1808, he studied with Luis Proust in Madrid and notably among his published works were: *On various minerals in the island* and *Opinion on the azucarometro invented by Larnier.*

In 1783, Carlos III, King of Spain, decreed for New Spain the creation of the *Real Seminario de Mineria*, which opened in 1795. It was founded with the aim of carrying out metallurgical studies. Its first director and first chemistry professor in

Table 13.1 Characteristic examples and comments of the first chemical revolution

Characteristics	Examples and comments
Historical highlights	Phlogiston; pneumatic trough; balance; atomic theory; gases (oxygen, hydrogen, nitrogen, carbon dioxide) quantitative chemistry, chemical language
Philosophical highlights	Chemistry was recognized as an independent quantitative discipline with atoms as its distinctive entity
World protagonists	J. Black; H. Cavendish; J. Dalton; A. Lavoisier; J. Priestley
Latin American protagonist	Brazil: J. Bonifacio; J. Manso
	Mexico: F. de Elhuyar; A.M. del Rio

Table 13.2 Characteristic examples and comments of the second chemical revolution

Characteristics	Examples and comments
Historical highlights	Kaliapparat, polarimeter spectroscope, Karlsruhe's Congress, valency, isomers, periodic table, organic chemistry
Philosophical highlights	Molecules, as a spatial specific atomic conglomerate with particular properties, become the *quintessential* chemistry entity. Organic chemistry emerges as a sub-discipline
World protagonists	R. Bunsen; S. Cannizaro; A. Kekule; J. Liebig; D. Mendeleiev; L. Pasteur; F. Wöhler; E. Frankland; M. Faraday
Latin American protagonists	Brazil: C. Alves Serrao; T. Peckolt
	Mexico: V. Ortigosa; L. Posselt

Table 13.3 Characteristic examples and comments of the third chemical revolution

Characteristics	Examples and comments
Historical highlights	Cathode ray tube, mass spectrometer, X-rays, radioactivity, polymers, Haber–Bosch ammonia synthesis, physical chemistry
Philosophical highlights	Atoms and molecules are composed of electrons and nuclei. Those two new entities were considered fundamental in the explanation of chemical bonds. Physical chemistry became one sub-discipline
World protagonists	F. Aston; G.N. Lewis; J.J. Thomson; Ostwald; S. Arrhenius; Vant Hoff; M. Curie
Latin American protagonists	Brazil: M. Saraiva
	Mexico: J. S. Agraz

Mexico and throughout the whole of America was Fausto de Elhúyar (1755–1833), eminent chemist who in 1783 with his brother discovered tungsten. The Spanish chemist and mineralogist Andrés Manuel del Río (1764–1849) worked with Antoine Lavoisier (1734–1794). After fleeing from the French Revolution, he came to New Spain where he established the exploitation and metallurgy of iron, in addition to the official teaching of mineralogy. As a professor of *Real Seminario de Mineria* he discovered a new chemical element: *eritronio*. The metal was rediscovered in 1830

Table 13.4 Characteristic examples and comments of the fourth chemical revolution

Characteristics	Examples and comments
Historical highlights	Massive incorporation of instruments in chemistry laboratories (IR, UV, NMR, chromatography), instrumental chemistry
Philosophical highlights	Despite spin chemistry being known before, it was in this period that it broke through significantly. Computers begin to be used in theoretical chemistry and is set as a sub-discipline in its ability to explain the chemical bond. Using NMR instruments, spin chemistry is "enthroned" in laboratories
World protagonists	L. Pauling; R. Woodward; R. Hoffmann; A.J.P. Martin
Latin American protagonists	Argentina: L. Leloir
	Mexico: E. Miramontes

Table 13.5 Characteristic examples and comments of the fifth chemical revolution

Characteristics	Examples and comments
Historical highlights	Electron capture detector, scanning tunneling microscope
	Montreal Treaty, organometallic, green, supra, nano, and femtochemistry
Philosophical highlights	A deep transformation in the very heart of chemistry. That is too say, related to substances, the size and the type of objects, the way in which they are produced, and the time in which they are transformed have changed. In one way or another, chemistry' limits had been set out
World protagonists	O. Fisher; G. Wilkinson; J.E. Lovelock; E. Ruska; G. Binning; H. Rohrer; M. Molina; F.S. Rowland; D.J. Cram; J.M. Lehn; C.J. Pedersen; R.F. Curl; H.W. Kroto; R.E. Smalley; A. Zewail
Latin American protagonist	Mexico: M. Molina

by the Swedish Nils Sefstrom (1787–1845), who named it vanadium. It was recognized as such by the European chemical community and it is still called this today. Apart from teaching chemistry, Vicente Cervantes (1755–1829) translated into Spanish Lavoisier's *Elementary treatise of chemistry,* before being translated Spain (Chamizo 2004).

In 1790, the Brazilian Jose Bonifacio de Andrada e Silva (1763–1838) was given the task, by the government of Portugal, of studying chemistry and metallurgy in some of the European countries to bring knowledge back to Portugal and Brazil. The trip lasted 10 years and took him from Paris, where he studied with Antoine Fourcroy (1755–1809), to Freiberg where he worked with Alexander von Humboldt (1769–1869), and to Sweden where he discovered the petalite and mineral spodumene. Later on the element lithium was identified in petalite. In Brazil, he built the first blast furnace in his home state Minas Gerais (Alfonso-Goldfarb and Ferraz 1990; Fernandes et al. 2004).

Joao Manso (1750?–1820) was an important Brazilian chemist. Born before 1750 in Minas Gerais, he was an important example of an autodidact. Manso did not receive higher education. He did not leave the country, but by the end of his life he could read Portuguese, Latin, Greek, German, and French. He published five books in Lisbon on practical chemistry, which includes many different topics, such as distillation, industrial production of salt, and chemical analysis of the ashes resulting from burning Brazilian plants.

13.4.2 Second Chemical Revolution (1828–1869)

In 1833, Custodio Alves Serrao (1800–1873) wrote *Lessons of chemistry and mineralogy*, the first treatise on chemistry published in Brazil; previous chemistry books by other authors such as Manso had been published in Lisbon because there were no official printers in Brazil before 1808. In 1845, he published a work on the purification of palladium.

The German pharmacist Theodor Peckolt (1822–1912) arrived in Brazil in 1848. He already had a great scientific reputation before being hired to reorganize the chemistry section of the National Museum. After that, he devoted himself to the chemical analysis of plants, with the purpose of discovering and commercializing new medical remedies. His scientific work was published in the best European journals, and gave him the highest academic distinctions. This research was mainly carried out at his pharmacies in Rio de Janeiro.

A few years after the independence of Chile, a project on the *New organization of the College of Coquimbo* was approved. Among other things, it established that,

> [...] a professorship in chemistry will be created and another of mineralogy, whose skills are necessary for this province that contains the main minerals of Chile, being the common practice of its inhabitants, mining.

In 1835, an industrial owner travelled to Europe to buy laboratory equipment, and he hired an expert, the Polish Ignacio Domeyko (1802–1889). He trained the largest group of chemists in the country, and published the first treatise on chemistry and mineralogy in Chile.

The German Luis Posselt (1817–1880) appeared as the only Latin American participant (for Mexico) at the Karlsruhe Congress (Chamizo 2015). He was born in Heidelberg in 1817, where he studied pharmacy, and in the 1840s travelled and worked in Giessen, under the direction of Justus von Liebig. At the University of Giessen, he met the first chemistry PhD student of the whole of American continent, Vicente Ortigosa, a Mexican who was working on an analysis of nicotine. Posselt travelled to Mexico in 1848. In an intense and solitary pilgrimage, he devoted himself to the benefit of silver and gold in Zacatecas, Nuevo Leon, and North Carolina. He never went to Mexico City, nor did he participate in any scientific associations. In 1860, he returned to Germany and attended the Karlsruhe Congress.

13.4.3 Third Chemical Revolution (1887–1923)

In Argentina, in 1896, pharmaceutical chemists established, at the University of Buenos Aires, the first title of chemistry in the country: a PhD in chemistry. This lasted for 4 years and was practically oriented toward analytical chemistry. Half a century later, more than 220 doctors had graduated from that institution. At the beginning of the twentieth century, other Argentinian universities started offering similar titles, but only in pharmacy and chemical engineering. In 1912, there were enough professors and PhD graduates to create the Sociedad Química Argentina, which in 1920 would be named Asociación Química Argentina. The following year, this institution began publishing its *Annals*, and organized the National Congress of Chemistry (Matharan 2016).

In 1918, in Rio de Janeiro, the Institute of Chemistry was established. Designed and directed for the following 20 years by Mario Saraiva (1815–1950), its main task was to analyze the French butter commercialized in Brazil. Another purpose of the Institute was the training of professional workers for the emergent chemical industry. In 1922, the Brazilian Society of Chemistry was established, in the midst of the celebrations of the first centenary of Brazil's independence.

Professional chemistry began in Mexico in 1916, when Juan S. Agraz (1881–1949), by presidential decree, was appointed director of the National School of Industrial Chemistry. A fragment of the document that gave rise to this institution shows its intentions:

> So chemistry, one of the mothers of the industry, teach the people how to get broad match beneath the sole and then, aware of scientific truths, it will feel more attachment to the piece of land that was fortunate to possess in the World and who has the right to call homeland.

Agraz studied at the *Institut de Chimie Appliquée* at the University of Paris, in the late nineteenth century. The institute, with a new pedagogical approach at that time, trained chemists for the development of science and industry at the same institution, with a single program lasting 3 years and focusing exclusively on chemistry.

13.4.4 Fourth Chemical Revolution (1945–1965)

In 1946, the Chilean Society of Chemistry was created with two purposes:

1. To promote scientific research and dissemination of the different branches of chemistry and related sciences, both theoretical and applied
2. To provide cooperation with the authorities, agencies, and institutions dedicated to teaching scientific and technological research. A few years later, the first Bulletin of the Chilean Society of Chemistry was published, now known as the *Journal of the Chilean Chemical Society*.

L. Leloir (1909–1987) was born in France to Argentinian parents, and graduated from the University of Buenos Aires medical school in 1932. One year later, he met Bernardo Houssay (1887–1971), the 1947 Physiology Nobel Prize winner, who pointed Leloir toward investigating in his doctoral thesis the suprarenal glands and carbohydrate metabolism. Leloir spent a few years working in England and in the United States. In those years, he studied chemistry and physics by himself. In 1947, he raised private financial support to set up the small *Instituto de Investigaciones Bioquímicas Fundación Campomar* in Buenos Aires, where he worked until he died. There, he and his group studied how sugar milk is processed by the human body. This in turn led to the breakthrough observation that sugar nucleotides help bodies to store certain sugars to be transformed into energy. In 1970, Leloir won the Nobel Prize for Chemistry "for his discovery of sugar nucleotides and their role in the biosynthesis of carbohydrates" (http://nobelprize.org).

In July 1951, *LIFE* magazine published an article entitled "Cortisone from giant yam." In this short article, one of the most remarkable pictures showed young researchers from the Mexican company Syntex next to barbasco, the plant from which cortisone was synthetized in a small laboratory in Mexico. Later, the same laboratory announced one of the most important advances of twentieth century chemistry, the synthesis and production of the birth control pill. The social implications of this discovery revolutionized the view that women had of themselves. In the mid-twentieth century it was inconceivable to imagine that cutting-edge research such as on steroids could be done in a country like Mexico (Soto 2005), but specific circumstances led to the creation of the Mexican hormone steroid industry that had a huge impact, both scientifically and socially worldwide. Thus, in 1999, the Marker degradation (the first practical synthesis of the pregnancy hormone, progesterone) and the Mexican hormone steroid industry were designated jointly by the American Chemical Society and the Chemical Society of Mexico as an International Historic Chemical Landmark (Raber 1999). One of the most important patents registered in the USA by Syntex is patent number 2744122. It is the synthesis of the main contraceptive pill compound developed by the young Mexican chemist E. Miramontes (1925–2004) with C. Djerassi (1923–2015) and G. Rosenkranz (1916–) (Syntex's research director).

13.4.5 Fifth Chemical Revolution (1973–1999)

In 1974, Sherwood Rowland (1927–2012) and his PhD Mexican student Mario Molina (1943–) published the results of his research into the effect of chlorofluoroalkanes in the ozone layer. It was not the first time that chemical companies and governments around the world faced difficulties regarding the ability to pollute the environment, but this time, unlike all previous times, the damage and the risk were unequivocally global. A couple of years earlier the use of DDT had already been banned in the USA. In 1995, Paul J. Crutzen (1933–), Mario J. Molina, and F. Sherwood Rowland were awarded the Nobel Prize for Chemistry "for their work in atmospheric chemistry, particularly concerning the formation and decomposition of ozone" (http://nobelprize.org).

After a strong struggle with the chlorofluoroalkane industry, where Molina and Rowland played an active role, the political response to ozone layer depletion was the Montreal Protocol. Signed in the 1980s by more than 200 countries, it is the first universally ratified treaty in United Nations history. Thus, after the publication of Rachel Carson's *Silent Spring* in the 1960s and the foundation of the US Environmental Protection Agency in the 1970s, green chemistry with its 12 principles was born, together with new concepts, such as "atom economy," "life cycle analysis," and also in its chemical context, "sustainable." About this, Roald Hoffmann, winner of the 1981 Nobel Prize for Chemistry said:

> The ecological imperative has crept down much more slowly to inventive yet unconcerned academia. However, I see its formative events there –in the interest in atmospheric chemistry and the ingenious construction of novel organic processes to avoid the use of organic solvents. A government carrot in the form of new research funds for green chemistry would, in my unpopular opinion, be just what is needed to channel the ingenuity of my colleagues, who love to say that they just want to do what is interesting, but…The same nicely obsessive penchant for control as that which is used to make molecules can do acrobatics and is being turned to the attainment of a necessary balance between our given imperative to create and our love for the world. (Hoffman 2000, p. xi)

13.5 Conclusions

> In the late 1940s the steroid production process in Mexico was constrained almost in its entirety by a single individual and scientist, (the American chemist) Marker (who founded Syntex). By contrast, in the 1970s, Mexican peasants, wholly unaware of chemical processes, controlled steroid hormone production by regulating the amount and type of wild yams that were extracted. (Soto 2005, p. 751)

The centennial pollution from copper extraction in Chile and Peru; the rich oil fields and its transformation into fuels in Colombia, Ecuador or Venezuela; the appropriation of Amazonian latex (rubber) and curare (anticoagulants); the strong analgesic epibatidine patent, whose active ingredient was obtained from a secretion of Ecuadorian frogs; the green revolution in Mexico; the International Potato Center in the Andean region; transgenic soybeans, which are used in Argentina and Brazil, are all examples of Latin American chemistry. These are treated not from the perspective of hagiography used in the previous section, but from that of environmental history (McNeill 2000), the so-called environmentalism of the poor (Martinez-Alier 2002) or cultural control (to be understood as a system and as an asymmetric process that considers resistance, appropriation, innovation, imposition, suppression, and disposal (Bonfil-Batalla 1988). We should study these stories that also became history. Scientific pluralism as a philosophical position recognized different approaches to the same phenomena (here, history of chemistry in Latin America). These approaches have the right to differ. That is to say, there are many different systems of representation for scientific use in understanding the world, none of them complete, and perhaps irreconcilable (Chang 2012).

Two brief examples show the diversity of participants and the difficulty in recognizing them, but pointing out how to pass from invisibility to visibility, from monism to pluralism, our pluralism in Latin America. The cinchona tree can be found in the jungle of Peru. In the seventeenth century, the wife of the Spanish viceroy accepted the knowledge of the natives of the place and was cured from a fever by eating the bark of the tree. For centuries, the Spanish crown fruitlessly tried to monopolize its cultivation. Its bark, dried, ground and frequently dissolved in wine, was used worldwide against malaria, until shortly before the fourth chemical revolution, when R.W. Woodward synthesized its active ingredient, quinine. For many years, it was considered one of the greatest treasures of Latin America. In the twentieth century the Mexican indigenous peasants already knew some of the properties of barbasco before this yam became the main raw material for the synthesis of cortisone and the contraceptive pill. Moreover, in recent research, we have identified a significant number of chemistry students from Mexican universities who related to this work (Hernández et al. 2016). Besides, as Soto showed:

> The merging of science and peasant life reshuffled social hierarchies in the countryside, granted monetary values to an erstwhile weed, and gave a novel reinterpretation to laboratory knowledge and its (social) uses. (Soto 2005, p. 743)

The chemistry that is now produced in Latin America is a very small part of that produced throughout the world. At the end of the fifth revolution, Latin American countries published 13,651 papers that were indexed in *Chemical Abstracts* (RICYT). Taking into account Spain and Portugal (Ibero-America), this number grows to 28,277. When compared with publications from all over the world (757,444), it represents less than 4%. This is the extent of chemistry in our region. We should reflect on the meaning of researching, discovering, and teaching our discipline, from our world region. The fifth chemical revolution calls for global responsibility. The signing of the *Montreal Protocol* and the *Treaty on Biological Diversity* (not yet ratified by the USA) indicates a possible way. Philosophical pluralism provides answers.

Finally, it would be useful to defend the value of the history of chemistry as one of the prime places for understanding chemistry. When a community, particularly an educational community, denies recognizing history, its own past, refusing to acknowledge that those events that should be part of the collective memory of the community, the image of the past, and the present and the future is built by others.

Acknowledgments To DGAPA-Universidad Nacional Autónoma de México for its support in a research stay at Centre d'historia de la ciència in Universitat Autònoma de Barcelona.

References

Alfonso-Goldfarb, A. M., & Ferraz, M. H. M. (1990). A recepção da química moderna no Brasil. *Quipu, Revista Latinoamericana de Historia de las Ciencias y la Tecnología, 7*, 73–92.

Aristizábal-Fúquene, A. (2015). El Platino: Contribuciones sociohistóricas y científicas desde el siglo XVIII. *Educación Química, 26*, 146–151.

Bond, J. (1993). *PreColumbian chemistry: A descriptive chemistry and history of technology from natural resources*. Medford: Tufts University.

Bonfil-Batalla, G. (1988). La teoría del control cultural en el estudio de procesos étnicos. *Anuario Antropológico, 86*, 13–53.

Chamizo, J. A. (2004). Apuntes sobre la historia de la química en América Latina. *Revista de la Sociedad Química de México, 48*, 165–171.

Chamizo, J. A. (2013). Technochemistry: One of the chemists' ways of knowing. *Foundations of Chemistry, 15*, 157–170.

Chamizo, J. A. (2014). The role of instruments in three chemical' revolutions. *Science & Education, 23*, 955–982.

Chamizo, J. A. (2015). Novohispanos y mexicanos en las revoluciones químicas: Aportes recientes a la historia de la química en México. México: UNAM-CEIICH.

Chamizo, J. A. (2016). How chemistry teachers, using history of chemistry could teach chemistry. In L. Jari, K. Juuti, J. Lampiselkä, A. Uitto, & K. Hahl (Eds.), *Science education research: Engaging learners for a sustainable future* (pp. 806–813). Helsinki: ESERA.

Chamizo, J. A. (2017). The fifth chemical revolution: 1973–1999. *Foundations of Chemistry, 19*, 157–179.

Chang, H. (2012). *Is water H_2O? Evidence, realism and pluralism*. Dordrecht: Springer.

Del Busto, J. (1970). *Historia general del Perú: Perú Antiguo*. Lima: Studium.

Fernandes, R., Seixas, L., & Domingos, D. (2004). *Historia da química no Brasil*. Campinas: Editora Átomo.

García-Canclini, N. (2002). *Latinoamericanos buscando lugar en este siglo*. Buenos Aires: Paídos.

Gilbert, J. (2006). On the nature of "context" in chemical education. *International Journal of Science Education, 28*, 957–976.

Golinski, J. (2005). *Making natural knowledge: Constructivism and the history of science*. Chicago: The University of Chicago Press.

Gónzalez-Mora, M. A. (1983). Apuntes sobre la historia de la química en Cuba (1820–1888). *Revista Cubana de Educación Superior, 3*, 101–114.

Hacking, I. (1983). *Representing and intervening*. Cambridge: Cambridge University Press.

Hernández, Y., Chamizo, J. A., Kleiche-Dray, M., & Russell, J. M. (2016). The scientific impact of Mexican steroid research 1935–1965: A bibliometric and historiographic analysis. *Journal of the Association for Information Science and Technology, 67*, 1245–1256.

Hoffman, R. (2000). Introduction. In N. Hall (Ed.), *The new chemistry* (pp. ix–xi). Cambridge: Cambridge University Press.

Holmes, F., & Levere, T. (Eds.). (2000). *Instruments and experimentation in the history of chemistry*. Cambridge: The MIT Press.

Höttecke, D., & Silva, C. C. (2011). Why implementing history and philosophy in school science education is a challenge: An analysis of obstacles. *Science & Education, 20*, 293–316.

Jensen, W. (1998). One chemical revolution or three? *Journal of Chemical Education, 75*, 961–969.

Kindi, V., & Arabatzis, T. (2012). *Kuhn's the structure of scientific revolutions revisited*. New York: Routledge.

Kragh, H. (1987). *An introduction to the historiography of science*. Cambridge: Cambridge University Press.

Kuhn, T. (1970). *The structure of scientific revolutions*. Chicago: The University of Chicago Press.

Kuhn, T. (1977). *La tensión esencial: Estudios selectos sobre la tradición y el cambio en el ámbito de la ciencia*. México: Fondo de Cultura Económica.

Kuhn, T. (1992). The trouble with the historical philosophy of science. In *Robert and Maurine Rothschild distinguished lecture, an occasional publication of the Department of History*. Cambridge: Harvard University.
Latour, B. (2012). *Investigación sobre los modos de existencia*. Buenos Aires: Paídos.
Martinez-Alier, J. (2002). *The environmentalism of the poor: A study of ecological conflict and valuation*. Cheltenham: Edward Elgar.
Matharan, G. (2016). La constitución de la química como disciplina en Argentina, México y Colombia: Un estudio comparado. *Educación Química, 27*, 67–73.
Matthews, M. (Ed.). (2014). *International handbook of research in history, philosophy and science teaching*. Dordrecht: Springer.
McNeill, J. (2000). *Something new under the sun. An environmental history of the twentieth century*. London: Penguin.
O'Gorman, E. (1958). *La invención de América*. México: Fondo de Cultura Económica.
Pickstone, J. (2000). *Ways of knowing*. Manchester: Manchester University Press.
Raber, L. (1999). Steroids industry honored: International historic chemical landmark acclaims success of Mexican steroid industry and a U.S. chemist who made it possible. *Journal of the Mexican Chemical Society, 43*, 235–237.
RICYT, Red Latinoamericana de Indicadores de Ciencia y Tecnología. http://ricyt.org. Accessed 13 July 2016.
Schwab, J. (1962). *The teaching of science as inquiry*. Cambridge: Harvard University Press.
Shapin, S., & Schaffer, S. (2011). *Leviathan and the air-pump: Hobbes, Boyle, and the experimental life*. Princeton: Princeton University Press.
Soto, G. (2005). Uncommon trajectories: Steroid hormones, Mexican peasants, and the search for a wild yam. *Studies in History and Philosophy of Biological and Biomedical Sciences, 36*, 743–760.
Tala, S. (2011). Enculturation into technoscience: Analysis of the views of novices and experts on modelling and learning in nanophysics. *Science & Education, 20*, 733–760.
Van Aalsvoort, J. (2004). Logical positivism as a tool to analyze the problem of chemistry's lack of relevance in secondary school chemical education. *International Journal of Science Education, 26*, 1151–1168.
Van Berkel, B., de Vos, W., & Pilot, A. (2000). Normal science education and its dangers. *Science & Education, 9*, 123–159.

Chapter 14
Natural History, Chemistry, and Teaching in Modern Scientific Culture

Luciana Zaterka and Ronei Clécio Mocellin

14.1 Introduction

The importance of history and philosophy of science to scientific teaching has been widely discussed in the last few decades (Matthews 1994; Paixão and Cachapuz 2003; Niaz 2001). In Brazil, we noticed, on the one hand, practical–political actions aimed at inserting history and philosophy of science into the curriculum; recall, for instance, the current PCNs (abbreviation in Portuguese that stands for National Curriculum Guidelines for High School Education) (Brasil 2000), and the New Curriculum Guidelines (Brasil 2006) for undergraduate courses; on the other hand, a theoretical concern reflected in an increasing number of scientific articles in the field in the last decades in specialized journals (*Ciência & Ensino*, *Caderno Brasileiro de Ensino de Física*, *Química Nova na Escola*, *Ciência & Educação*). Undoubtedly, the reasons for this discussion embrace different perspectives that are often complementary: expansion of the conceptual network of science students, criticism of the positivist view of science; collapse of the myth of scientific neutrality; contextualization of science teaching, etc.

The history and philosophy of chemistry allow us to understand the historical–conceptual genesis of several concepts approached in the current chemistry teaching. For instance, one of these concepts refers to the dichotomy between nature and history, or even between history and artifact. Nature was within the scope of laws, under the domain of necessity, namely what can only be exactly as it is, impossible to become different from what actually it is – or according to Aristotle (384–322 BC) what can not exist in a different way and, therefore, what exists in only one way

L. Zaterka (✉)
Federal University of ABC (UFABC), Santo André, SP, Brazil

R. C. Mocellin
Federal University of Paraná (UFPR), Curitiba, PR, Brazil
e-mail: roneimocellin@ufpr.br

M. E. B. Prestes, C. C. Silva (eds.), *Teaching Science with Context*, Science: Philosophy, History and Education,
https://doi.org/10.1007/978-3-319-74036-2_14

is needed (Aristotle 1984). History, conversely, was under the scope of contingency, choice, deliberation, and place, that "it was like this but it could be different." Thus, modern chemistry which began in the early years of the seventeenth century with leading names such as Robert Boyle (1627–1691), would stand side by side with nature, laws, experiments, and protocol. Correct? Not exactly.

This chapter addresses issues related to chemical knowledge and its material and cognitive products, from a pedagogical, philosophical, and historical perspective. We consider authors and contexts from the seventeenth and eighteenth centuries, approaching them as case studies and concrete examples that characterize this historical period. Francis Bacon's (1561–1626) new program for natural history, more specifically his view of *history of arts*, was adopted as the core idea of our work. Our first goal is to describe the reasons that made chemistry the key knowledge of Bacon's new program.

Investigating the development of the concept of modern natural history – essential to our understanding of modern chemistry origins –shows otherwise. We will see that the division between the disciplines that study the history of objects, facts, and events on the one side and those that deal with the sciences of nature on the other, though still present, lacked support back then. Through this study, we will see the beginning of an experimental philosophy of nature, richer and more complex than we could imagine at first sight. An experimental philosophy that is concerned with the natural history of things devised by art. Here, we shall watch the creation of key concepts for its further development, such as *a posteriori* science (which, therefore, operates from the effects to the causes), strongly empiric, more concerned with the study of bodies and their operability than with their atomic constituents.

Hereafter, we intend to approach the convergence between manipulations aimed at producing artifacts and those aimed at intervening in living matter that took place in an originally chemical epistemic place: the laboratory. Our purpose is to explain the identity of technical and conceptual operations in the manipulation of both living and inert matter. The chemical laboratory is indeed a theater of material manipulations and operations. As such, chemical knowledge can surely be characterized as a long-term *style of reasoning,* in which the laboratory is a place for the production and creation of theories and artifacts (Bensaude-Vincent and Simon 2008, pp. 55–74).

This means that there is no chemical science without a chemical laboratory. To a chemist, the laboratory is not a cabinet of curiosities or a place to demonstrate natural laws, but a place where transformations happen, where inner components matter and several different levels of social structure interact and relate. From those spaces, emerged a scientific culture that identified a community of practitioners. Such a culture comprised practical attitudes, gestures, theories, instruments, products, teaching manuals and methods, translations, correspondences, symbolic forms, and was also of interest to a wide-ranging audience. Apart from a particular style of reasoning originating in laboratories, the values of chemistry spread throughout society, generating practical results to complex philosophical debates.

Hence, we intend to draw attention to the centrality of chemical knowledge in the scientific culture of that time and how chemistry teaching is a manifestation of that social interest. Our purpose here is to point out the common objectives between the *natural history of arts* and chemistry teaching, as laid out in the manuals and courses inspired by the lessons on Baconian chemical philosophy of the physician–chemist Herman Boerhaave (1668–1738). We give two examples of this centrality, one directly related to philosophical issues and the other related to the teaching of chemistry itself. Based on Bacon's ideas and the *concepts* and *operations* of French chemistry in the middle of the eighteenth century, Denis Diderot (1713–1784) developed his concept of *experimental philosophy*. Among the several different ways of learning chemistry at the end of eighteenth century, the course on chemistry and scientific culture taught by the group led by Louis-Bernard Guyton de Morveau (1787–1816) at Dijon Academy is considered a unique example, especially because of the importance of this chemist for the conceptual and linguistic changes during the so-called "chemical revolution".

Finally, we limit the historic period studied herein to the time when the *Encyclopédie méthodique* (1786) coordinated by Guyton and the *Traité élémentaire de chimie* (1789) by Antoine de Lavoisier (1743–1794) were published. Our choice here is not based on traditionally claimed reasons, such as for his discovery of the role that oxygen plays in combustion, but because that was the moment when the history of chemistry and, therefore, part of the natural history of its products was set aside during the process of education of new chemists. However, even with the deconstruction of the "Lavoisier myth," the fact remains that his *Traité éléméntaire* was a didactic–pedagogical model for the teaching of chemistry. In this *method* of teaching there was no place for the natural history of chemical artifacts. Thus, we would like to point out that, paradoxically, the *manual* that for many characterizes a scientific revolution and the beginning of modern chemistry also marks an important pedagogical break, that of ending the interest in the history of this science in its own teaching–learning process.

14.2 Bacon: Chemistry and Natural History

The "Baconian program"[1] of science can be an important route of access to the studies of modern chemistry. Thus, to comprehend aspects of this knowledge program, developed in the seventeenth and eighteenth centuries, with ethical and epistemic outcomes until now, it is necessary to understand some of the concepts from the

[1] We are aware of the complexity of defining what the "Baconian program" of knowledge is. In the present work, we use the traditional definition by T. Kuhn. The Baconian program can be understood as "a new cluster of research areas that owed their status as sciences to the seventeenth century's characteristic insistence upon experimentation and upon the compilation of natural histories, including histories of the crafts. To this second group belong particularly the study of heat, of electricity, of magnetism, and of chemistry" (Kuhn 1961, p. 186). Thus, this "program" of knowledge includes the quantification of aspects, construction of technical instruments, reproducibility, communities of people of science working cooperatively, new techniques and methods, and natural philosophers working in laboratories (Sukopp 2013, p. 58).

philosophy of Lord Verulam, mainly, the notions that allowed chemistry to slowly advance toward the field of the new experimental philosophy of nature and thus to acquire a nuclear *locus* within this new program. Amongst these, we come across the concept of natural history.

In the several works in which Bacon addresses his concept of natural history, we highlight his *Parasceve ad historiam naturalem et experimentalem* (Preparative toward a Natural and Experimental History), published in 1620. There, we can, without a doubt, identify the main concepts of his new understanding. Bacon immediately dissociates himself from the old concepts of natural history, such as those formulated by Aristotle, Theophrastus (372–287 BC) or Plinius (23–79 AD), because, according to him, none of them understood what a "true natural and experimental history of nature" is, narrowing their methods to mere descriptions of natural phenomena. Thus, Bacon aims to establish a new route to knowledge – his *Instauratio Magna's project* (Great Instauration) – in which he intends to review the concept of knowledge as known in ancient times and the late scholastic, based on syllogisms and meaningless abstractions.

In this review, Bacon points out the importance of the operational scope of nature, hence introducing the natural and experimental history of nature as a basis for the entire system. "For knowledge are as pyramides, whereof history is the basis: so of Natural Philosophy the basis is Natural History" (Bacon 1963a, v. 3, p. 356). In other words, we should start the search for knowledge in its basis, that is, "prepare a Natural and Experimental History, appropriate and good." Thus, everything that is related to bodies, its properties, qualities, and materials should be, as far as possible, recorded, numbered, weighed, measured, and then defined. Descriptions of new technical inventions should also be cataloged. Bacon, thus, attachs history to the experimental method. As a result, we see the rise of a concept of protocol science. Because of its own programmatic nature, natural histories gain unprecedented importance (Sukopp 2013, p. 57).

If, on the one hand, we notice the meticulous accumulation of knowledge provided by those histories, on the other hand, we also notice the production of practical effects that they necessarily generate. Based on such an operational perspective, according to Bacon, histories can neither be a simple record of empirical facts, mere memories or recollections, nor should they be restricted to immediately useful things. Mainly, natural history should, first and foremost, shed light on the discoveries of causes. According to him, this goal can only be achieved if we dissect, change, torment nature through experiments, so that humans can get closer to the hidden causes through the evident effects observed in nature. That is why natural history is the place of causes and not of effects. In other words, natural history objects are the bodies inspected at their several levels of ontological texture, that is, from minute to manifest bodies. In his *Historia densi et rare* (History of dense and rare), Bacon prioritizes the study of bodies in their changes – rarefaction, condensation, expansion, and contraction – aiming to get closer to the causes of the material processes through their effects and, thus, make the research more productive (Giglioni 2012, p. 68).

Bacon, in *Historia densi et rare*, develops a history of dilatations by the envelopment and contact with a favorable body, that is, a history of bodies capable of losing (or not) their resistance when combined with other bodies:

1. Sugar and some gums, infused in liquors, dissolve; for (like sponges) they willingly slacken their parts to take the liquor in.
2. Paper, animal hair, wool and suchlike porous bodies so open themselves up when immersed in liquors, or otherwise moistened, that they become softer, more easily torn and practically rotten.
3. Sudden joy, as you have with good news, on seeing something you have been longing for, and so forth, although they embrace not the body, but only the fantasy, dilate the spirits of animals strikingly, and they sometimes do it enough to risk sudden fainting fits or death. The imagination does the same in sexual love. (Bacon 2007, v. XIII, p. 117)

Bacon believes that these processes happen in the scope of microscopic bodies and in a more complex ambit such as the human body, where the spirits – material bodies, very subtle and weightless, that have appetites and impulses and are, therefore, responsible for the matter activity –join each other or volatilize, causing, for instance, diseases, aging, and even death. This Baconian perspective clearly shows a concern for materiality and the operationality of substances.

We are not in the metaphysical domain of a general substance, but of particular substances, which will be analyzed according to the *a posteriori method* that, as we have seen, essentially arises from an experimental history of things. As a result of this new knowledge program, we notice the rise of a strongly empirical epistemology and an anti-essentialist metaphysics. For this reason, the experimental method is based on rules and protocols of laboratory research, the true scientist has to interfere in nature and be "its minister" and, first and foremost, this scientist must undertake a science directed at the welfare of society. This philosophy is less focused on the knowledge of forms and universal essences and more on the process of this operational science.

This methodological position as proposed by Bacon can be noticed in the work of one of his main heirs, the chemist R. Boyle. In *Some considerations touching the usefulnesse of experimental natural philosophy*, where Boyle presents the reasons for the importance and usefulness of the new experimental philosophy of nature, the chemist writes nearly two pages about antimony, actually putting into practice a natural history of antimoy. First, he presents a wide range of treatises about this "mineral abstruse": *Anatomia antimonii* (Anatomy of antimony, 1617), of Angelo Sala, and the *The triumphant chariot of antimony de Basil Valentine* (English version of 1660), *Basilica antimonii de Hamerus Poppius* (Basilica of antimony, 1618), and *Liber unus de secretis antimonii* (Of the secrets of antimony, 1570) by Alexander von Suchten, among several others. Next, Boyle presents the different processes of preparation, purification, medicinal use, etc. Finally, "the shortness of life makes it impossible for one man thoroughly to learn Antimony, in which every day something of new is discovered" (Boyle 1999, v. 3, p. 208).

However, such an approach is precisely the one that the experimental philosopher is expected to adopt. To an experimenter, a laboratory worker who believes in *a posteriori* science, the truth is never complete and finished; the help and testimony of peers, the memory and, therefore, the natural history are essential. Herein, we clearly notice the importance of historical knowledge as a description and report of things and their productions and of whom produces and describes such things as well.

This history plays a significant role in the scope of teaching, because teaching something means, above all, teaching its natural history, which includes the transmission of everything that is known about it until now by means of reliable testimonies, reports, and experiments.[2] Ultimately, one only knows in fact nature after counting, recounting, and recounting again its history. Perhaps that is why when Bacon writes about the disciplines he believes are missing at his time (*desiderata*) and, therefore, are to achieve in his new "scientific–pedagogical," the inductive history, that is, natural history appears between the first (Bacon 1963b, v. 1, p. 838). Here, we observe the temporary, dynamic, changeable, and public character of this new knowledge.

Thus, the transmission of chemical knowledge, at that time, pointed directly to the imbrication between the natural history of materials manipulated in laboratories and manufactories in the professional training of chemists, doctors or apothecaries. Allan Debus underlined the role of creating chemistry professors in medical schools across Europe in the social and institutional recognition of chemistry as a science. In fact, in the seventeenth century, authors such as Jean Baptist van Helmont (1580–1644) reformulated Paracelsianism, emphasizing quantification, techniques of observation, and, above all, chemical explanations of physiological processes, which enabled chemistry to develop in a university environment (Debus 1977). In addition, this transmission was accelerated by the creation of a large book market, which besides constituting an audience for texts on chemistry, also structured didactic models for learning this science. According to Owen Hannaway, a striking feature of the emergence of chemistry as an autonomous science is directly linked to the publication of specialized manuals. According to him, the printed book introduced a new regime of knowledge, breaking away from the hermetic tradition, so that chemistry became a public science. Two works marked the entrance of chemistry onto the scene of the modern scientific revolution, *Alchemia* by Andreas Libavius, published in Frankfurt in 1597, and *Basilica chymica* by Oswald Croll, published in the same city in 1609 (Hannaway 1975).

[2] "When Bacon uses History in these experimental and natural histories it includes, for instance, names of winds, what people have said about them, when they blow, where they come from, and what they do to trees and plants [...]. The history of the longevity of humans begins with a collection of textual evidence. The history of the circulation of the blood not only tells how blood might move, but also facts relevant to circulation and heating" (Miller 2005, pp. 358–359). Thus, Miller argues that history, in Baconian terms, indicates, on the one hand, a preliminary work in which the facts are reported and informed numerous times and thus, taught, and presents subsequent elaborations, including the experimental ones; and, on the other, this history must include the structure and functioning of nature.

The author of *Novum Organum* innovates once again by classifying nature into three states: natural processes (or generations), monsters of nature (or preter-generation), and the nature modified by human's domain (or arts) (Bacon 1963c, v. 4, p. 253). Thereby, as history refers both to what is created by nature and by men, it includes, according to Bacon, what the nature generates itself and what it generates on the action of man.

It is important to emphasize that this conception of natural history comprises the effect of mankind, that is, we are not in the scope of a merely descriptive history of nature itself. We are in the scope of an "active" history of human operations and techniques on nature, dominating it for the benefit of human beings:

> The most noble end of Natural History is: being the base and raw material of the truth and legitimate induction, and sufficiently extract from the senses to provide to the intellect. (Bacon 1963c, v. 4, pp. 251–252)

Therefore, natural history deals with the nature of things, whether those things are "free," as in natural species, or "disturbed," as in monsters or beauties, or "confined" as in experiments. The latter is the nature troubled and modified by human experiments. Traditionally, it was outside the scope of natural history because it was seen and understood as defective, fragmented, and careless.

This perspective represents an important turn toward modernity. From now on, men are not only capable of accelerating the ordinary course of nature, but are also capable of producing new natures – artificial nature ontologically similar to original nature. This is why chemistry can acquire a privileged place in this new program. Solar heat can be compared to fire, natural gold in sand is identical to the artificial one produced in furnaces. Thus, men of science no longer need to fear divine cholera by manipulating, and altering nature because manipulating the divine work is one of their designs. As we know, this statement of a nonsubstantial difference between the products of art and nature – present in some of the greatest exponents of the new natural philosophy – contrasts radically with the Aristotelian view of art or technique that only completes the work of nature or imitates its production. Herein, arts are seen as the adulteration and falsification of nature.

Hence, European culture was compelled to abandon the inferiority thesis of the technique compared with science or "theoretical knowledge," and of manual labor compared with intellectual, very much alive during classic and medieval civilization. We noticed a change in the very conception of nature in the genesis of modernity. In such a change, chemistry reaches a fundamental place among human knowledge. For chemistry, based upon natural and experimental history of things and artifacts produced by man, is a vital tool in the reconstruction of knowledge, as proposed by the Baconian program.

14.3 The Laboratory: Experimental Knowledge of Living Bodies and Artifacts

Perhaps Bacon, in none of his histories, has been so clear about the goal of the philosophical enterprise (i.e., the proposal for a restoration of lost knowledge by original sin through a new conception of experimental and operational science, directed to the welfare of most people) than in his thought-provoking *Historia vitae et mortis* (History of life and death) (Bacon 2006, v. XII). In this work, Bacon uses his matter theory to approach the question of continuance and longevity of man and animals. At the end of the book, in which longevity is effectively discussed, the genesis of the modern empiricism is observed, in addition to its foundation and realization.

Usually, science textbooks introduce Baconianism in a caricatured way. This confuses the professor of science in the presentation of the origins of scientific modernity, by emphasizing the empiricism/rationalism dichotomy. Such a dichotomy disguises the essential concepts of the constitution of modern methodology. As currently defined in textbooks on science teaching, empiricism uses experience as a fundamental methodological tool, as opposed to rationalism. In addition, the goal of empiricism is to combine logical, psychological, and historical arguments that were traditionally addressed separately and, therefore, elaborate an experimental history of nature. Therefore, we must note the nuclear importance of understanding the knowledge and its practices from its own context, thus examining the primary sources. This avoids anachronism and reductionism.

Our focus here is that Bacon, based on the complex matter theory, proposes possible scenarios of physical regeneration. Bacon stated:

> […] if a man could arrange to put into an old body spirits of the kind characteristic of a young one, it is likely that this mighty wheel might put the other, lesser wheels into reverse, and turn back the course of nature. (Bacon 2006, v. XII, p. 245)

In other words, Bacon believes that if the spirits that compose the human body can be properly managed, the body's senescence can be postponed and a long-lasting life can be achieved. In this scenario, Bacon introduced the possibility, unimagined until then, of man exercising dominion over nature; this empire could not only be practiced over external nature (i.e., the natural world), but also over men themselves, and, ultimately, over their own body. Thus, Bacon anticipated some of the fundamental ideas that would lead some members of the Royal Society to test experiments on blood transfusions. Thereafter, the doors opened to the modern dream of introducing several and effective individual "rejuvenation" or "restoration" techniques.

Chemists have consolidated an experimental style of investigating natural and artificial materials in their laboratories and manufactories. *Chemical operations* performed on those epistemic spaces corresponded to all chemical means to act on the materials to achieve through art the purpose of chemistry, which is to perform *separations* and *unions*. Chemists had chemical and physical instruments to perform such operations. In the period studied herein, the notion of *chemical instruments* developed by Boerhaave had a great impact. Boerhaave's notion, which is based on the Baconian methodology of discovery, had the purpose of collecting the highest number of chemical facts to establish a *natural history* of chemistry. For Boerhaave,

> Chemistry being employed in the examination of all sensible bodies, 'tis obvious of how much importance it is in science of natural philosophy, and extends itself through all the parts thereof: besides, making great assistance to natural philosophy, in regard fire is the most general instrument which nature makes use of for the accomplishing of almost all her works. (Boerhaave 1753, p. 172)

Boerhaave's view of the instruments:

> In all arts which direct bodies to be chang'd, the name instruments is given to certain things capable of a particular motion; which being thence apply'd to the body intended to be chang'd, produces the requisite change therein: thus, in the present art, there are certain bodies by whose means the requisite actions are produced. These, with the best chemists, we usually reduce to six principal ones: fire, water, air, earth, menstruums, and utensils; which are to be well understood, in order to have a just notion of the operation performed by them. (Boerhaave 1753, p. 205)

This concept also indicated the indifference to whether the material was obtained in the chemist's laboratory or extracted directly from nature, given that what characterized the material were the properties known solely by its chemical behavior. Thus, it would be possible to produce in the laboratory materials analogous to the ones found in nature, which led to a breach of the classical contrasting division between the natural and the artifact (Bensaude-Vincent 2003, pp. 155–74). In addition, chemists also aspired to build a theory that could overcome the classical division of nature in the three kingdoms: mineral, plant, and animal. Overcoming the kingdom division neither means the reduction of each kingdom's specificity into chemistry nor the disregard of its operative characteristics, but represents the transversality of chemical operations among natural kingdoms, which would allow the material to flow among them (Simon 1999, pp. 65–80).

Therefore, the experimentalism and operationality of chemistry led to a scientific practice whose concepts were meaningful only if they could be reconstructed in the laboratory. Chemical concepts were neither purely theoretical, nor merely empirical, in fact, they were a *mixture* of theory and practice. The common trait between chemistry and other *arts* was that they were constructed based on a technique, an experimental method organized to produce a stable artifact. Chemistry from that period was effectively a *science of materials*, the only science focused on investigating and manipulating different substances, so that the technicality was intrinsic to this field of knowledge (Klein and Lefèvre 2007, pp. 1–3).

This technicality that had begun at the lowest level of materiality was linked with other levels of material organization because chemical artifacts were often produced in large amounts. However, regardless of the order of magnitude, controlled and methodically organized material production was a socially structured practice involving workers from different occupations, such as trained chemists and industrial workers, in addition to State interests.

Ultimately, chemistry left aside the hierarchical distinction between theory and practice, rationality, and experience, because knowing the material world meant mastering operatorial techniques used by chemists. Chemist rationality emerged from a community culture, a way of thinking and taking action created from a specific materiality level. This rationality was absolutely tied to its place of origin, the physical and epistemic space essential to do chemistry. The image of the laboratory as a theater of chemical and instrumental operations, of qualitative and quantitative changes that brought about new knowledge of materials have been consolidated as the genesis of a chemical style of thinking.

14.4 Chemistry Teaching As a New Scientific-Philosophical Culture

As an experimental science, chemistry participated in a culture that allowed the rise of an enlightened public. For instance, in Paris chemical demonstrations performed on the chemistry course of the apothecary Guillaume-François Rouelle (1703–1770) at the *Jardin du Roi* in 1742 and 1764 were carefully followed by great names of the French Enlightenment such as d'Holbach, Diderot, Turgot, Rousseau, and Condorcet, in addition to great future chemists such as Pierre-Joseph Macquer (1718–1784), Antoine-Laurent de Lavoisier, and the Irish Richard Kirwan (1733–1812).

Indeed, experimental philosophy developed by Diderot found strong support in the chemical science presented in the *Encyclopedia* by his friend Gabriel-François Venel (1723–1775). According to Bacon, Diderot also considered that experimental philosophy was a way of asking what the practical knowledge was and where it came from. Diderot distinguishes:

> [...] two types of philosophy: experimental and rational. One is blindfolded, always trying to walk, collecting all that falls into their hand and eventually finds pretentious things. The other collects those pretentious materials and turns them into a guide. However, until today, this pretentious guide was less useful than its rival's attempts and this should be so. (Diderot [1875] 2005, p. 73)

That is the reason why Diderot paid great attention to *arts*, *experimental science*, and *know-how*. Chemistry was not a simple external model to a philosophy created beyond its domains. In fact, convergence between experimental philosophy defended by Diderot and chemistry was an attempt to develop an authentic theory based on practical and operational knowledge of nature (Pépin 2012).

A substantial part of this social and intellectual acknowledgment of chemistry in the eighteenth century was due to Boerhaave, professor at the University of Leiden from 1701 to 1737. If historians and philosophers who describe and analyze the rise of modern scientific culture from the physical and mathematical sciences considered Bacon's contributions of marginal importance, the chemical works of Boerhaave suggest that this negative image should be widely minimized (Powers

2012). This is not only because of the centrality of his work in Bacon's chemical philosophy, but also the importance of Bacon's philosophy in developing a speech that allowed chemistry to enter the listing of socially respected knowledge (Peterschmitt 2005).

Boerhaave raised central elements of the baconian program, such as ordinated experimentation, the utility of chemical knowledge, and, mostly, the alliance between the senses and reason, refuting the symmetric opposition between rationalism and empiricism in the narrative construction of his *Elementa chemiae* (1732). For Boerhaave,

> Such are the noble fruits which chemistry, duly cultivated, holds no natural philosophers; and from this will arise such a system of physical knowledge as the great Lord Bacon wish'd for, and begun; and which, in pursuance of his design the immortal Mr. Boyle considerably promoted. (Boerhaave 1753, p. 174)

This social respectability acquired by chemistry was the result of a variety of factors; however, adoption of a philosophically well-founded speech and the interest in the learning of this science by a broad and eclectic audience were particularly important (Principe 2007). We give as an example a case that strikes at the same time a typical event in the cultural environment around chemistry teaching in a French academy (Dijon), and the transition to a new type of education, whose characteristic is to neglect part of the natural history that has been considered of great importance so far.

The chemistry course at Dijon Academy was created in 1776 and directed by Guyton de Morveau until 1791. Professor, encyclopedist, translator, and correspondent with Torbern Bergman (1735–1784), Kirwan, and Franz Karl Achard (1753–1807), and close to Lavoisier and Claude-Louis Berthollet (1748–1822), Guyton was one of the most respected French chemists, coordinator of a network that allowed a continuous debate among important centers for scientific research such as Uppsala, London, Berlin, and Paris (Bret 2016).

Guyton had a self-taught education based on Boerhaave's work, and on that of his countryman Georges-Louis de Buffon (1707–1788), but mainly, Macquer's work. At Dijon Academy, which had a well-equipped laboratory, Guyton brought together a group of scientists and amateurs working to develop and publish original research in addition to the translation of books and foreign articles on chemistry. The main goal of the course was to train people interested in applying chemistry in a particular productive activity. However, beyond this practical side, an environment of intense social activity was organized around the course. One of the most important activities was a translation group organized by Guyton and Claudine Picardet (1735–1820), who translated into French the work of Swedish, British, German, and Italian chemists, and hence, contributing to spreading foreign Enlightenment.

Guyton's chemical conception was structured around the investigation of the dissolutions and the processes derived from it, such as precipitation and crystallization, where all chemical operations were understood within the framework of Newtonian affinities. The figure of the chemical bodies, not in the geometrical sense, but as the product of their surface texture, would be the cause of the different relationships between the bodies, as they altered the distance between them. In relation to Buffon and Macquer, the originality of his work was in the proposition that crystallization could serve the discovery of these figures and that the affinities could be estimated by the measure of the force of adhesion between surfaces. In fact, as Mi Gyung Kim points out, Bergman, Guyton, and Kirwan represented the frontier of theoretical chemistry in the 1780s (Kim 2003, p. 268).

Another fundamental contribution originated in the Guyton's laboratory consisting in the first systematic nomenclature reform. In 1782, he published a dissertation that we may consider to be the first wholly devoted to chemical language (Guyton de Morveau 1782). The main representatives of the "Republic of Chemists," such as Macquer, Bergman, Kirwan, and Lavoisier, welcomed with enthusiasm the linguistic proposals of Guyton. This method of nomenclature began to be used in the French translations and the articles published in scientific journals such as the *Observations sur la physique* and the *Journal de Sçavans* also began to standardize a chemical language (Crosland 1978).

Finally, it should be remembered that, for the lusophone public, the chemistry practiced and taught by Guyton was of great importance in the formation of Luso-Brazilian chemists. Both the new method of nomenclature that Guyton and the group led by Lavoisier proposed in 1787, and the extensive entry "*Affinité*", written by Guyton and published in the *Encyclopédie méthodique* (1786), were soon translated into Portuguese. The Brazilian Vicente de Seabra Telles (1764–1804) proposed a Portuguese version of the chemical nomenclature of Guyton and Lavoisier, first in his *Elements of chemistry* (2 vols, 1788–1789), then in a specific dissertation on the new nomenclature (1801), while the Portuguese Thomé Rodrigues Sobral (1759–1829) translated "*Affinité*" (1793) (Luna 2013).

The French Revolution in 1789 offered Guyton the opportunity to apply his pedagogical and chemistry teaching ideas on a national scale. Guyton was one of the coordinators of the group of scientists involved in military projects designed to "save the Republic," he also organized the *revolutionary courses*, and, more importantly, he was one of the founders of the great educational institution created by the Revolution, the *Polytechnical School*.

14.5 Final Remarks: Natural History, Chemical Revolution, and the End of the History

Two key aspects of the so called "Chemical Revolution" at the end of the eighteenth century were the new concept of simple bodies – the ultimate level of materiality obtained through analysis – and the renewal of chemical nomenclature. Both of them are directly derived from investigations inspired by Boerhaave's Chemistry Natural History and developed by the Swedish chemists Axel Cronstedt (1722–1765) and Bergman. The investigations gained fundamental importance in theoretical and linguistic changes made by Lavoisier and Guyton in 1770–1790 (Llana 1985, pp. 71–91).

Nevertheless, both Guyton and Lavoisier set aside an important part of the natural history of the arts, the history of its authors and the origins of chemical knowledge. In the *Preface* in the first volume of *Encyclopédie méthodique* (1786), Guyton considers that the history of chemistry should be an independent work because "it is not connected to the dogmatic portion, therefore, one can study it separately" (Guyton de Morveau 1786, p. 1). Similarly, Lavoisier excluded the history of chemistry from *Traité élémentaire de chimie*, and also presents the chemical knowledge in a dogmatic order, that is, as a logical sequence of rationally justified propositions (Bensaude-Vincent et al. 2003, p. 223).

If natural history inspired by Bacon and chemistry forged solid bonds during the period studied herein, the so-called Chemical Revolution represents the abandonment of a part of the scientific culture in the chemists' education that was previously considered essential. Although the first part of *Elementa* by Boerhaave was entirely dedicated to the history of chemists and their art, the post-revolutionary manuals abolished this scope from chemists' culture.

However, some of the central issues arising from techno-conceptual overlaps between Bacon's inspiration and chemistry remain relevant both from the philosophical point of view and from the pedagogical. Eventually, although during the last two centuries, chemistry and the fundamentals of the Baconian program continued to have a common history, its perceptions and implications may still remain hidden by the lack of history and philosophy beyond *dogmatic* exhibitions of chemical knowledge. Here, we observed, once again, the consequences of a reductionist learning in which history is neglected in favor of purely dogmatic knowledge, that is, exclusive to science, as it was possible to disentangle epistemic and scientific knowledge from its history.

References

Aristotle. (1984). Metaphysics. In J. Barnes (Ed.), *The complete works of Aristotle* (Vol. 2, pp. 1552–1728). Princeton: Princeton University Press.

Bacon, F. (1963a). Advancement of learning. In J. Spedding, R. Leslie, & D. D. Heath (Eds.), *The works of Francis Bacon* (Vol. III, pp. 253–491). London: Longman.

Bacon, F. (1963b). Novus Orbis Scientiarum, sive Desiderata. In J. Spedding, R. Leslie, & D. D. Heath (Eds.), *The works of Francis Bacon* (Vol. I, pp. 838–840). London: Longman.

Bacon, F. (1963c). Description of a natural and experimental history. In J. Spedding, R. Leslie, & D. D. Heath (Eds.), *The works of Francis Bacon* (Vol. IV, pp. 251–263). London: Longman.

Bacon, F. (2006). Historia vitae & mortis. In G. Rees (Ed.), *The Oxford Francis Bacon* (Vol. 12, pp. 142–377). Oxford: Oxford University Press.

Bacon, F. (2007). Historia densi & rare. In G. Rees (Ed.), *The Oxford Francis Bacon* (Vol. 13, pp. 36–169). Oxford: Oxford University Press.

Bensaude-Vincent, B. (2003). La nature laboratoire. In I. B. Bensaude-Vincent & B. Bernardi (Eds.), *Rousseau et les sciences* (pp. 155–174). Paris: L'Harmattan.

Bensaude-Vincent, B., & Simon, J. (2008). *Chemistry: The impure science*. London: Imperial College Press.

Bensaude-Vincent, B., Belmar, A., & Bertomeu, J. R. (2003). *L'émergence d'une science des manuels*. Paris: Éditions des Archives Contemporaines.

Boerhaave, H. (1753). *A new method of chemistry.* Translated from the original Latin of Dr. Boerhaave *Elementa Chemiae* by Peter Shaw. London: Longman.

Boyle, R. (1999). *The works of Robert Boyle* (Vol. 3, M. Hunter &, E. Davis, Eds.). London: Pickering & Chatto.

Brasil, Ministério da Educação. (2000). *Parâmetros Curriculares para o Ensino Médio*. Brasília: Ministério da Educação.

Brasil, Ministério da Educação. (2006). *Orientações Curriculares para o Ensino Médio* (Vol. 3). Brasília: Ministério da Educação.

Bret, P. (Org.). (2016). Louis-Bernard Guyton, "l'illustre chimiste de la République". *Annales historiques de la Révolution française*, 383. Paris: Armand Colin.

Crosland, M. P. (1978). *Historical studies in the language of chemistry*. New York: Dover Publications.

Debus, A. G. (1977). *The chemical philosophy*. Mineola: Dover Publications.

Diderot, D. ([1875] 2005). *De l'interprétation de la nature*. Paris: Flammarion.

Giglioni, G. (2012). Historia and materia: The implications of Francis Bacon's natural history. *Early Science and Medicine, 17*, 62–86.

Guyton de Morveau, L.-B. (1782). Mémoire sur les dénominations chimiques, la nécessité d'en perfectionner le système, les règles pour y parvenir, suivi d'un tableau d'une nomenclature chimique. *Observations sur la Physique, 19*, 370–382.

Guyton de Morveau, L.-B. (1786). *Encyclopédie méthodique: Chymie, pharmacie et métallurgie, t.1*. Paris: Panckoucke.

Hannaway, O. (1975). *The chemists & the word: The didactic origins of chemistry*. Baltimore: Johns Hopkins University Press.

Kim, M. G. (2003). *Affinity, that elusive dream: A genealogy of chemical revolution*. Cambridge, MA: MIT Press.

Klein, U., & Lefèvre, W. (2007). *Materials in eighteenth-century science: A historical ontology*. Cambridge, MA: MIT Press.

Kuhn, T. S. (1961). The function of measurement in modern Physical Science. *Isis, 52*, 161–193.

Lavoisier, A. ([1789] 1937). *Traité élémentaire de chimie*. Paris: Guther-Villar Éditeur.

Llana, J. W. (1985). A contribution of natural history to the chemical revolution in France. *Ambix, 32*, 71–91.

Luna, F. (2013). Vicente de Seabra Telles e a criação da nomenclatura em português para a química "nova" de Lavoisier. *Química Nova, 36*, 921–926.

Matthews, M. R. (1994). *Science teaching: The role of history and philosophy of science*. New York: Routledge.

Miller, P. (2005). Description terminable and interminable: Looking at the past, nature and peoples in Peiresc's archive. In G. Pomata & N. G. Siraisi (Eds.), *Historia, empiricism and erudition in early modern Europe* (pp. 355–397). MIT Press: Cambridge, MA.

Niaz, M. (2001). How important are the laws of definite and multiple proportions in chemistry and teaching chemistry? A history and philosophy of science perspective. Science & Education, 10, 243–266.

Paixão, F., & Cachapuz, A. (2003). Mudança na prática de ensino da química pela formação dos professores em história e filosofia das ciências. *Química Nova na Escola, 18*, 31–36.

Pépin, F. (2012). *La philosophie expérimentale de Diderot et la chimie: Philosophie, sciences et arts*. Paris: Garnier.

Peterschmitt, L. (2005). Bacon et la chimie: a propôs de la réception de la philosophie naturelle de Francis Bacon aux XVIIe et XVIIIe siècles. *Methodos, 5*, 1–22. https://methodos.revues.org/385. Accessed 20 July 2016.

Powers, J. C. (2012). *Inventing chemistry: Herman Boerhaave and the reform of the chemical arts*. Chicago: The University of Chicago Press.

Principe, L. (Ed.). (2007). New narratives in eighteenth-century chemistry. *Contributions from the First Francis Bacon Workshop*, 21–23 April 2005, California Institute of Technology, Pasadena, California. Dordrecht: Springer.

Simon, J. (1999). L'Homme de verre? Les trois regnes et la promiscuité de la nature. *Corpus, 36*, 65–80.

Sukopp, T. (2013). Robert Boyle, Baconian science and the rise of chemistry in the seventeenth century. *Society and Politics, 7*, 54–73.

Chapter 15
The History of Science in Teacher Training Programs: A Series of Contributions and Debates for the Teaching of Electrochemistry

Johanna Camacho González, Mercé Izquierdo, and Núria Solsona

15.1 Introduction

Nowadays, there is a consensus on assuming as imperative that teacher formation should be oriented toward the development of a professional profile that encourages future teachers to make decisions to carry out their professional practice. Nevertheless, teachers' traditional conceptions of science, its methods and its nature are seen as obstacles that consequently lead teachers to think of science as decontextualized matter that rarely focuses on its social, cultural, ethical, and creative dimensions (Matthews 1994).

In this regard, here we propose a teacher intervention–development model based on the history of science (HOS) for the teaching of electrochemistry. Our research problem is the construction of the knowledge of chemistry seeking new didactical strategies that could contribute positively to changing the teachers' misconceptions of science, and promote the development of scientific competencies. This intervention model was developed together with two chemistry teachers and it was implemented while secondary students were taught theoretical electrochemistry. In addition, it was included in an in-depth case-study research [program] (Camacho 2010).

J. Camacho González (✉)
University of Chile, Santiago, Chile
e-mail: jpcamacho@uchile.cl

M. Izquierdo · N. Solsona
Autonomous University of Barcelona, Barcelona, Spain
e-mail: merce.izquierdo@uab.cat; nsolsona@xtec.cat

M. E. B. Prestes, C. C. Silva (eds.), *Teaching Science with Context*, Science: Philosophy, History and Education,
https://doi.org/10.1007/978-3-319-74036-2_15

15.2 The History of Science in the Chemistry Class

Teaching chemistry at school promotes the development of skills and the construction of knowledge by people, ideas *of* chemistry, and ideas *about chemistry*. That is to say, this approach seeks to establish a less dogmatic concept of science and relate it to individuals' daily life activities in which men and women work to solve problems. Therefore, the teaching of chemistry in schools can be developed through Scientific Problematic Situations (SPS) (Camacho 2010), and open to then be evaluated and discussed in the chemistry class. Such situations are similar to those that often occur in real scientific settings (Camacho and Quintanilla 2008) and these activities must be meaningful for students learning chemistry; be relevant to the subject being taught; and promote reflective processes that students can later discuss through scientific modeling (De Jong and Treagust 2002).

One of the purposes of incorporating the HOS in chemistry class activities is not to transform students and teachers into historians of science, but rather to promote and develop competencies of scientific reasoning that foster a wider and deeper knowledge and understanding of chemistry (Camacho 2010). Reading and interpreting primary sources, reproducing experiments, engaging students in debate about the HOS, among other activities will enable them to understand how scientific knowledge is constructed or how scientific explanations are elaborated; to be able to explain why one scientific approach was developed more successfully than another; what made it possible for a theory to evolve, in what context it evolved or how it was influenced by the development of a particular piece of knowledge; how people dealt with theoretical, experimental, and professional issues; how scientific theories were sustained; what equipment was available at that time and what made it possible to develop and consolidate certain beliefs and concepts; what such beliefs and concepts led scientists to conclude; and how these conclusions have been evaluated by specialists and taken into account from other perspectives such as science and gender, material culture, socialization, and popularization of science. According to Gooday et al. (2008), the aforementioned approaches tend to be more effective than simply learning and repeating content from science textbooks and laboratory routines, as they stimulate the development of students' awareness of the incompleteness and fallibility of scientific models, and the complex nature of science.

15.3 Trends and Obstacles in Teaching and Learning Electrochemistry Theory

Electrochemistry has been considered one of the most difficult and challenging subjects to teach in the classroom. According to De Jong and Treagust (2002), curriculum and textbooks have traditionally divided this concept into two main parts: oxide reduction and electrochemical batteries or galvanic cells. Compared with other chemistry-related subjects, there is little empirical evidence that comprehensively addresses and relates the students' learning difficulties, the main obstacles in teaching electrochemistry, and the new proposals of innovative teaching. De Jong and Treagust (2002) point out that the difficulties of redox processes can be grouped into two issues: *conceptual* and *procedural*. Although the first is concerned with the mutual dependence of the reactions in the oxidation reduction process, the meaning of the oxidation number, and the cathode and anode charge in electrochemical batteries, the latter refers to the identification of reactants as oxidizing or reducing agents, and the chemical equations in terms of the oxidation-reduction couple. Regarding the difficulties found in electrochemical batteries from a *procedural* perspective, the authors make mention of the issues in predicting products and the magnitude of potential in galvanic cells. From a *conceptual* point of view, they address the obstacles present in the understanding of electrical conduction, electrical neutrality in electricity, and in the identification of a battery cathode and anode and their corresponding charges.

The complexity of teaching and learning of electrochemistry is mostly based upon three main issues: first, the multiple definitions and models that changed as different scientific concepts evolve; second, the experimental practices associated with other areas in science, such as physiology, physics, chemistry, and mathematics; and finally, the instrumental and algorithmic components whose teaching is often reduced, with prioritization of memorization of oxidation numbers, balance of chemical equations, and the resolution of quantitative exercises, thus weakening conceptual understanding.

15.4 A Proposal for the Teaching of Electrochemistry from a History of the Science Perspective

Taking into account all the above, it is evident that teaching electrochemistry from a historiographical and didactic perspective can contribute to the design of an innovative, authentic proposal. As a consequence, an intervention–development model was suggested. This focused mostly on reflection and practice to modify teachers' schemes and misconceptions that underlie their own views of science and teaching (Astudillo et al. 2008). Throughout a diverse sequence of activities, a number of metacognitive situations were promoted in which teachers thought alternatively as an individual who teaches, but who also learns. The tasks incorporate a historical perspective so as to develop a series of varied strategies through which it is possible to identify the following: engagement of students in reading-based research and innovation articles that connect the HOS and classroom activities; stimulation of discussion about the advantages and disadvantages of taking the said approach into account by considering both bibliographical sources and the teachers' own experience in the classroom; the design and implementation of a didactical sequence for teaching electrochemistry; assessment of students' output; and the design and application of a self-assessment task consisting of the analysis and reflection of teachers' practice.

This didactical intervention was developed throughout four phases that were based on/relied on/drew from a dynamic perspective and comprised a macro and micro vision related to the different stages of development of scientific thinking (Labarrere and Quintanilla 2002).

15.4.1 Stage I: Data Collection and Preliminary Information

During the first stage, the objective was to identify and characterize the notions and concepts held by chemistry teachers regarding the HOS. Thirty-two secondary school chemistry teachers from Santiago, Chile, participated from the same school over a period of 5 years. Data collection consisted of two parts: first, 32 volunteer teachers had to fill in a Likert-type scale. Subsequently, 18 teachers were selected through a theoretical sampling stemming from the results/responses of the Likert scale. Only eight were requested to answer an open questionnaire about the strategies, advantages, and disadvantages of incorporating the HOS into the teaching of chemistry. Some of the results obtained at this stage showed that teachers' theoretical formation lacked relevant aspects concerning the HOS, and revealed a scant understanding of the relationship between the history and didactics of science (Camacho and Quintanilla 2009). A theoretical sampling was applied, and as a result, four teachers were chosen (three women and one man) to be part of stage II.

15.4.2 Stage II: Theoretical Background

Although at this stage, four teachers were requested to participate, only two of them were considered for the case study analysis. Emilio was a 40-year-old teacher of chemistry and chemical engineer who had more than 10 years of teaching experience. He used to teach chemistry, biology, physics, and natural sciences at a privately owned but publicly subsidized school. He was the only science teacher at his workplace. Caroline, who was aged 50, was a teacher of chemistry with a master's degree in education and a specialization in evaluation. She worked for more than 20 years at a private school where she was head of the science department. These teachers are a representative sample of Chile's current educational system, specifically, in the area of science. According to records provided by the OECD (2006), science teachers in Chile are aged 40 – and even older – and have more than 10 years of experience.

Given the results above, a *Course on the History of Science and Teacher Development* was proposed, which was divided into ten sessions or Workshops of Teaching Reflection (WTR) of 120 min each.

Theoretical grounds were covered from WTR01 to WTR04. The main purpose of these sessions was to offer teachers a historical point of view for the teaching of chemistry at school in addition to understanding the contributions of the HOS from a conceptual, procedural, and contextual perspective. The workshops were centered on the goals and obstacles of incorporating the HOS into the chemistry class with regard to the following aspects: the teachers' perceptions and the data provided by research in the field of didactics of science teaching; the theoretical framework sustaining the HOS, and the different historiographical views (anachronistic, diachronic, and recurrent). Furthermore, the debate addressed the relationship between the history and nature of science; the role of the materials and activities designed under this approach; and finally, the influence of female scientists in the development of the HOS.

15.4.3 Stage III: Design of a Didactic Unit from a Didactics Approach

The objective of the third stage was to engage teachers in the design of a teaching–learning sequence for the teaching of electrochemistry based on the didactics of science and on HOS approaches. Throughout WTR05 and WTR06, the discussion focused on the notions of teaching electrochemistry. To do so, data needed to be taken from Jane Marcet (1817–1857) so as to offer a review of the history of electrochemistry that covered the period between 1800 and 1853. Moreover, the examination was enriched by the conceptual framework of teaching electrochemistry provided by Grapí (2006). Based on the information mentioned above, it was possible to design a chart that shows the evolution of concepts in the field of electrochemistry through time (Chart 15.1).

In the next sessions – WTR07 and WTR08 – both the discussion and analysis are devoted to the design of the didactic unit called "A theory of electrochemistry for secondary education students: A proposal of didactics based on the history of science" (Camacho 2010) which offers a didactic sequence of activities mostly influenced by the constructivist learning theory. This unit is summarized in Chart 15.2.

15.4.4 Stage IV: Implementation and Assessment

In the last two sessions – WTR09 and WTR10 – the analysis focused on the implementation of the didactic unit throughout classroom observation (CO) and the examination of some of the students' products with the purpose of encouraging teachers to discuss the advantages and disadvantages of incorporating the HOS into the teaching and learning of electrochemistry.

Time Period	Landmark	Relevant Concepts in Electrochemistry	Relevant Concepts in Chemistry
1791	Galvanism	Electricity is generated by means of natural processes	Chemistry relates to other disciplines (Physiology)
1800	Volta's electrical battery invention	Electricity is generated by means of artificial processes	Value and importance of scientific instruments
1806	First edition of *Conversations on Chemistry*	Relationship between electrical energy and chemical reactions	Spreading and education of chemistry
1806	Conference by H. Davy *On Some Chemical Agencies of Electricity*	Polarization of substances Chemical decomposition caused by electricity	Evolution and development of the concept of chemical element
1810	Faraday first reads *Conversations on Chemistry*	Study of chemical reactions	The construction of knowledge integrates the social and personal context
1811	The Dual Theory by J.J. Berzelius	Coexistence of a negative and positive pole in every atom Attraction and repulsion of electricalcharges on chemical substances (chemical affinity)	Distinctions and debates between organic and inorganic chemistry
1817	Chemical agents in electricity in *Conversations on Chemistry* is first read	Volta's electrical battery and the relationship between Galvanism and Electrolysis	Female scientists take part in the construction of scientific knowledge
1832	A proposal of Faraday's law first appears	Oxidation and Reduction Charge conservation The amount of chemical compound in state of decomposition is proportional to the amount of electricity used in the process	Chemistry relates to other disciplines (Mathematics-Physics)
1836	Daniell cell	Oxidation and Reduction properties in substances Redox reactions	Value and importance of scientific instruments
1853	Last version of *Conversations on Chemistry*	Quantitative relationship between the amount of substance and electricity Relationship between electricity and magnetism	Female scientists take part in/become part of the construction of scientific knowledge

Chart 15.1 Conceptual evolution in the theory of electrochemistry between 1800 and 1853

Stages	Activity	Objective	Description
Exploring previous ideas and knowledge	The chemical aspects in electricity	Identifying students' previous knowledge on the relationship between electricity and chemical reactions	Reading of "Conversations on Chemistry" by Marcet (1853) in order to engage students in a discussion that explores how electricity is generated and what the role performed by women is in the construction of scientific knowledge (Solsona 2006)
Introduction of new concepts	Debate Berzelius vs Davy	Explaining how chemical reactions occur in order to generate electricity, and how the process of obtaining new substances from electricity works	A debate between two groups in which one party first advocates for one position, and later, reaches an agreement with the contrary group in order to understand how science is built up upon the individuals' capacity to negotiate
Systematization and conclusions	What competencies were developed?	Identifying and characterizing the scientific competencies developed, and revising the concepts that are sustaining the main contributions to the theory of electrochemistry	Assessing the students' output so as to develop a self-regulation criterion to be applied in their own process of learning scientific knowledge Designing a timeline in order to demonstrate the complexity of scientific knowledge, since its evolution is intrinsically bound to a series of internal and external factors
Implementation	Construction of Daniell cell	Students will explain how the Daniellcell works and will also refer to the use of the different types of electrical batteries	Reflecting on the value of experimental replication in teaching in order to understand redox reactions in electrical batteries, as well as becoming aware of the relevance of experiments in the history of electrochemistry (Grapí 2006)

Chart 15.2 Description of the activities in the didactic unit of electrochemistry theory

15.5 Data Analysis Plan

The data analysis plan used for this study took into account the proposals of Miles and Huberman (1994) and Rodríguez et al. (1999), which focus on the complexity rather than on the linearity of the tasks to be carried out: *data reduction; organization and display of data; and to draw, interpret, and verify the conclusions* (Fig. 15.1). Regarding the processes followed in this research, *data preparation* was part of the preliminary activities, whereas *methodological triangulation* was used in the last stage of the study to enhance the interpretation and verification of the conclusions, and therefore to obtain a more accurate and trustworthy version of them.

Data interpretation was based on a *conceptual profile* model, which presents in a scheme format the different ways of thinking and talking by teachers in different contexts, as pointed out by Mortimer (1995). The conceptual profiles describe the changes in the thought process of individuals as a result of development processes (learning), and contemplate epistemological and ontological dimensions that are strongly linked to culture and context. This dynamic process can be considered to be a construction that is based on new facts and experiences, and it was on these grounds that development of the case study was proposed through a process of teacher training.

The *interpretation* and *conceptualization* processes were framed by the content types (Wang and Cox-Petersen 2002); the stages of development of scientific thinking (Labarrere and Quintanilla 2002); and the conceptual aspects from a moderate rational perspective (Toulmin 1977; Izquierdo 2005). This analysis was pivotal to performing a conceptual interpretation that gains nourishment from the interaction of different interrelated levels, hence revealing the teachers' conceptual profile that changes depending on the stage that they may be at throughout the intervention process.

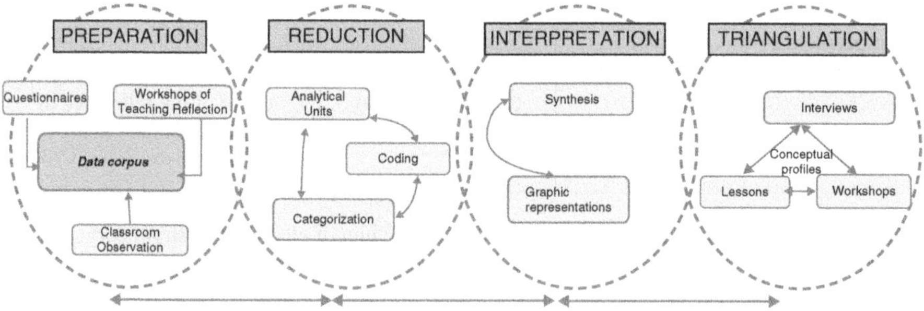

Fig. 15.1 Data analysis

15.6 Results

15.6.1 *Caroline's Conceptual Profile*

During the analysis phase, the conceptual profile obtained shows the importance of incorporating HOS to understand theoretical models from a sociocultural context and its impact on the chemistry class. Although teachers were provided with a theoretical framework and centered the discussion on one of the notions of the HOS, there is evidence that the theoretical models refer particularly to the conceptual aspect, which in turn, may contribute positively to the understanding of the scientific knowledge and how it is developed. In addition, Caroline states that it is of paramount importance that the national curriculum of chemistry includes the HOS perspective, to raise awareness of the fact that scientific knowledge emerges in sociocultural contexts. In fact, this aspect is part of the first section of the didactic unit.

From this point of view, and taking into account the approach by Wang and Marsh (2002), the conceptual profile defined by Caroline mostly views concepts in specific sociocultural contexts; in other words, incorporating the HOS perspective into the teaching of chemistry allows both teachers and students to become aware of how concepts vary according to sociocultural contexts. Although the HOS and the construction of scientific knowledge are barely related, there seems to be no evidence sustaining the assumption that this construction is highly complex in relation to the nature of science.

15.6.2 *Emilio's Conceptual Profile: How It Is Used?*

Regarding the conceptual profile defined by Emilio, it basically shows his interest in the HOS approach in terms of the procedures to be followed within the classroom.

According to his view, the contributions of incorporating the HOS approach to teaching electrochemistry and teacher development had a procedural purpose. In other words, he mostly paid attention to the procedural strategies, from an operative instrumental point of view, the know-how-to do perspective, and the representation techniques; in brief, the aim was to incorporate new aspects/approaches to his teaching strategies while discarding the reflective dimension. According to Emilio's conceptual profile, the importance of incorporating the HOS approach into the national curriculum of chemistry, and the strategies offered by this perspective are reduced to its procedural and technical aspects; that is to say, to find out what strategies and materials can be used in the classroom in addition to learning how to use them, and to knowing when these can be applied.

Although Emilio mostly centered his attention on the procedural aspects of the HOS approach during the last stage of the study, he decided to take into account aspects of the HOS that promote the understanding of the conceptual and contextual dimensions, so as to enhance the students' learning of chemistry. It was observed while Emilio was teaching in the classroom that he can analyze their own performance, as long as metacognitive spaces are provided that foster the reflection on their own practice while teaching, which will consequently lead them one step further, according to Sánchez and Varcarcel (2000). So far, the present research is aimed at engaging teachers in the problems of the notions of the HOS; still, it is necessary to generate new strategies that allow teachers to get more involved in the reflective process involved in their own conceptual profiles.

15.6.3 Results of the Collective Case Study

Although Emilio's profile showed a particular interest in the procedural aspects of the HOS perspective for the teaching of electrochemistry, Caroline's assumptions and concepts moved toward a more contextual perspective, considering these aspects essential to the construction of the theoretical model, and to the understanding of conceptual contents that have been largely seen as obstacles in the teaching and learning of electrochemistry. Another purpose of the present study was to generate links that promote a relationship between the national curriculum and the HOS, as the current trends in the teaching of science often reduce the scientific activity at school to the design and use of materials that scarcely incorporate a historiographical perspective.

In summary, there is evidence that suggests that incorporating the historical aspect into chemistry teaching – based on the present case study – has been done with partial awareness by teachers, although special attention has been paid to how this aspect is to be included in the lesson planning. Nevertheless, there is little understanding of how to relate the historical aspect of chemistry to the didactics of science, which means that most of the conceptual profiles will keep revealing a procedure-oriented interest from teachers adopting the historical approach.

15.7 Conclusions

Regarding chemistry teachers' assumptions and beliefs on the HOS, findings have revealed that these are related to contextual content and often occur in the communicative dimension, as has been addressed in previous research (Wang and Marsh 2002; Wang and Cox-Petersen 2002). Teachers participating in the preliminary stage of this study value the contributions made by incorporating the HOS perspective, in addition to recognizing its positive impact in the chemistry class. However, these beliefs are frequently overshadowed by Monk and Osborne's perspective (1997), which reduces the role of a historical perspective to the procedural and instrumental dimensions of the scientific activity.

Evidence has demonstrated that teachers, who participated in the early stages of this study, incorporated the historical component in the teaching of chemistry with little awareness and understanding. However, they manifested a great interest in taking into account the aspects provided by the HOS approach. Furthermore, the progressive adoption of this perspective by teachers showed their limited notions of didactics and its theoretical basis.

To explain this phenomenon, two views are revised. In the first place, and according to specialized literature, teachers of chemistry are unfamiliar with didactic materials, especially those designed for students at school, as pointed out by Álvarez (2006) and Chamizo (2007). Second, teacher training programs on chemistry rarely focus on the area of didactics; an observation which is sustained by the findings presented in research by Izquierdo et al. (2006), Matthews (2009), Monk and Osborne (1997), and Rudge and Howe (2009).

During the development of the intervention model, particularly when designing the didactic unit, Emilio and Caroline's discourses reveal the struggles in elaborating material that incorporates the HOS, which consequently turns out to be of great relevance when it comes to designing materials that are based on a history of the science approach to the teaching of electrochemistry. Therefore, it is evident that their assumptions and beliefs are oriented toward the instrumental and procedural dimensions of the scientific activity, as referred to by Monk and Osborne (1997), focusing teaching on the products derived from scientific activity rather than engaging students in the construction of scientific knowledge. The latter responds, according to their reflections, to the fact that electrochemistry has been always taught with the purpose of describing the process of reduction of oxide, the memorization of oxidation numbers, the balancing of redox equations, and problem-solving according to Faraday's laws, a common issue that has been also addressed by De Jong and Treagust (2002).

Despite the difficulties mentioned above, important changes were observed in the conceptions of Emilio and Caroline, the key concepts of electrochemistry were better understood. Predictably, this evidence coherently matches that presented previously by Níaz (2009) and García (2009).

In conclusion, training of science teachers in services that focus on the development of scientific competencies, and the construction of scientific knowledge from the HOS approach, provide a new insight into the strategies that incorporate the historiographical aspect into the activities designed for and applied in the chemistry class. Furthermore, it promotes new relations that enrich the activities in the field of chemistry that take place in school, as pointed out by Emilio and Caroline during stages 1 and 4. In these phases, both demonstrated the possibility of integrating curriculum content, carrying out an interdisciplinary work, promoting the development of scientific reasoning, among other aspects referred to by specialized literature (Matthews 1994; Solbes and Traver 2001; Monk and Osborne 1997; Rudge and Howe 2009) and that promote a new perspective in teacher development requiring more research, and seeks to innovate and transform chemistry teaching in schools.

References

Álvarez, M. (2006). La historia de la ciencia en la formación del profesorado de ciencias naturales. In M. Quintanilla & A. Adúriz-Bravo (Eds.), *Enseñar ciencias en el nuevo milenio: Retos y desafíos* (pp. 239–256). Santiago: Ediciones Pontificia Universidad Católica de Chile.

Astudillo, C., Rivarosa, A., & Ortiz, F. (2008). El discurso en la formación de docentes de ciencias: Un modelo de intervención. *Revista Iberoamérica de Educaciónarchive*. http://rieoei.org/2107.htm. Accessed 18 Oct 2016.

Camacho, J. (2010). *Concepciones del profesorado y promoción de la explicación científica en la actividad química escolar: Aportes de un modelo de intervención desde la historia de la ciencia para la enseñanza de la electroquímica*. Tesis de Doctorado en Ciencias de la Educación, Pontificia Universidad Católica de Chile, Santiago, Chile.

Camacho, J., & Quintanilla, M. (2008). Resolución de problemas científicos desde la historia de la ciencia: Retos y desafíos para promover competencias cognitivolingüísticas en la química escolar. *Ciência y Educação, 14*, 197–212. https://doi.org/10.1590/S1516-73132008000200002.

Camacho, J., & Quintanilla, M. (2009). Concepciones de los profesores de química sobre naturaleza de la ciencia e historia de la ciencia. *Enseñanza de las Ciencias, número extra*, 458–461.

Chamizo, J. (2007). Teaching modern chemistry through "Recurrent Historical Teaching Models". *Science & Education, 16*, 197–216. https://doi.org/10.1007/s11191-005-4784-4.

De Jong, O., & Treagust, D. F. (2002). The teaching and learning of electrochemical. In J. K. Gilbert, O. De Jong, R. Justi, D. F. Treagust, & J. Van Driel (Eds.), *Chemical education: Towards research based practice* (pp. 317–337). New York: Kluwer Academic Publishers.

García, Á. (2009). Aportes de la historia de la ciencia a la formación permanente del profesorado universitario: Un caso en el área de la fisicoquímica. *Enseñanza de las Ciencias, número extra*, 1685–1688.

Gooday, G., Lynch, J., Wilson, K., & Barsky, C. (2008). Does science education need the history of science? *Isis, 99*, 322–330. https://doi.org/10.1086/588690.

Grapí, P. (2006). L'electrificació de la química al Segle XVIII: Una xarxa de guspires. *Nova Época, 1*, 65–74.

Izquierdo, M. (2005). Com fer problemàtiques els problemes que no en són proa: Noves temàtiques per als problemes de química. In M. Izquierdo (Ed.), *Resoldre problemes per aprendre* (pp. 45–52). Bellatera: Universitat Autònoma de Barcelona.

Izquierdo, M., Vallverdú, J., Quintanilla, M., & Merino, C. (2006). Relación entre la historia y la filosofía de las ciencias II. *Alambique, 48*, 78–91.

Labarrere, A., & Quintanilla, M. (2002). La solución de problemas científicos en el aula: Reflexiones desde los planos de análisis y desarrollo. *Pensamiento Educativo, 30*, 121–137.

Marcet, J. (1853). *Conversations on chemistry in which the elements of that science are familiarly explained and illustrated by experiments*. London: Longman, Brown, Green & Longmans.

Matthews, M. (1994). *Science teaching: The role of history and philosophy science*. New York: Routledge.

Matthews, M. (2009). Science and worldviews in the classroom: Joseph Priestley and photosynthesis. *Science & Education, 18*, 929–660. https://doi.org/10.1007/s11191-009-9184-8.

Miles, M., & Michael Huberman, A. (1994). *An expanded sourcebook qualitative data analysis*. Thousand Oaks: Sage Publications.

Monk, M., & Osborne, J. (1997). Placing the history and philosophy of science on the curriculum: A model for the development of pedagogy. *Science Education, 81*, 405–424. https://doi.org/10.1002/(SICI)1098-237X(199707)81:4<405::AID-SCE3>3.0.CO;2-G.

Mortimer, E. F. (1995). Conceptual change or conceptual profile change? *Science & Education, 4*, 267–285. https://doi.org/10.1007/BF00486624.

Níaz, M. (2009). Progressive transitions in chemistry teachers' understanding of nature of science based on historical controversies. *Science & Education, 19*, 43–65. https://doi.org/10.1007/s11191-007-9082-x.

OECD. PISA. (2006). *Marco de la evaluación: conocimientos y habilidades en ciencias, matemática y lectura*. Paris: OECD.
Rudge, D. W., & Howe, E. M. (2009). An explicit and reflective approach to the use of history to promote understanding of the nature of science. *Science & Education, 18*, 561–580. https://doi.org/10.1007/s11191-007-9088-4.
Sánchez, G., & Valcárcel, M. V. (2000). ¿Qué tienen en cuenta los profesores cuando seleccionan el contenido de enseñanza? Cambios y dificultades tras un programa de formación. *Enseñanza de las Ciencias, 18*, 423–437.
Solbes, J., & Traver, M. (2001). Resultados obtenidos introduciendo historia de las ciencias en las clases de física y química: mejora de la imagen de la ciencia y desarrollo de actitudes positivas. *Enseñanza de las Ciencias, 19*, 151–162.
Solsona, N. (2006). Las mujeres en la historia de la ciencia. In M. Quintanilla & A. Adúriz-Bravo (Eds.), *Enseñar ciencias en el nuevo milenio: Retos y desafíos* (pp. 37–63). Santiago: Ediciones Pontificia Universidad Católica de Chile.
Toulmin, S. (1977). *La comprensión humana: El uso colectivo y la evolución de conceptos*. Madrid: Alianza Editorial.
Wang, H. A., & Cox-Petersen, A. M. (2002). Comparison of elementary, secondary and student teachers' perceptions and practices related to history of science instruction. *Science & Education, 11*, 69–81. https://doi.org/10.1023/A:1013057006644.
Wang, H. A., & Marsh, D. D. (2002). Science instruction with a humanistic twist: Teachers' perception and practice in using the history of science in their classrooms. *Science & Education, 11*, 169–189. https://doi.org/10.1023/A:1014455918130.
Rodríguez, G., et al. (1999). *Metodología de la investigación cualitativa*. Málaga: Ediciones Aljibe.

Chapter 16
An Empirical View of the Teaching of the Chemical Element Concept

Dulce María López-Valentín

16.1 Theoretical Background

The concept of a chemical element is considered a prerequisite for the study of chemistry, essential for understanding John Dalton's (1766–1844) idea of chemical change, and therefore, understanding subsequent and more complex concepts, such as: chemical reactions, substance quantity, and all the stoichiometric problems that derive from it. The concept of a chemical element (CCE) is an elementary concept because it allows the diversity of existing ordinary materials to be explained with a few chemical elements, constituting a unitary structure, and on the other hand, the search for this unitary structure makes it possible to explain the material changes that occur in chemical reactions. Therefore, it represents a model structure that supposes the conservation or permanence of those elements in the materials' chemical transformations (López-Valentín 2008).

This concept has been reviewed from several standpoints: philosophical, historical, and conceptual. Historically, several authors have studied the *development* of this concept over time.[1]

From the philosophical perspective, Paneth was the first to review the epistemological status of the CCE in his article titled "The Epistemological Status of the Element's Chemical Concept" (Paneth 2003). This work was considered as the starting point for many studies in the contemporary philosophy of chemistry (Scerri 2009). Paneth argues for the need for a dual conception of the term "element" that distinguishes two different meanings: "basic substance, indestructible" (microscopic

[1] Partington (1948); Holton and Roller (1963); Rocke (1986); Mierzecki (1991); Bensaude-Vincent and Stengers (1997); Brock (1998); Paneth (2003); Hendry (2005); Scerri (2007); Ruthenberg (2009).

D. M. López-Valentín (✉)
National Pedagogical University, Mexico City, Mexico
e-mail: dvalentin@upn.mx

M. E. B. Prestes, C. C. Silva (eds.), *Teaching Science with Context*,
Science: Philosophy, History and Education,
https://doi.org/10.1007/978-3-319-74036-2_16

meaning) present in compounds and elemental substances and "elemental substance" (with macroscopic meaning).

The other angle, starting in the 1980s, incorporating some of the difficulties that students have regarding understanding the CCE, was studied and reported in the literature.[2]

On the other hand, the generalized lack of significant learning about any scientific concept, as has been seen in research, must be related to a critical analysis of how it is being taught. It is believed that one of the many reasons why this significant lack of learning in students exists could be because of a lack of knowledge of teachers of the history of science (the *development* of historical models) and the existence of distorted epistemological versions of the nature of science and scientific activity (Furió 1994; McComas et al. 1998; Fernández et al. 2002). Transmission of these distorted visions, specifically in the teaching of chemistry, often manifests itself implicitly in the organization and sequencing of content in textbooks and in the faculty, and it is supposed that they will also be transmitted in the teaching of the CCE. In this work, the focus is on the empiric–inductivist and atheoretical vision. According to Fernández, the empiric–inductivist vision is understood as:

> The role of neutral observation and experimentation, without a priori ideas, is highlighted, forgetting the essential role of hypotheses and the construction of a coherent body of theoretical knowledge. In addition, despite the verbal importance given to observation and experimentation, teaching is generally based on books with little experimental work. Particular emphasis is placed on the atheoretic vision in presenting the learning of science as a matter of discovery or is reduced to practice to processes, forgetting the contents. (Fernández et al. 2002, 479)

For the purposes of this work, the empiric–inductivist idea for CCE is assumed if the operational definition of a simple substance (an elemental substance, which cannot be decomposed) is identified with the theoretical definition of chemical elements at the macroscopic level of interpretation. It should be borne in mind that the idea of a chemical element, from the macroscopic point of view, can be understood as a basic material system designed to explain the composition of various substances that are the empirical referents (elemental and compound substances). The procedural definitions are based on empirical knowledge accepted as evidence in a determined historical model that is incommensurable (Kuhn 1971) with ontological definitions belonging to another historical model that tries to explain that evidence causally.

Thus, it is assumed that the chemistry faculty presents an empirical view of science if he/she identifies the definition of a simple substance with the theoretical definition of a chemical element at the macroscopic level of representation. This error is fundamentally based on the faculty's empirical views, who do not distinguish the procedural definition of a simple substance, which is considered empirical evidence, and the ontological definition of a chemical element proposed in Dalton's model. This identification of reality with the model is usually pedagogically justified

---

[2] Caamaño et al. (1983); Holding (1985); Llorens (1991); Pozo et al. (1991); Solsona and Izquierdo (1998); Laugier and Dumon (2003); López-Valentín (2008).

when the professor tries to bring the students closer to the transformation of substances into others in the macroscopic world, but then it is only adequate to talk about simple and compound substances. The explanation of these changes is carried out by conceiving a unique structure as in the introduction of the CCE as a group of atoms that are equal in mass and that belong to the microscopic world (López-Valentín 2008).

On the other hand, if we do not contemplate the development of knowledge, that is, if we do not take into account the history of science, we will not know about its difficulties, and the epistemological obstacles that were necessary to solve them, issues that are fundamental to understanding the difficulties of students (Saltiel and Viennot 1985). This is why it is important to have knowledge of the history of chemistry, because teachers can use this as a tool to define structural concepts (in this case, the CCE), and also as a topic in class where they can analyze or point out the difficulties in its development, conceptual problems, and the obstacles that needed to be solved.

After this brief introduction of the CCE, the hypothesis from which it stems (to detect didactic deficiencies in the teaching of the CCE) is based on the fact that the usual teaching of physics and chemistry follows conventional methods of verbal transmission of scientific knowledge, with the limitations that this implies and that do not facilitate learning. For the purpose of this work, only the empirical vision of science in the introduction of the CCE will be analyzed.

As the teaching received is the external factor that influences learning more, it is important to analyze how this process of teaching is carried out to be able to solve possible didactical deficiencies that could block learning. This is why this work is intended to solve the following question: Do faculty and textbooks present an empirical view of the CCE?

16.2 Methodology

This study was carried out with the participation of 48 Mexican teachers. These individuals had 6–34 years of service and teaching experience at the secondary and pre-university levels. Most had a Bachelor's degree in chemistry. Regarding the chemistry textbooks (CTBs) evaluated, 30 books were reviewed (13 from the pre-university level and 17 from university-level general chemistry).

With the aim of determining if the research question was valid or not, two convergent experimental designs were proposed. The first was oriented to show an empirical view of science of the CCE in teachers and the second one was used to confirm the empirical view of the CCE in chemistry textbooks (a total of five).

The analysis was performed by two independent researchers, and if there was a discrepancy on any item, it was reviewed again, and if it persisted, it was eliminated or an intervention by a third researcher was requested.

16.3 Presentation and Analysis of Results

The objective of this analysis is to corroborate if the operational definition of a simple substance is identified with the ontological definition of a chemical element in the macroscopic level of representation.

The results of the analysis are presented below. First are the results of the empirical view that correspond to the faculty, and second, those related to chemistry textbooks.

16.3.1 Does the Faculty Present an Empirical View of the CCE?

The results obtained by the faculty view of the CCE are shown in Table 16.1.

Question 1 had the greatest number of correct answers. An example of this case is Teacher #21 who distinguished the properties attributed to the CCE and those corresponding to the simple substance. "*The atomic number, atomic weight and valence are properties of the atom itself, but the bonding of atoms forms the "substance called hydrogen." The former characteristics are macroscopic and the latter are microscopic*" (Teacher #21).

In the case of *Question 2*, the results show that half of the teachers (50%) do not distinguish between the CCE and that of a simple substance on the assumption that simple substances originate a chemical element. For example, the following is a correct answer: "*The element oxygen is a group of atoms that have the same atomic number." Chemical elements are the components of substances; these can be present both in simple as well as compound substances. For example, the simple substance oxygen is represented by O_2, while the element oxygen is present in ozone (O_3); water (H_2O), etc.*" (Professor #31).

Finally, another remarkable answer selected shows the confusion (unusual, although existing) between two terms, allotropes and isotopes, mentioned in the following example: "*Because they are different isotopes*" (Teacher #19).

Regarding *Question 3*, only four teachers (13%) did not have an empirical view of the text, as seen in the following answer: "*Presenting an element in its pure state implies the generalization that all elements are substances. Since it introduces the phrase with the word 'some,' it is assumed that the rest of the elements are found in the form of mixtures; in other words, it associates the idea of a chemical compound with that of a mixture and that of an element with a pure substance, as if a compound could not form a pure substance. Only an element always forms a pure substance, therefore it doesn't make sense to say it is present in a pure state. Here it associates the idea of an element with that of a substance*" (Teacher #14).

As seen in this answer, Teacher #14 solves the *empirical vision of chemistry*, as in the first two lines he identifies that in the cited text, the word *substances* should be mentioned instead of *elements*. On the other hand, this educator also reflects on

Table 16.1 Objectives, questions, and results obtained for the empirical view of the teachers regarding the teaching of the concept of a chemical element (CCE)

Objective	Question	Correct responses (%) ($N = 48$)
Identify whether teachers differentiate the properties attributed to the CCE and those corresponding to the simple substance	1. The following figure shows some of the information of the element hydrogen that is usually found on the periodic table 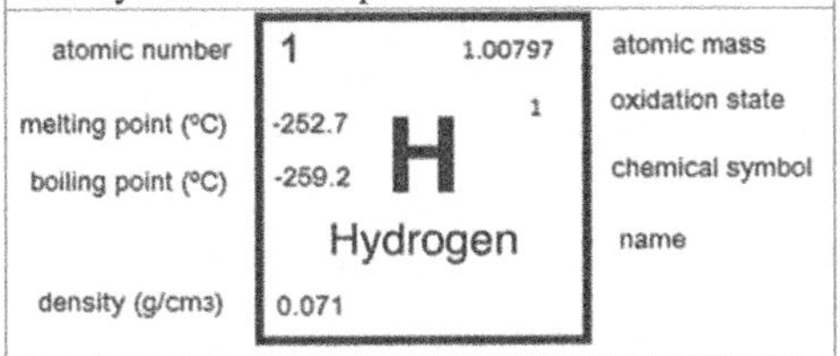 Regarding this information, a teacher says: "*The density, boiling and melting point are properties of the simple substance called hydrogen; while the atomic number, the atomic mass and the valence are properties of its element*." Argue whether you agree or disagree with this statement	58
Confirm the teacher defines oxygen as a set of atoms with the same atomic number, and therefore distinguishes between the chemical element and a simple substance	2. Atmospheric oxygen and stratospheric ozone are simple substances that have different properties. How do you explain to your students what the element oxygen is?	50
Verify that the teacher identifies chemical elements as elemental substances that may occur in nature	3. The concept of chemical elements, like all scientific concepts, has a validity range that depends on the definition given and the theoretical context in which it is introduced. Comment on the validity of the following expression found in a chemistry textbook: "*Some of the elements have been in a pure state in nature for thousands of years*"	13

the implicit association that "could be made in the text" of the association between the compound and mixture; and finally, comments on the erroneous redundancy made when the pureness of substances or of chemical elements is mentioned.

On the other hand, 23% of the sample considers the text invalid, but the faculty do not express practical reasons to justify them, such as: "*I consider that this statement does not have all the validity since the native elements are found in nature in pure state, not thousands of years ago, but since our planet was formed*" (Teacher # 4).

This teacher focuses his comment on the time factor as an error, as it is considered that the expression "for thousands of years" should be greater, that is to say, "since our planet was formed," so this expression is not valid philosophically. He explicitly identifies *native elements* with *pure simple substances*.

When reviewing the results of the three questions together, it is confirmed that the faculty has an *empirical view* by identifying the ontological definition of a chemical element with the operational definition of a simple substance, as the maximum number of correct answers obtained of 58% for Question 1 was similar to that of Question 2 (50%). This can be explained by the degree of difficulty being the same, as question 1 (differentiation between the properties attributed to the CCE and those corresponding to the simple substance) and question 2 (defining the element oxygen as a set of atoms with the same atomic number) are apparently easier to answer than question 3 in which the percentage obtained was 13%, because in the latter, teachers needed to criticize the error when identifying a chemical element with a simple substance in a given text. This can indicate that the faculty, in general, does not have a clear idea of these concepts, as when they answer the direct question (question 1) they are "careful" of their answer, but when they have to identify the error, they seem not to see it.

To close with the analysis of this section, the great abundance of teachers who do not solve this empiricist vision is striking, because 12% of teachers leave the answer without responding.

16.3.2 Do Chemistry Textbooks Show an Empirical View of the CCE?

To continue the analysis of the *empirical view of science*, two questions were applied to the CTB. The results are shown in Table 16.2.

Regarding *Question 1*, most textbooks do not solve the *empirical and atheoretical view of science* or in another way, only four texts (13.3%) "warn" the reader about the possible confusion (very common) of understanding the idea of a chemical element as a basic or fundamental substance with which "real" substances are formed and that can be considered empirical referents. The rest of the texts do not

Table 16.2 Objectives, questions, and results obtained on the empirical view of the CTB in the introduction of the CCE

Objective	Question	Correct responses (%) ($N = 30$)
Identify whether the text makes the emphasis needed to avoid confusion when using the operational definition of a simple substance as the theoretical definition of a chemical element at the macroscopic level of interpretation	1. The text explicitly or implicitly presents that a simple (or elemental) substance is not the same as a chemical element	13.3
Discover if an element is identified with a simple substance as a mixture of two simple substances	2. The textbook presents, at least, some of the possible difficulties that can occur when identifying a simple substance	10

advise about this detail, and what is worse, superimpose the concept of chemical element with that of elemental (or simple) substance.

The following is an example of a positive valuation where the idea of an element as a material formed by a particular class of atoms (Daltonian model) and the elemental substance is implicitly differentiated:

> When a species of matter is represented by a particular class of atoms it is called an *element.* All pure substances can be divided into two classes: elemental and compound substances. *An elemental substance is that consisting of atoms of a single class. A compound is a substance consisting of atoms of two or more different kinds* […] Thus, an elemental substance is composed of an element; a compound of two or more […]. (Pauling 1961, 76)

However, strictly speaking, in this example a macro–micro overlap can be seen as it needs to be added that the (compound) substance is composed of "particles" or "molecules" all the same, because if they cannot be confused at a microscopic level (mixture and compound concepts). This identification between element and elementary substance was very frequent when the Periodic System was introduced as an empirical synthesis of the results obtained by chemistry in 1864. An example is the following paragraph of a text:

> The Russian Mendeleev obtained the most impressive result in 1869, when he generated the periodic table. You can find a modern version of the periodic diagram on the back cover of this text. In it, you can see photos of the elements that make it clear that those in the same column have a similar appearance. (Garritz and Chamizo 1994, 144)

In short, the textual quotation suggests that the chemical element is the elemental substance, when in fact, it is not, transmitting a certain epistemological obstacle by not being able to differentiate between the atomic–molecular model conceptualization of the chemical element and, the interpretation of the operational definitions of a simple substance and compound introduced in the Empiricist Model (eighteenth century).

To conclude the analysis of this question, I refer to the paragraph shown in Table 16.3 and discuss some of its erroneous ideas:

Table 16.3 Quoted paragraph by Hein (1992, 5)

1	*"Modern chemistry developed slower than astronomy and physics,*
2	*it started at the beginning of the XVII and XVIII centuries when Joseph Priestley (1773*–1804) who*
3	*Discovered oxygen in 1774, and Robert Boyle (1627–1691), started to record and*
4	*publish the results of their experiments, and to openly present their theories.*
5	*Boyle, who has been called the founder of modern chemistry, was one of the*
6	*first to practice it as a true science. He believed in the*
7	*experimental method. In his most important book. The Skeptical Chemist, he clearly*
8	*distinguished between an element and a compound or mixture. Boyle is better known today*
9	*because of the law of gases that has his name. A French chemist*
10	*Antoine Lavoisier (1743–1794), put chemistry on a firm foundation with*
11	*experiments in which he used a balance to perform quantitative measures of the*
12	*the weight of substances that formed part of chemical reactions"*

In this text, Hein presents diverse aspects about the nature of chemistry that show the empirical–inductivist view of the author as follows:

- One of these clues is when the author cites in line 5 that "Boyle was one of the first to practice chemistry as a true science." In this paragraph, he states that in this era the theoretical body of chemistry was established, when actually, historians indicate that this is when chemistry began as a science (Holton and Roller 1963). Therefore, we can doubt the existence of a theoretical body in Boyle's time. What this author did was contribute the introduction of the empirical conceptual model that allowed the classification of material systems using an operational definition into "mixtures, not perfectly mixed bodies (simple substances) and perfectly mixed bodies (compounds)" (López Valentín 2008).
- On the other hand, lines 7 and 8 seem to use the terms "compounds" and "mixtures" as synonyms, which induce the reader to a conceptual error (although this could also be attributed to the translator of the text). Coherently, in lines 7 and 8, the concept of *element* is explicitly identified with the empirical idea of a *simple substance*, which is in opposition to the concepts of "mixture or compound."
- The author states his empirical vision explicitly (lines 6 and 7) when he associates science with the experimental method, as if the scientific method could be reduced to only the performance of experiments (Fernández et al. 2002).
- Last, the birth date of Joseph Pristley "(1773*–1804)" is incorrect, as it is not possible that he discovered oxygen at 1 year of age. The correct date should be 1733–1804 (Biografias y vidas).

Regarding the results that correspond to *Question 2*, only 10% of the CTB make a comment about the difficulties that resulted from the introduction of various definitions of the chemical element and, specifically, identify an element with a simple substance. Table 16.4 shows an example:

When the author mentions that "compound substances are those formed by two or more simple substances (lines 2 and 3), the conceptual error of identifying an element as a simple substance, and later, defining a compound formed by two or more elements or simple substances, that is, as a mixture of substances, can be

Table 16.4 Quoted paragraph of Feo and Izquierdo (1976, 211–212)

1	*"Simple substances are those that cannot be divided into simpler ones.*
2	*Hydrogen, sulfur, oxygen, zinc, etc., are simple substances.*
3	*Compound substances are those formed by two or more simple substances, with these*
4	*being divided forming simple substances, and obtaining, when grouped, simple substances*
5	*that form them. Water, hydrochloric acid, calcium carbonate, etc., are compound substances.*
6	*Since matter is formed by atoms, which are individual units*
7	*of matter, whether a substance is simple or compound will depend*
8	*Only if it is composed of equal or different atoms"*

introduced. This is the confusion that must be avoided in teaching, even more so when we consider that it is much generalized among students who are starting to study chemistry (Furió and Domínguez 2007).

On the other hand, the text does not introduce Dalton's concept of element as it implicitly identifies it with a simple substance when it states "since matter is formed by atoms, which are individual units of matter, whether a substance is simple or compound will depend only if it is composed of equal or different atoms" (lines 6–8). This empirical idea eliminates the idea of an element as it is identical to that of a simple substance. It is convenient to remember, historically, that Dalton introduces in his proposition the concept of an element as a group of equal atoms and that according to the rule of maximum simplicity, the simple substance is formed of equal atoms because "compound atoms" (molecules) cannot be formed. This is why Dalton was opposed to the results obtained by Joseph Louis Gay-Lussac (1778–1850) especially, the explanation of the experiments of gas reactions performed by Amedeo Avogadro (1776–1856), in which the hypothesis that molecules of simple substances were composed of more than one atom was introduced.

To complete with the analysis of the CTB, most textbooks (86.7%) do not go beyond an empiricist view of science, because they do not "alert" the reader about the possible confusion of understanding the chemical element and the simple substance (very common). Only 10% of the books comment on the possible difficulties when introducing several definitions of the chemical element and, in particular, the identification of an element with a simple substance.

16.4 Conclusions

The teaching of the CCE is actually empirical. Teachers do not know the philosophy and history of science and do not have a critical view of the development of chemistry. On the other hand, in textbooks there is an absence of topics concerning the history and development of the CCE.

The development of historical models plays an important role in the origin and development of chemical knowledge (Justi and Gilbert 2000), and especially in the construction of the concepts of chemical and elemental substances. Similarly, knowing the history and philosophy of chemistry helps to understand better the nature of science (Matthews 1994). Regarding the didactic implications that this may have for chemical education, I consider that this is a good topic that teachers should know and that should be taken into account in training courses for teachers. It is well known that teachers must not only have a good knowledge of the subject to do their job well, but they also need to know the history and epistemology of scientific constructs beyond the contents of chemistry. In addition, knowing chemical epistemology provides the teachers with the necessary tools to learn, to discuss, and to reason in chemistry, which means being able to establish suitable relationships between, for example, the macroscopic and microscopic levels of representation in chemistry. In this case, it means being able to explain the specific properties of substances, the characteristics of the macroscopic level of representation from the properties of atoms and molecules, entities from the microscopic or submicroscopic model of matter (Gabel 1998; Erduran and Scerri 2003; Treagust et al. 2003).

References

Bensaude-Vincent, B., & Stengers, I. (1997). *Historia de la química*. Madrid: Addison-Wesley.

Biografias y vidas: La Enciclopedia Biográfica en Línea. Joseph Pristley. http://www.biografiasyvidas.com/biografia/p/pristley.htm. Accessed 21 December 2016.

Brock, W. (1998). *Historia de la química*. Madrid: Alianza.

Caamaño, A., Mayós, C., Mestre, G., & Ventura, T. (1983). Consideraciones sobre algunos errores conceptuales en el aprendizaje de la química en el bachillerato. *Enseñanza de las Ciencias, 1*(3), 198–200.

Erduran, S., & Scerri, E. (2003). The nature of chemical knowledge and chemical education. In J. Gilbert (Ed.), *Chemical education: Towards research-based practice* (pp. 7–27). Riverwoods: Kluwer Academic Publishers.

Feo, R., & Izquierdo, M. (1976). *Física y química 2° BUP*. Valencia: Bello.

Fernández, I., Gil-Pérez, D., Carrascosa, J., Cachapuz, A., & Praia, J. (2002). Visiones deformadas de la ciencia transmitidas por la enseñanza. *Enseñanza de las Ciencias, 20*(3), 477–488.

Furió, C. (1994). Tendencias actuales en la formación del profesorado de ciencias. *Enseñanza de las Ciencias, 12*(2), 188–199.

Furió, C., & Domínguez, C. (2007). Problemas históricos y dificultades de los estudiantes en la conceptualización de sustancia y compuesto químico. *Enseñanza de las Ciencias, 25*(2), 241–258.

Gabel, D. (1998). The complexity of chemistry and implications for teaching. In B. Fraser & K. Tobin (Eds.), *International handbook of science education* (pp. 233–248). London: Kluwer Academic Publishers.

Garritz, A., & Chamizo, J. A. (1994). *Química*. México: Addison-Wesley Iberoamericana.

Hein, M. (1992). *Química*. México: Grupo Editorial Iberoamérica.

Hendry, R. (2005). Lavoisiser and Mendeleev on the elements. *Foundations of Chemistry, 7*, 31–48.

Holding, B. (1985). *Aspects of secondary students' understanding of elementary ideas in chemistry: Summary report. Children's learning in science project*. Leeds: Centre for Studies in Science and Mathematics Education of the University of Leeds.

Holton, G., & Roller, D. (1963). *Introducción a la física moderna*. Barcelona: Reverté.

Justi, R., & Gilbert, J. (2000). History and philosophy of science through models: Some challenges in the case of "the atom". *International Journal of Science Education, 22*(9), 993–1009.

Kuhn, T. (1971). *La estructura de las revoluciones científicas*. México: Fondo de Cultura Económica.

Laugier, A., & Dumon, A. (2003). Obstacles épistémologiques et didactiques à la construction du concept d'elément chimique: quelles convergences? *Didaskalia, 22*, 69–97.

Llorens, J. A. (1991). *Comenzando a aprender química: Ideas para el diseño curricular*. Madrid: Visor.

López-Valentín, D. M. (2008). *La enseñanza y el aprendizaje del concepto de elemento químico en la educación secundaria y el bachillerato: Análisis crítico y propuesta de mejora*. Tesis doctoral, Departamento de Didáctica de las Ciencias Experimentales y Sociales, Universidad de Valencia, Valencia, España.

Matthews, M. (1994). Historia, filosofia y enseñanza de las ciencias: La aproximación actual. *Enseñanza de las Ciencias, 12*(2), 255–277.

McComas, W., Clough, M., & Almazroa, H. (1998). The role and character of the nature of science in science education. In W. McComas (Ed.), *The nature of science in education, rationales and strategies* (pp. 3–39). Dordrecht: Kluwer Academic Publishers.

Mierzecki, R. (1991). *The historical development of chemical concepts*. Dordrecht: Kluwer Academic Publishers.
Paneth, F. (2003). The epistemological status of the chemical concept of element. *Foundations of Chemistry, 5*, 113–145.
Partington, J. (1948). The concepts of substance and chemical element. *Chymia, 1*, 109–121.
Pauling, L. (1961). *Química general: Una introducción a la química descriptiva y a la moderna teoría química*. Madrid: Aguilar.
Pozo, J. I., Gómez, M. Á., Limón, M., & Sanz, Á. (1991). *Procesos cognitivos en la comprensión de la ciencia: las ideas de los adolescentes sobre la química*. Madrid: CIDE.
Rocke, A. (1986). *Chemical atomism in the nineteenth century*. Columbus: Ohio State University Press.
Ruthenberg, K. (2009). Paneth, Kant and the philosophy of chemistry. *Foundations of Chemistry, 11*, 79–91.
Saltiel, E., & Viennot, L. (1985). ¿Qué aprendemos de las semejanzas entre las ideas históricas y el razonamiento espontáneo de los estudiantes? *Enseñanza de las Ciencias, 3*(2), 137–144.
Scerri, E. (2007). *The periodic table: Its story and its significance*. Oxford: Oxford University Press.
Scerri, E. (2009). Chemistry goes abstract. *Nature Chemistry, 1*, 679–680.
Solsona, N., & Izquierdo, M. (1998). La conservación del elemento, una idea inexistente en el alumnado de secundaria. *Alambique, 17*, 76–84.
Treagust, D., Chittleborough, G., & Mamiala, T. (2003). The role of submicroscopic and symbolic representations in chemical explanations. *International Journal of Science Education, 25*(11), 1353–1368.

Chapter 17
An Educational Blend of Pseudohistory and History of Science and Its Application in the Study of the Discovery of Electromagnetism

Roberto de Andrade Martins

17.1 Introduction

For four decades I have been researching and teaching history of physics to undergraduate students, in Brazil. The educational approach I used throughout these years underwent many changes, of course. Although I have always strived to present what seemed to be an adequate view of the nature of science (NOS) (Hodson 2014) via historic episodes, sometimes I was dissatisfied noticing that many students kept their previous, naïve and inadequate opinions about this subject and they were unable to distinguish a well-grounded historical narrative from an unreliable one. In recent years, while working as a visitor professor at the State University of Paraíba (UEPB, Campina Grande, Brazil), I developed a new strategy of combining pseudohistory and history of science (HOS) that seemed to me very useful for changing the naïve views of the students.

This chapter describes my recent line of attack, presenting one specific instance: the history of the discovery of electromagnetism.

The author is a retired professor from the State University of Campinas (Unicamp), now collaborating with the Municipal Foundation of Higher Education of Bragança Paulista (FESB), Bragança Paulista, Brazil. Part of this research was developed while he was a visiting researcher at the São Carlos Physics Institute (IFSC) of the University of São Paulo (USP) sponsored by the São Paulo State Research Foundation (FAPESP).

R. d. A. Martins (✉)
Municipal Foundation of Higher Education of Bragança Paulista (FESB),
Bragança Paulista, Brazil
e-mail: roberto.andrade.martins@gmail.com

M. E. B. Prestes, C. C. Silva (eds.), *Teaching Science with Context*,
Science: Philosophy, History and Education,
https://doi.org/10.1007/978-3-319-74036-2_17

17.2 The Teaching Approach

At the State University of Paraíba, history of physics is mandatory for all students. It is a 60-h discipline, taught after the general physics courses but before modern physics. Instead of attempting to cover the whole history of physics, I chose to address a few specific case studies of classical physics, to present a deeper and exemplary study of the HOS.

The topics addressed during the history of physics course varied, but this is a typical list of the subjects:

1. Archimedes and the crown of king Hero
2. The legend of Newton's apple
3. Aristotelian physics: natural and violent motion
4. Medieval dynamics and Buridan's impetus theory
5. Galileo and the principle of circular inertia
6. Descartes' laws of nature
7. Newton's principle of inertia
8. The study of magnetism in Antiquity and in the Middle Ages
9. Electricity in the eighteenth century and Volta's pile
10. The early development of electromagnetism
11. Newton's optics
12. The law of energy conservation.

In this example, topics 3–7 were chosen to provide an instance of a long history of changes of scientific concepts. Alternatively, sometimes I dealt at great length with the history of astronomy, from Antiquity to Newton – another illustration of long-term historical changes that provided similar historical insights.

There are many different approaches to the use of HOS in science education (McComas 2010). My classroom activities included a discussion of short primary and secondary historical sources (in Portuguese), and explanations using computer presentations. Some of the secondary sources used in the course were papers written by historians of science. Other, however, were reproductions of web pages or extracts from popular books concerning the historic topics that were studied.

Each new topic was introduced using a carefully selected faulty (pseudohistorical) account, usually found on the internet. Douglas Allchin (2004) has provided a detailed analysis of the meaning and dangers of the pseudohistory of science. I use "pseudohistory" as a broad label, including many types of carelessly written historical accounts that can be easily challenged by an experienced historian. Pseudohistorical narratives are written by amateurs; they usually repeat ungrounded myths as truth, they are uncritical of the sources they employ, they seldom make use of primary sources and they always present an oversimplified version of history. In the case of the pseudohistory of science, besides presenting erroneous information and interpretation of history, those writings are also responsible for the perpetuation of wrong views concerning scientific endeavor.

It is very easy to find deeply flawed accounts of Archimedes' study of king Hero's gold crown, or of Newton's apple episode, for instance. I usually chose web pages that appeared among the first results of a Google search for the topic and that contained lots of historical mistakes. The students received a paper copy of the web page and they were first asked to mark any odd detail of the text while I read aloud the whole text without making any comments. At the beginning of the course, the usual reaction of the students was complete acceptance and agreement with the faulty historical description. Indeed, the common initial attitude of the undergraduates was the uncritical acceptance of everything that they read or were told concerning the history of physics.

After the first inspection of the inaccurate historic account, I called the attention of the students to some strange, inconsistent or unbelievable features of the text. Could Archimedes build machines that were able turn the Roman ships upside down? What would be the power required to pull an ancient war ship from the water fast enough? It is easy to make an approximate estimation, and the power would be enormous. How strong should the wooden beam of the machine be that would extend from the shore or harbor to the ships (before they landed) to sustain the weight of the ship? It was probably impossible have such strong beams. Was the whole shore full of those machines, or would the Roman sailors just go in the direction of the few available machines? If Archimedes' marvelous inventions were so powerful, how did the Romans ultimately defeat those machines and conquer Syracuse?

Using simple physics and common knowledge it is possible to highlight that some points of the popular historical accounts are indeed strange or unbelievable.

Was an apple the very first thing that Newton ever observed to fall? Hadn't everybody, since Antiquity, known that heavy things fall? Why should the fall of an apple (and not that of a stone or a fork) lead Newton to think about gravity? And why nobody supposedly did never think about gravity before? Did Newton discover gravity, as stated in many web pages and books? But didn't Galileo study freefall and the motion of projectiles? Yes, and Galileo died before Newton was born. Isn't this inconsistent?

Some of the pseudohistorical texts I used were so flawed that not a single sentence was free of problems. Of course, some of the mistakes were not obvious and spotting them required historical information that was not available to the students. It was not difficult, however, to show that the text did not take into account much relevant information.

After demolishing the first pseudohistorical accounts, the students were shocked and confused. They became aware that they had read and heard stories that were wrong – and that they had trusted them. Some of them even declared that they had formerly accepted and passed on those erroneous versions to other people.

The question then arose: Is there any trustful account of the HOS? How is it possible to distinguish an acceptable historical version from a made up story? This was discussed over and over again, during the whole course, providing an elucidation of the basic techniques of historical research.

The discussion of the flawed account was followed by the study of some sound historical version of the same topic, accompanied by excerpts from primary sources.[1] While studying the secondary sources, the attention of the students was drawn to several relevant features, such as the complexity of the historic narratives, the bibliographic references used by the authors, and the messages conveyed about scientists and NOS – contrasting them with the pseudohistorical tales.

The same strategy, contrasting pseudohistory and history of physics, was used for each new topic of the course. Some of the students soon began to become critical of the flawed accounts and to point out relevant problems of those texts. The smartest ones even began to have reservations about the accounts written by historians of science, pointing out possible gaps and some relevant uncertainties of those narratives. Repeating the comparison between inadequate stories and serious HOS again and again, and stimulating the students to be critical of the information they received or found out, was intended to produce a change of attitude – not a mere learning of HOS and its epistemological and sociological messages.

In this approach, besides dealing with the HOS and discussing the NOS, the very meaning of history and its research methodology were also dealt with. This scheme stimulated the students to be careful when reading about HOS and to choose carefully the secondary sources they could reasonably trust. As those students would become physics teachers in the future, I expected them to use this acquired attitude to avoid transmitting flawed stories to their own students. This approach was used at university level; it would be more difficult to apply it in high-school teaching.

17.3 Case Study: The Discovery of Electromagnetism

17.3.1 The Faulty Secondary Source

As an illustration of the line of attack described above, I present an instance of analysis of a faulty historical account, such as those used during my history of physics course. However, I will not reproduce a text I actually used with my students, because in my teaching practice I chose passages in Portuguese, and here I must use an example in English. The topic, however, is one of those I presented in my courses: the discovery of electromagnetism by Hans Christian Ørsted (1777–1851). The extract transcribed below is part of Kendall Haven's book *100 Greatest Science Discoveries of All Time* and it also appears on some web pages, sometimes without due acknowledgement of the source.[2]

[1] The comparison of primary sources with the flawed historical account presented in textbooks was used by Silvana Galdabini and Ornella Rossi (1993), for instance.

[2] See, for instance: http://superbeefy.com/who-discovered-electromagnetism-and-how-does-an-electric-current-creates-a-magnetic-field/ and http://www.dadazi.net/new/diff/100science/book04/scbk04-01.html

> In the spring of 1820, Hans Oersted was giving a lecture to one of his classes when an amazing thing happened. He made a grand discovery – the only major scientific discovery made in front of a class of students. It was a simple demonstration for graduate-level students of how electric current heats a platinum wire. Oersted had not focused his research on either electricity or magnetism. Neither was of particular interest to him. Still, he happened to have a needle magnet (a compass needle) nearby on the table when he conducted his demonstration.
>
> As soon as Oersted connected battery power to his wire, the compass needle twitched and twisted to a point perpendicular to the platinum wire. When he disconnected the battery, the needle drifted back to its original position.
>
> Each time he ran an electric current through that platinum wire, the needle snapped back to its perpendicular position. Oersted's students were fascinated. Oersted seemed flustered and shifted the talk to another topic.
>
> Oersted did not return to this amazing occurrence for three months, until the summer of 1820. He then began a series of experiments to discover if his electric current created a force that *attracted* the compass needle, or *repelled* it. He also wanted to try to relate this strange force to *Urkraft*.
>
> He moved the wire above, beside, and below the compass needle. He reversed the current through his platinum wire. He tried two wires instead of one. With every change in the wire and current, he watched for the effect these changes would produce on the compass needle.
>
> Oersted finally realized his electric current created *both* an attractive and a repulsive force at the same time. After months of study, he concluded that an electric current created a magnetic force and that this force was a whole new type of force radically different than any of the forces Newton had described. This force acted not along straight lines, but in a circle *around* the wire carrying an electric current. Clearly, he wrote, wires carrying an electric current showed magnetic properties. The concept of electromagnetism had been discovered. (Haven 2007, p. 66)

The HOS teacher who intends using the approach described in this chapter should be able to select a plausible but defective version of the topic chosen for study. This entails, of course, a deep historical knowledge of the subject.

17.3.2 Some Odd Details

After the initial reading of the text reproduced above, the students would usually notice a few peculiar elements. If Ørsted was not interested in electricity or magnetism, why was he lecturing about those subjects, and how did he manage to study this new phenomenon? Why did he return to this subject only after 3 months? What is the *Urkraft* that Ørsted was trying to study, and what did he decide about it? Why did he conclude that the electric current both attracts and repels the compass needle – as we know that this cannot happen? Which forces did Newton study, besides gravitation?

Notice that, at this stage, the students could not be sure that there was anything *wrong* with the account they had read. Because they had already studied electromagnetism, they knew that the electric current does not attract and repel the magnetic needle, but perhaps in 1820 Ørsted had simply reached a conclusion that was later discarded.

There are, of course, other particularities of the text that could arouse the attention of someone who had already studied the history of the discovery of electromagnetism. They will be dealt with later on.

17.3.3 The Author

Depending on the phase of the course, after the initial reading of the extract I asked the students to look for information about the author and to analyze the bibliography he/she used. The HOS teacher should not bring to the classroom all the answers; he/she should show the students how the relevant questions can be solved. There are many ways of obtaining data about an author. In the specific case of Brazil, there is a national database containing the detailed curricula of graduate students, professors, and researchers.[3] It is very easy to check the qualifications of Brazilian authors.

At the back of Kendall Haven's book, we find this information about the author: "The only West Point graduate and only senior oceanographer to become a professional storyteller, Haven has performed for four million." So, he is not a historian of science. He is a prolific writer and according to the *WorldCat*[4] he published two other books in 2007, two books in 2006 and one in 2008 – just to show his production at the time he wrote about Ørsted. Supposing that he devoted a whole year to the preparation of his *100 Greatest Science Discoveries of All Time* (and that is a generous, unlikely assumption), he would have had about 3 days to study each of those episodes – not enough time to become an expert on the discovery of electromagnetism.

Students should become aware that HOS is a specific discipline and that writing good essays on this subject requires specific skills, effort, and time. A sufficiently serious and devoted amateur can write an acceptable historical essay, but this seldom happens. Checking the background of the author of a historical piece and the circumstances of its composition is an important step for the evaluation of its merit.

17.3.4 The Bibliography

Students should become aware that the analysis of the bibliography used by an author is an essential step in trying to evaluate the merit of his/her historic account.

In this specific case, the author supplied a bibliography for each chapter of his book, and provided these references for his account of the discovery of electromagnetism (Haven 2007, p. 66):

[3] "Plataforma Lattes," maintained by the Brazilian National Council for Scientific and Technological Development (CNPq): http://lattes.cnpq.br

[4] http://www.worldcat.org

Beaumont, L. (1997). *Memoir of Oersted*. Washington, DC: Smithsonian Institution.
Brain, R. M. (Ed.). (2006). *Hans Christian Oersted and the Romantic Quest for Unity*. New York: Springer.
Cohen, B. (1995). *Revolutions in science*. Cambridge, MA: Harvard University Press.
Dibner, B. (1995). *Oersted and the discovery of electromagnetism*. Cambridge, MA: Burndy Library.
Oersted, H. (1994). *The discovery of electromagnetism made in the year 1820*. London: H. H. Theiles.

He seemingly consulted five books to write that chapter and his bibliography included both secondary and primary sources (the last one being authored by Ørsted himself). There is, however, a bad sign that could be noticed at this point: the bibliography does not include any academic papers published in a HOS journal.

Students should become aware that there are different types of bibliographic sources and that a work with a good bibliography may be more creditable than a composition with a bad bibliography or no references at all. However, even a very large and impressive bibliography can be misleading, because the author could have merely copied the references from some place, without studying them.

Did Kendall Haven really read them… in about 3 days? Are they relevant books, written by competent historians of science? Those are questions that the students should become used to asking about any historic text. Sometimes it is not easy to answer the questions, but the teacher should show how that can be done, checking the references and their authors. In the case of academic papers, the journal in which it was published is also a relevant piece of evidence for or against its merit.

The first title of the bibliography drew my attention because, although I have read a lot about Ørsted, I had never come across that work. A quick internet search provided very few occurrences of that book. One of them was the Hungarian *Wikipedia* notice about Ørsted,[5] which provided exactly the same five references that appear in Kendall Haven's book. However, the *WorldCat* did not include that book, and this was very odd. At last, I found the reference of an article published in 1868:

Beaumont, L. E. (1868). Memoir of Oersted. *Annual Report of the Board of Regents of the Smithsonian Institution… for the year 1868*: 166–184.

We can conclude that Haven's first bibliographic reference does not exist at all. There is no book called *Memoir of Oersted* published by Lelonce Beaumont in 1997, but there is a short paper with the same title published by Léonce Elie Beaumont in 1868. This circumstance strongly suggests that Kendall Haven did not read this work (otherwise he would have provided the correct bibliographic reference) and that he copied it from somewhere else.

This conjecture is confirmed by analyzing the other references. The second book does not exist either. *Hans Christian Oersted and the romantic quest for unity* was

[5] https://hu.wikipedia.org/wiki/Hans_Christian_Ørsted

the preliminary title of a book that was finally published as *Hans Christian Ørsted and the romantic legacy in science: Ideas, disciplines, practices*. The editors were Robert M. Brain (the only name cited by Haven), Robert S. Cohen, and Ole Knudsen. It was published in 2007, not 2006, and the city of publication was Dordrecht, not New York. I suppose that Kendall Haven never saw this book.

The third book does exist, with a slightly different title: *Revolution in science* (nor *Revolutions*). However, it does not deal with Ørsted's work at all. The fourth reference is *almost* correct – Dibner's book was published in 1961, not 1995, and it is outdated. The last one is wrong. There is a book called *The Discovery of Electromagnetism Made in the Year 1820*, but it was not published in London, in 1994. It was published in Copenhagen, in 1920, to commemorate the first centennial of Ørsted's discovery, containing a reproduction of his original Latin publication, accompanied by translations in French, English, German, Italian and Danish. This is its full reference:

Oersted, H. C. (1920). *The discovery of electromagnetism made in the year 1820 by H.C. Oersted;* Published for the Oersted Committee at the expense of the State, by Absalon Larsen. Copenhagen: [H. H. Thieles bogtrykkeri].

Again, it seems that Kendall Haven never consulted this publication. His bibliography does not add credit to his work – on the contrary, it suggests that he did not consult a relevant bibliography to write this chapter of his book.

17.3.5 The Relevant Primary Sources

When reading a historical account, students should learn to ask a few fundamental questions: How can the author know that this did really happen? Are there any contemporary documents or other vestiges of the past that could allow us to check the reliability of that story? Which are the relevant primary sources available for the study of this episode?

In this specific case, the author did not elucidate those points. Did Ørsted describe how his discovery happened? Was there any witness, besides Ørsted, who wrote about what happened in the classroom? At one particular passage, Haven seems to imply that he was quoting Ørsted himself: "Clearly, he wrote, wires carrying an electric current showed magnetic properties".

Of course, the physics undergraduate students will be unable to answer those questions. However, they should think about them and try to notice the primary sources described in historical descriptions. The HOS teacher, on the other hand, should have pondered these questions well in advance, and he/she should be prepared to offer adequate answers and to provide additional reading materials containing or referring to the primary sources.

Concerning the discovery of electromagnetism, there are two relevant contemporary authors: Ørsted (both in his first publication about the electromagnetic effect and in later accounts[6]) and Christopher Hansteen, an acquaintance of Ørsted who wrote a famous letter to Faraday about this subject.

Those primary sources are not too voluminous and they can be delivered to the students, for analysis and discussion. In my actual teaching practice, I handed to the students a historical paper in Portuguese that contained translations of those passages (Martins 1986). For other historical episodes, I provided both copies of primary sources and one or two secondary sources separately. The second alternative seems to me the best one, because it calls the attention of the students to the different natures of those writings.

17.3.6 Hansteen's Version of the Discovery

Hansteen wrote a letter to Faraday on 30 December 1857 telling him about the discovery of electromagnetism. The relevant part of the letter is this:

> Professor Oersted was a man of genius, but he was a very unhappy experimenter; he could not manipulate instruments. He must always have an assistant, or one of his auditors who had easy hands, to arrange the experiment; I have often in this way assisted him as his auditor. Already in the former century there was a general idea that there was a great conformity, and perhaps identity, between the electrical and magnetical force; it was only the question how to demonstrate it by experiments. Oersted tried to place the wire of his galvanic battery perpendicular (at right angles) over the magnetic needle, but remarked no sensible motion. Once, after the end of his lecture, as he had used a strong galvanic battery in other experiments, he said, "Let us now, while the battery is in activity, try to place the wire parallel with the needle." When this was done, he was quite struck with perplexity by seeing the needle making a great oscillation (almost at right angles with the magnetic meridian). Then he said, "Let us now invert the direction of the current," and the needle deviated in the opposite direction. Thus the great discovery was made; and it has been said, not without reason, that "he tumbled over it by accident." He had not before had any more idea than any other person that the force should be transversal. But as Lagrange said of Newton on a similar occasion, "such accidents only meet persons who deserve them." (Williams 1971, vol. 2, p. 892)

The attention of the students should be called to the possibility that even primary sources might transmit misleading information. The style of the letter suggests that Hansteen might have been an eye witness of the incident. There is, however, no clear evidence that he attended Ørsted's lectures during this time and he may have been reporting secondhand information (Stauffer 1953). Even if he were really present during the event, his report might be unfaithful for several reasons, including the delay between the fact (1820) and the correspondence with Faraday (1857). Although this letter has been taken into account by historians of science and does deserve attention, it should not be regarded as the "true version" of the discovery.

After reading Hansteen's report the students should be able to compare it with Kendall Haven's story and notice the many points of disagreement between them.

---

[6] One paper published in 1821, a short autobiography published in 1828 and an article on thermo-electricity published in *The Edinburgh Encyclopaedia* in 1830 (Stauffer 1957).

Comparison of conflicting accounts is an important task, and the students should acquire the skill of finding out similarities and differences.

Remark that Hansteen's letter states that Ørsted believed before 1820 that there was a strong connection between electricity and magnetism and that he had tried to find the influence of the electric current upon a magnetic needle, but had found no effect because of the relative position of the wire and needle (perpendicular to one another). What was new during the lecture was the direction of the wire (parallel to the magnetic needle) and the discovery that the effect was transversal. There are many other relevant differences between the letter and Kendall Haven's version, of course.

17.3.7 Ørsted's Descriptions

At the beginning of Ørsted's 1820 paper we find a short historical account of the discovery:

> The first experiments respecting the subject which I mean at present to explain, were made by me last winter, while lecturing on electricity, galvanism, and magnetism, in the University. It seemed demonstrated by these experiments that the magnetic needle was moved from its position by the galvanic apparatus, but that the galvanic circle must be complete, and not open, which last method was tried in vain some years ago by very celebrated philosophers. But as these experiments were made with a feeble apparatus, and were not, therefore, sufficiently conclusive, considering the importance of the subject, I associated myself with my friend Esmarck to repeat and extend them by means of a very powerful galvanic battery, provided by us in common. Mr. Wleugel, a Knight of the Order of Danneborg, and at the head of the Pilots, was present at, and assisted in, the experiments. There were present likewise Mr. Hauch, a man very well skilled in the Natural Sciences, Mr. Reinhardt, Professor of Natural History, Mr. Jacobsen, Professor of Medicine, and that very skillful chemist, Mr. Zeise, Doctor of Philosophy. I had often made experiments by myself; but every fact which I had observed was repeated in the presence of these gentlemen. (Ørsted 1820, pp. 273–274)

Notice that according to Ørsted the first observation was made during the winter, not in the spring of 1820, as stated by Haven. He states that the first results were not conclusive (contrary to Hansteen's version) because the galvanic apparatus was weak, and that after some time he carried out new experiments, together with his friend Esmarck [sic] and others, using a stronger piece of equipment. Lauritz Esmarch was a lawyer with strong scientific interests, who published works on mineralogy. In a paper published in 1817, Ørsted described that they had been collaborating for 1 year in the construction and use of a strong galvanic apparatus (Ørsted 1998, p. 402). This cooperative component of Ørsted's research is seldom cited. We should also remark that Ørsted referred to other researchers who had attempted to find an influence of the battery upon the magnetic needle with an open circuit (without producing an electric current).[7] Also, contrary to Haven's and Hansteen's

[7] Early attempts to produce effects upon a magnetic needle using a Voltaic pile are described by de Andrade Martins (2001).

descriptions, there is no evidence that the effect observed in the first experiment was strong and striking.

There are slight differences between Ørsted's several descriptions of his discoveries that are highly instructive. Because of the limits for this chapter, I only reproduce here his account published in 1830, written in the third person:

> Electromagnetism itself, was discovered in the year 1820, by Professor *Hans Christian Oersted*, of the university of Copenhagen. Throughout his literary career, he adhered to the opinion, that the magnetical effects are produced by the same powers as the electrical. He was not so much led to this, by the reasons commonly alleged for this opinion, as by the philosophical principle, that all phenomena are produced by the same original power. In a treatise upon the chemical law of nature, published in Germany in 1812, under the title *Ansichten der chemischen Naturgesetze*, and translated into French, under the title of *Recherches sur l'identité des forces électriques et chimiques*, 1813, he endeavoured to establish a general chemical theory, in harmony with this principle. In this work, he proved that not only chemical affinities, but also heat and light are produced by the same two powers, which probably might be only two different forms of one primordial power. He stated also, that the magnetical effects were produced by the same powers; but he was well aware, that nothing in the whole work was less satisfactory, than the reasons he alleged for this. His researches upon this subject, were still fruitless, until the year 1820. In the winter of 1819–20, he delivered a course of lectures upon electricity, galvanism, and magnetism, before an audience that had been previously acquainted with the principles of natural philosophy. In composing the lecture, in which he was to treat of the analogy between magnetism and electricity, he conjectured, that if it were possible to produce any magnetical effect by electricity, this could not be in the direction of the current, since this had been so often tried in vain, but that it must be produced by a lateral action. This was strictly connected with his other ideas; for he did not consider the transmission of electricity through a conductor as an uniform stream, but as a succession of interruptions and re-establishments of equilibrium, in such a manner, that the electrical powers in the current were not in quiet equilibrium, but in a state of continual conflict. As the luminous and heating effect of the electrical current, goes out in all directions from a conductor, which transmits a great quantity of electricity; so he thought it possible that the magnetical effect could likewise eradiate. The observations above recorded, of magnetical effects produced by lightning, in steel-needles not immediately struck, confirmed him in his opinion. He was nevertheless far from expecting a great magnetical effect of the galvanical pile; and still he supposed that a power, sufficient to make the conducting wire glowing, might be required. The plan of the first experiment was, to make the current of a little galvanic trough apparatus, commonly used in his lectures, pass through a very thin platina wire, which was placed over a compass covered with glass. The preparations for the experiment were made, but some accident having hindered him from trying it before the lecture, he intended to defer it to another opportunity; yet during the lecture, the probability of its success appeared stronger, so that he made the first experiment in the presence of the audience. The magnetical needle, though included in a box, was disturbed; but as the effect was very feeble, and must, before its law was discovered, seem very irregular, the experiment made no strong impression on the audience. It may appear strange, that the discoverer made no further experiments upon the subject during three months; he himself finds it difficult enough to conceive it; but the extreme feebleness and seeming confusion of the phenomena in the first experiment, the remembrance of the numerous errors committed upon this subject by earlier philosophers, and particularly by his friend *Ritter*, the claim such a matter has to be treated with earnest attention, may have determined him to delay his researches to a more convenient time. In the month of July 1820, he again resumed the experiment, making use of a much-more considerable galvani-

> cal apparatus. The success was now evident, yet the effects were still feeble in the first repetitions of the experiment, because he employed only very thin wires, supposing that the magnetical effect would not take place, when heat and light were not produced by the galvanical current; but he soon found that conductors of a greater diameter give much more effect; and he then discovered, by continued experiments during a few days, the fundamental law of electromagnetism, viz. *That the magnetical effect of the electrical current has a circular motion around it*. (Ørsted 1830, p. 575)

In this report, Ørsted clearly stated (and provided a relevant reference) that he had been thinking about the relationship between electricity and magnetism since 1812 and that he was aware of the influence of lightning (an electric discharge phenomenon) upon the compass. Observe that according to Ørsted's description, the discovery was not made by chance, as he had predicted the influence of the electric current upon the magnetic needle and had prepared the experiment before the lecture. He described the first effect as weak and irregular and stated that the experiment did not impress the audience – contrary to Hansteen's and Haven's accounts. The experiment performed during the lecture did not allow him to reach definite conclusions and he associates the discovery of electromagnetism with his further experiments and observations of July 1820, when he established that the electric current produced a circular magnetic effect around the wire.

Notice that Ørsted described a highly complex trend of ideas he had about the relationship among electricity, magnetism, and other physical effects. In addition, it is relevant to remark that he pondered various hypotheses concerning the direction of the magnetic effect produced by the electric current, before arriving at the final conclusion.

Of course, Ørsted's reports could be incorrect and the discovery could have happened in a completely different way. Again, it is necessary to draw the students' the attention to the difficulty in ascertaining whether a primary source should be trusted. In this specific case, I believe that Ørsted did present a plausible and faithful account, and that there are no serious inconsistencies among his reports (Martins 2003).

17.3.8 Secondary Historical Sources

The use of primary sources is an important step, but it is not sufficient. The students must also read at least one sound historical paper on the subject. The secondary historical sources add much relevant information concerning the broader context and significance of each episode. Suitable secondary sources will moreover be instrumental in conveying important messages about the NOS – and those messages should be explicitly stressed by the teacher.

The study and discussion of the secondary sources in the classroom should draw the students' attention to the adequate use of sources and arguments by the historian, in contrast to pseudohistorical narratives.

In the specific case of the discovery of electromagnetism, historians of science have studied the precedents of Ørsted's work (Martins 2001), the strong influence of philosophical doctrines (especially Kant and the *Naturphilosophie*) upon his scientific research (Christensen 1995; Friedman 2007; Martins 2007; Shanahan 1989), the complexity of his ideas and how different they are from the interpretation of electromagnetism that is conveyed by current textbooks (Gérard 1961; Martins 1988), the difficulties of acceptance of his work (de Andrade Martins 2003), and many other interesting topics. The comparison between Ørsted's work and the investigations of Ampère, for instance, are also highly elucidative, showing the difficulties of interpretation of the phenomena they studied (Caneva 1980). Of course, the choice of the secondary sources depends on the particular characteristics of the NOS that the teacher intends to convey.

The use of two or more essays with conflicting interpretations can be very instructive, showing that historical research is not straightforward and that documents should be *interpreted*, and that the selection and analysis of historical sources by different historians can lead to diverging conclusions.

The teacher should draw the students' attention to the parallel between the historical research and research in the natural sciences, in which experiments and observations do not lead by themselves to definite and unquestionable conclusions, but must be clarified and can produce different interpretations. The teacher should also explain that this does not mean that in history or in the natural sciences everything is doubtful and that one opinion is equivalent to any other: radical relativism is incompatible with significant research, and the students should perceive that historians and natural scientists are seriously engaged in the production of meaningful knowledge using the best data and procedures they have.

Disagreement between historians should not be a disappointing finding, but a stimulating realization that there is further historical research to be done.

In the case of most episodes of the HOS that will be dealt with in the teaching practice, there will be an overall agreement of historians concerning the main points. Of course, agreement does not mean that those historians have reached the final and true understanding of the episode.

17.3.9 Assessment of Haven's Account

After presenting and discussing some primary and secondary sources, the HOS teacher should return to the first (faulty) version to analyze what part of its content could be regarded as acceptable. Let us now review Haven's story.

"In the spring of 1820, Hans Oersted was giving a lecture to one of his classes when an amazing thing happened. He made a grand discovery – the only major scientific discovery made in front of a class of students." In the spring – but perhaps in winter – of that year Ørsted made a relevant experiment, but the discovery of electromagnetism did not occur at that time, because he did not *understand* the effect at that time.

"It was a simple demonstration for graduate-level students of how electric current heats a platinum wire. Oersted had not focused his research on either electricity or magnetism. Neither was of particular interest to him. Still, he happened to have a needle magnet (a compass needle) nearby on the table when he conducted his demonstration." This is completely wrong. Ørsted had performed electrical and magnetic researches and he had been thinking about the relationship between electricity and magnetism for years. The experiment he carried out was planned to find the influence of the electric current upon the magnetic needle.

"As soon as Oersted connected battery power to his wire, the compass needle twitched and twisted to a point perpendicular to the platinum wire. When he disconnected the battery, the needle drifted back to its original position. Each time he ran an electric current through that platinum wire, the needle snapped back to its perpendicular position. Oersted's students were fascinated. Oersted seemed flustered and shifted the talk to another topic." No, the effect was weak and irregular. The students were not impressed. Ørsted was not flustered, he was probably happy and excited by observing the effect he was searching for.

"Oersted did not return to this amazing occurrence for three months, until the summer of 1820. He then began a series of experiments to discover if his electric current created a force that *attracted* the compass needle, or *repelled* it. He also wanted to try to relate this strange force to *Urkraft*." There was, indeed, a 3-month interval between the classroom experiment and Ørsted's further investigation of the effect; but there is no evidence that he was looking for an attraction or repulsion of the magnetic needle (it could only turn, therefore he could only notice a torque) and the idea of a universal force (*Urkraft*) was not tested in his experiments: it was a basic philosophical assumption he had accepted since he was very young.

"He moved the wire above, beside, and below the compass needle. He reversed the current through his platinum wire. He tried two wires instead of one. With every change in the wire and current, he watched for the effect these changes would produce on the compass needle." This is correct and is described in Ørsted's 1820 paper, except for the use of two wires.

"Oersted finally realized his electric current created *both* an attractive and a repulsive force at the same time. After months of study, he concluded that an electric current created a magnetic force and that this force was a whole new type of force radically different than any of the forces Newton had described. This force acted not along straight lines, but in a circle *around* the wire carrying an electric current. Clearly, he wrote, wires carrying an electric current showed magnetic properties. The concept of electromagnetism had been discovered." No, Ørsted never concluded that the electric current produced both an attractive and a repulsive force on the compass needle at the same time. He did conclude that the current produced a magnetic effect circling around the wire, and that was his most important (and revolutionary) conclusion. The only forces that Newton investigated were gravitational; other people studied electric and magnetic attraction and repulsion. The electromagnetic effect (not concept) discovered by Ørsted was indeed different than these, because it did not act along straight lines – it was not an attraction or repulsion at all, and it was not both an attractive and a repulsive force at the same time, as asserted by Haven. Perhaps this author did not understand the basic facts of electromagnetism at all.

17.4 Final Remarks

This chapter described a distinctive way of using different types of historical sources, including pieces of pseudohistory, to teach HOS. This approach was effectively used but no attempt was made to evaluate its results. My personal feeling is that the introduction of flawed historical accounts, their discussion and comparison with primary and sound secondary sources provides lessons and stimulates attitudes that are very difficult to convey using other strategies. Particularly, one of the aims of this approach is to teach the students how to distinguish a well-grounded historic narrative from an unreliable one and to develop a critical attitude concerning the flawed stories that proliferate on the internet and in popular science books.

Acknowledgements The author is grateful to the Brazilian National Council for Scientific and Technological Development (CNPq) and to the State of São Paulo Research Foundation (FAPESP) for sponsoring this research.

References

Allchin, D. (2004). Pseudohistory and pseudoscience. *Science & Education, 13*, 179–195.

Caneva, K. L. (1980). Ampère, the etherians, and the Ørsted connection. *British Journal for the History of Science, 13*, 121–138.

Christensen, D. C. (1995). The Ørsted-Ritter partnership and the birth of romantic natural philosophy. *Annals of Science, 52*, 153–185.

Friedman, M. (2007). Kant – *Naturphilosophie* – Electromagnetism. In R. M. Brain, R. Cohen, & O. Knudsen (Eds.), *Hans Christian Ørsted and the romantic legacy in science: Ideas, disciplines, practices* (pp. 135–158). Dordrecht: Springer.

Galdabini, S., & Rossi, O. (1993). Using historical papers in ordinary physics teaching at high school. *Science & Education, 2*(3), 239–242.

Gérard, J.-P. (1961). Sur quelques problèmes concernant l'œuvre d'Ørsted en électromagnétisme. *Revue d'Histoire des Sciences et de leurs Applications, 14*(3–4), 297–312.

Haven, K. (2007). *100 greatest science discoveries of all time*. Westport: Libraries Unlimited.

Hodson, D. (2014). Nature of science in the science curriculum: Origin, development, implications and shifting emphases. In M. R. Matthews (Ed.), *International handbook of research in history, philosophy and science teaching* (Vol. 2, pp. 911–970). Dordrecht: Springer.

Martins, R. A. (1986). Ørsted e a descoberta do eletromagnetismo. *Cadernos de História e Filosofia da Ciência, 10*, 89–122.

Martins, R. A. (1988). Contribuição do conhecimento histórico ao ensino do eletromagnetismo. *Caderno Catarinense de Ensino de Física, 5*, 49–57.

Martins, R. A. (2001). Romagnosi and Volta's pile: Early difficulties in the interpretation of voltaic electricity. In F. Bevilacqua & L. Fregonese (Eds.), *Nuova Voltiana: Studies on Volta and his times* (Vol. 3, pp. 81–102). Pavia: Università degli Studi di Pavia/Ulrico Hoepli.

Martins, R. A. (2003). Resistance to the discovery of electromagnetism: Ørsted and the symmetry of the magnetic field. In F. Bevilacqua & E. Giannetto (Eds.), *Volta and the history of electricity* (pp. 245–265). Pavia: Università degli Studi di Pavia/Editore Ulrico Hoepli.

Martins, R. A. (2007). Ørsted, Ritter and magnetochemistry. In R. M. Brain, R. Cohen, & O. Knudsen (Eds.), *Hans Christian Ørsted and the romantic legacy in science: Ideas, disciplines, practices* (pp. 339–385). Dordrecht: Springer.

McComas, W. F. (2010). The history of science and the future of science education: A typology of approaches to the history of science in science instruction. In P. V. Kokkotas, K. S. Malamitsa, & A. A. Rizaki (Eds.), *Adapting historical knowledge production to the classroom* (pp. 37–54). Rotterdam: Sense Publishers.

Ørsted, H. C. (1820). Experiments on the effect of a current of electricity on the magnetic needle. *Annals of Philosophy, 16*, 273–276.

Ørsted, H. C. (1830). Thermo-electricity. In D. Brewster (Ed.), *The Edinburgh Encyclopaedia* (Vol. 18, pp. 573–589). Edinburgh.

Ørsted, H. C. (1998). *Selected scientific works of Hans Christian Ørsted* (K. Jelved, A. D. Jackson, O. Knudsen, Trans., & Ed.). Princeton: Princeton University Press.

Shanahan, T. (1989). Kant, 'Naturphilosophie', and Oersted's discovery of electromagnetism: A reassessment. *Studies in History and Philosophy of Science, 20*, 287–305.

Stauffer, R. C. (1953). Persistent errors regarding Oersted's discovery of electromagnetism. *Isis, 44*, 307–310.

Stauffer, R. C. (1957). Speculation and experiment in the background of Oersted's discovery of electromagnetism. *Isis, 48*, 33–50.

Williams, L. P. (Ed.). (1971). *The selected correspondence of Michael Faraday* (Vol. 2). Cambridge: Cambridge University Press.

Chapter 18
Isaac Newton and "Hidden Forces" in Universal Gravitation: Delimiting an Approach for Teacher Training

Thaís Cyrino de Mello Forato

18.1 Introduction

In the last decades of the twentieth century, historical research has intensified on the relationship between Isaac Newton's (1642–1727)[1] science and his interests in other fields of knowledge, for example, alchemy, Kabbalah, certain mystical knowledge of a remote antiquity and the biblical prophecies (e.g., Manuel 1974; Westfall 1987; Rattansi 1988; Dobbs 1991; Forato 2006b). Such approaches take into account the ideas of the period in which he lived, the intellectual environment in Cambridge, his unique neo-Platonic worldview, and perspectives of other thinkers of his time.

This complex confluence of knowledge, which permeated the development of the Newtonian doctrine, has not been found in textbooks (Braga et al. 2008) and it is present in only a few studies focused on teaching, and on proposals implemented in classrooms, such as that held by Leal (2014). In general, when educational and outreach materials mention Newton's interests, they continue to propagate naive views about the period (Martins 2006) and some stereotypes, such as the aberrant "philosopher-magician" (McGuire and Rattansi 1966). In addition to disregarding Newton's specific personal and social context, they ignore ontological and epistemological aspects of problems of the period, for example, explaining actions at a

A previous version of this work was presented at the 2nd IHPST Latin American Regional Conference, October, 2012, Mendoza, Argentina (Forato, Thaís Cyrino de Mello. Isaac Newton: science and religion in the classroom. Book of Abstracts. 2nd Latin American Regional Conference of the International History, Philosophy, and Science Teaching Group, Mendoza-Argentina, October, 2012, pp. 8–9).

[1] Newton's birth date was 25 December 1642, following the Julian Calendar adopted in England until 1700, but may appear as 4 January 1643, following the Modern Gregorian Calendar.

T. C. d. M. Forato (✉)
Department of Exacts and Earth Sciences, Federal University of São Paulo, São Paulo, Brazil

M. E. B. Prestes, C. C. Silva (eds.), *Teaching Science with Context*,
Science: Philosophy, History and Education,
https://doi.org/10.1007/978-3-319-74036-2_18

distance and using inductivism as methodology (Martins 1993, 1998). In science education, the "myth of the apple" (Martins 2006) is still perpetuated, reinforcing naive conceptions of the nature of science (NOS).

These are examples of a naive historical approach that has been contributing to fostering dissonant views of a contextualized position on science, at different levels of education and in several countries (Gil-Perez et al. 2001; Abd-El-Khalick 2012). This scenario requires concrete actions to question stereotypes and naive interpretations of NOS.

Approaching Newton's peculiar neo-Platonist conception in his science involves elements to discuss physics concepts, epistemic and non-epistemic aspects of science in science education, especially in teacher training. This historical episode allows critical and explicit reflections on science development and it can promote a well-informed position regarding historical and epistemological approaches of the sciences (Rudge and Howe 2009). It also brings the subjects science and religion to teacher training to offer them subsidies to deal with controversial positions and fundamentalists, which usually occur in classes on Universe creation (Leal 2014). Several teachers mock students' religious values, which can cause their lack of interest in science.

A respectful approach to science and religion allows dialogue with the students' beliefs and it fosters greater understanding of the differences and similarities between science and other ways of trying to understand and explain the natural world (Bagdonas and Silva 2015).

The didactic use of history of science (HOS) presented in this work assumes the impossibility of neutral approaches to the practices of the scientists, researchers of any area, including the historians of science, and in the practices of the science teachers. Accordingly, this proposal requires explanation of the adopted perspective on NOS (Forato et al. 2012). Understanding science as a dynamic human enterprise, a member of numerous cultures throughout history, it cannot be imagined that there is one single correct or appropriate conception of NOS (Gil-Perez et al. 2001; Lederman 2007). Every period, every culture and every field of science has their own normative and axiological rules, allowing the specialized dialogue and validation of theories, although they are opened to refutation (Martins 1999).

Fruitful discussions among scientists, philosophers, sociologists, and historians of science on NOS suggest the importance of problematizing views about science and its nature as dogmatic, only objective, neutral, and decontextualized (Bagdonas and Silva 2015). In the education community, the current different approaches on NOS show a general concern with the critical and well-informed training of citizens about epistemic and non-epistemic aspects of sciences (e.g., Lederman 2007; Allchin 2011; Matthews 2012; Acevedo-Díaz and Garcia-Carmona 2016; Dagher and Erduran 2016). Thus, it is understood that this historical approach reveals peculiar aspects about NOS, which may encourage reflection on science development, but that it should not be extended to all sciences, in all periods (Forato et al. 2011; Bagdonas and Silva 2015).

Among the challenges in training autonomous and protagonist teachers in their learning about NOS is promoting the understanding that their choices are not neutral, political or epistemological (Freire 1996), and that their educational practice perpetuates, implicitly or explicitly, a conception of NOS (Carvalho and Gil Perez 2001). However, experts believe that historical approaches that have permeated training courses are not suitable for such purposes (Höttecke and Silva 2011). Within this scope, teacher training requires the experience of explicit reflections on NOS, founded by history, philosophy, and sociology of sciences (Adúriz-Bravo and Bejarano 2015) and knowledge of the intrinsic challenges to the uses of HOS in Basic School, as obstacles arising from social, family, and cultural environments of which the student is a part (Forato et al. 2011). Nor can the conflict be ignored that new epistemological approaches possibly find in teacher training compared with their school culture (Höttecke and Silva 2011).

From this perspective, this research is aimed at developing a historical approach, contemplating Newton's involvement with different fields of knowledge, to promote the learning of physics concepts, to problematize stereotypes in HOS, to discuss epistemic and non-epistemic aspects of science with the goal of offering elements for a critical and autonomous positioning considering approaches to NOS.

The delimitation of an approach for the context of teacher training endeavored to reconcile the historiographical emphasis contextualized to the school environment. As a starting point, the pedagogical purposes were established as the following:

1. Promoting contextualized teaching of Universal Gravitation, explicating some problems to describe falling bodies
2. Discussing differences between Newton's neo-Platonism and the classical mechanism, which would have influenced conceptions of forces by contact or at a distance
3. Exemplifying the influence of the context and personal values of a thinker in science construction
4. Problematizing the "apple myth"
5. Discussing the role of biblical prophecies in the Newtonian project to fight atheism, questioning stereotypes
6. Offering subsidies to Basic School teachers for them to position themselves critically and respectfully toward the debates between students' religious or atheistic options, regardless of their personal beliefs.

Next, a brief summary of the historical period is presented, followed by an analysis of the methodology adopted for its delimitation, and, finally, a possibility for using the historical episode in teacher training is discussed.

18.2 Newton and the "Hidden Forces" in the Development of Universal Gravitation

The relationships among natural philosophy, alchemy, religion, Renaissance Hermetic magic tradition, and the neo-Platonism of Cambridge thinkers, present in the world conception of Isaac Newton, have been analyzed and described by historians of science since at least mid-twentieth century (Manuel 1974; Copenhaver 1980; Westfall 1987; Rattansi 1988; Dobbs 1991; Forato 2006b).[2]

The socio-cultural context in which Newton lived is often portrayed as emblematic of a break with the magical thinking of the Renaissance and the emergence of other ways of investigating nature. In general, this period saw greater emphasis on mathematization and experimentation, both understood to be fundamental characteristics of modern science. But beyond that, the period was intense and complex with regard to methodological debates and the use of technical devices to investigate nature, besides the presence of neo-Platonisms in the work of several thinkers (Debus 1978; Alfonso-Goldfarb 2000).

The concern in explaining phenomena such as gravity and the attraction between magnets and iron, for example, were impregnated, for several natural philosophers, with an intrinsic relationship between the existence of immaterial agents in nature and the divine action on matter. Since ancient times, there were supporters of the idea that such actions would only occur by contact and of the idea that they were based on actions at a distance (Martins 1993).

To explain falling bodies, Newton developed some mechanical models of actions by contact, imagining that a down flow ether would "push" bodies toward the earth surface. But when questioning what would "push" the ether, he found limitations in such models, and he started to adopt alchemical ideas to support different explanations, based on the conception of action at a distance (Dobbs 1991). René Descartes (1596–1660), for example, did not accept the existence of these actions and he argued that the interactions between bodies that were not in direct contact, should be mediated by some kind of material medium (Martins 1998).

Neo-Platonic conceptions in the intellectual environment of several natural philosophers of the seventeenth century in Britain – for example, Robert Boyle (1627–1691) – were inspired by studies of secret literature, such as hermetic treatises and confidential craft manuals and texts influenced by the Baconian proposal for dissemination of the new science (Alfonso-Goldfarb 2000). In this ideology, there was the conception of a deity who acted in the natural world, and that alchemy would be a way of understanding this action (Dobbs 1991; Rattansi 1988). In the European continent, the growing acceptance of mechanism philosophies – explaining phenomena only by the mediated interaction between material objects – promoted atheism, leading some natural philosophers to engage in criticizing mechanism and defending the existence of God (Westfall 1987).

Newton, for example, performed alchemical experiments to understand the action of God in nature (Dobbs 1991) and he used biblical prophecies as an argument to fight atheism, believing that these achievements would be a convincing proof of the existence and work of God in the natural world:

---

[2]As the didactic proposal was developed for the Brazilian context, the suggested texts are in Portuguese. For example, here, Forato (2006b) offers a synthesis (developed for Brazilian science teachers) of relevant aspects discussed in the literature indicated in English. Whenever possible, several sources in English with similar content are also indicated.

He gave this and the Prophecies of the Old Testament, not to gratify men's curiosities by enabling them to foreknow things, but that after they were fulfilled they might be interpreted by the event, and his own Providence, not the Interpreters, be then manifested thereby to the world. For the event of things predicted many ages before, will then be a convincing argument that the world is governed by providence. [...] There is already so much of the Prophecy fulfilled, that as many as will take pains in this study, may see sufficient instances of God's providence [...]. (Newton 1733, pp. 251–252)

According to Westfall (1987), Newton's studies on religion started up soon after his student days and continued for the other 60 years of his life. We can identify passages about God, for example, both in the *Principia*, since its first edition of 1687, and in Question 28 of *Opticks*, since its first edition of 1704 (Newton [1713] 1952). However, in these published works, Newton does not assign the cause of gravity explicitly to God (Mcguire and Rattansi 1966).

In his theory of universal gravitation, presented in the first edition of the *Principia*, Newton described the behavior of the universe and the attraction between bodies, but he did not explain the cause of that attraction. In the second edition of the *Principia*, of 1713, at the end of Book III entitled "The system of the world (in mathematical treatment)" he added the "General Scholium" (Newton [1713] 1952, pp. 369–372). In this scholium, he included his well-known statement "hypotheses non fingo," stating that he does not invent hypotheses to explain the cause of gravity's power and reinforcing the experimental aspect of his natural philosophy (Martins 1989). According to Martins (1989, 1998) Newton was mainly concerned with responding to the disapproval received from the Dutch Christiaan Huygens (1629–1695), and other natural philosophers, who criticized him both for the methodology, which generalized the laws by induction, and for certain Renaissance principles that would be implied in his ideas of forces acting at a distance:

Hitherto we have explained the phenomena of the heavens and of our sea by the power of gravity, but **have not yet assigned the cause of this power**. This is certain, that it must proceed from a cause that penetrates to the very centers of the sun and planets, without suffering the least diminution of its force; [...] But hitherto **I have not been able to discover the cause of those properties of gravity from phenomena, and I frame no hypotheses**; for whatever is not deduced from the phenomena is to be called an hypothesis; and **hypotheses, whether metaphysical or physical, whether of occult qualities or mechanical, have no place in experimental philosophy**. In this philosophy particular propositions are inferred from the phenomena, and afterwards rendered general by induction. Thus it was that the impenetrability, the mobility, and the impulsive force of bodies, and the laws of motion and of gravitation, were discovered. And **to us it is enough that gravity does really exist, and act according to the laws which we have explained,** and abundantly serves to account for all the motions of the celestial bodies, and of our sea. (Newton [1713] 1952, pp. 371–372, no emphasis in the original)

Although Newton does not make the conjecture that God is the cause of planetary motion explicit, he mentions the divine presence several times in the material world, for example:

This most beautiful system of the sun, planets, and comets, could only proceed from the counsel and dominion of an intelligent and powerful Being. [...] This Being governs all things, not as the soul of the world, but as Lord over all; [...] He is omnipresent not *virtually*

> only, but also *substantially*; for virtue cannot subsist without substance. In him[3] are all things contained and moved; yet neither affects the other: God suffers nothing from the motion of bodies; bodies find no resistance from the omnipresence of God. (Newton [1713] 1952, pp. 369–370)

In an unpublished manuscript containing a draft for the same "General Scholium" and known as "The classical scholia," Newton writes that God is the cause of gravity. He also describes the divine action in the matter, mentions atoms moving in the void, states inversely proportional forces to the square of the distance between bodies, and attributes all these ideas to the Egyptian priests of remote antiquity (Mcguire and Rattansi 1966; Forato 2008). These views were amalgamated: it was not possible to separate the action of gravity from God's action in the natural world, known through alchemical activity. Today, such ideas belong to different fields of knowledge and epistemological approaches, but at the time, they expressed a way of understanding the behavior of the material world through a neo-Platonic interpretation of the Universe.

Newton, often presented as a model of rationality in HOS and in science teaching, makes explicit mentions to subjective factors, something that is currently considered as extrascientific. This example provides critical reflection on the idealized model of rationality that was assigned to him during the eighteenth century (Rattansi 1988; Braga et al. 2008).

This very brief presentation of some of the concerns that permeated Newton's life and work exemplify the complexity of his ideas, in addition to the most popular aspects of mathematics, optics, mechanics, astronomy, studies on heat, etc. In the didactic proposal presented here, texts will be indicated for a broader approach of this perspective.

The didactization of this proposal requires the educational context, the assumptions on science education adopted, and the teacher training to be assessed, in addition to considering the historiographical and epistemological consistency of the approach with the desired educational goals. In the next section, the analysis developed for this delimitation is presented.

18.3 Theoretical and Methodological Assumptions

This research has a theoretical scope on two fronts: the delimitation of the historical approach to the school environment and the development of a didactic proposal. These analyses considered that the reflections on the historical content cannot be dissociated from the didactic prerogatives of science and education (Forato 2009). Thus, these steps are based on references with resonant conceptions, as presented below.

[3] Here Newton added a long note mentioning several thinkers of Classical Antiquity and bible passages.

18.3.1 Delimiting the Historical Approach

The context for the construction of the didactic proposal considers its inclusion in a course in physics teacher training or related areas. The delimitation of the historical approach uses parameters to select contents to be emphasized or omitted from the training goals defined (Forato 2009; Forato et al. 2012). Twenty-two items are organized into four blocks of reflections, aiming at the consistency of the historical approach in the desired NOS view. Owing to space limitations, we present a synthesis of the reflection promoted by each item, to explain the points assessed.

I. Establishing the goals and conceptions of the didactic proposal:

   I.1 *Establishing the educational purposes for HOS in education.*
   I.2 *Explicating the science conception adopted and/or the epistemic and non-epistemic aspects intended.*

The aspects mobilized by items *I.1* and *I.2* are as follows:

- Physics conceptual learning: gravity; interaction forces in matter; universal gravitation
- Understanding of the debate "action at a distance versus action by contact" in the seventeenth century
- Knowledge of the complex relationship between science and religion in Newton, allowing to question the naive view of an eternal conflict between science and religion in HOS
- Criticizing the uniquely rational conception of science, showing the interlocution given by Newton to different fields of knowledge
- Understanding science as a historical construction, locally and temporally contextualized, mutable, also influenced by factors currently considered as unscientific
- Understanding that the cultural environment and personal beliefs of a thinker influence the construction of science
- Questioning naive or extreme relativistic positions
- Understanding some of the reasons why such content is not present in current textbooks
- Improvement of the critical reading of texts, argumentation, and integration for group work, aimed at teacher autonomy.

II. Defining the approach and focus for the historical episode

   II.1. *Selecting the subject and the appropriate historical content for the goals*

The confluence of aspects of natural philosophy, alchemy and Newton's religious thought to explain gravity, to elaborate universal gravitation and the relationship between such aspects and physical models of his time. Reasons to demonstrate the fulfillment of the bible prophecies in the Newtonian project.

II.2. *Selecting aspects to emphasize or to omit in the HOS approach*

Emphasize the harmony between the different ways of studying nature for Newton, God as the cause of gravity, and the relevance of this approach to the cultural environment. Present the limitations of mechanical models to explain falling bodies, introducing the alchemical vision to make action at a distance explicit and how God worked on matter, determining the laws governing nature. Newton's concern with atheism and his commitment to "prove" the existence of God through the fulfillment of bible prophecies.

Omit[4]: Without compromising the pedagogical goals, omit the different neo-Platonisms existing in the seventeenth century, or the controversy if Newton found or imposed God in/on nature. Omitting Kabbalistic aspects for God's *sensorium* and the methodological debate between Huygens and Newton, unless the teacher considers it appropriate to bring the debate "deduction *versus* induction" to his/her context. Omit Newton's further prestige as a possible influence for the acceptance of his doctrine, leading some natural philosophers of the eighteenth century to ignore some of its controversial aspects. Omit the issue of the great apostasy that Newton attributes to the Catholic Church and his interpretation of biblical prophecies. The omission of the contributions of medieval mathematics related to natural phenomena and mystical and religious issues also does not cause loss regarding the pedagogical purposes.

II.3. *Confronting omitted aspects with the objectified epistemic and non-epistemic contents*

The omitted aspects would allow a deepening of the subject and they could make significant contributions to teacher training, but they were excluded because of didactical time, content specificity, and the goals established, involving a small number of classes.

II.4. *Defining the detail level of the non-scientific context to be treated, avoiding promotion of extreme relativistic interpretations*

God's actions and mystical ways of interpreting the natural world are considered in a nonscientific context in this proposal, but in the seventeenth century, they were not consensually excluded from the science heuristics. Although this classification is anachronistic, it is needed for this analysis. The atheistic implications because of mechanism are approached to clarify Newton's and other natural philosophers' concern to fight them. For example, Robert Boyle bequeathed resources to fight atheism, which funded the lectures of the theologian Bentlley who explained the Newtonian world system, ensuring the presence of a creator (Westfall 1987). Although not currently considered as scientific, these ideas influenced science during the period and they were fundamental to the Law of Universal Gravitation. However, emphasizing the influence of personal beliefs and interests in the science

---

[4]This reflection is on the omission of closely related aspects to the adopted approach, which already excludes several subjects on Newton's life and work (Forato 2006a, 2006b, 2008).

construction can promote extreme relativistic interpretations. To oppose them, it is interesting to highlight objective aspects, such as the methodological debate in the construction of explanations and the relevant role of mathematics and experimentation in the Newtonian doctrine.

This tension between differentiating what is and what is not currently considered and in the seventeenth century context may favor reflections on the historical anachronism problem and make the existence of different NOS in the HOS explicit. This discussion is formative for teachers, allowing the reflection that goes beyond the specific content to a single historical episode (Forato et al. 2017).

III. Problematizing stereotypes, myths, and prejudices

III.1 *Pondering the use of primary sources: if, when, how many and how to introduce them*

The importance of the contact with the primary sources of HOS is admitted for future teachers, as they are discussed in diachronic perspective and with the support of secondary sources. Contextualized passages can aid the proposal goals, for example, Newton explaining the divine action in matter, in the "Classical scholia," and in the fulfillment of biblical prophecies as proof of God's existence.

III.2 *Diachronically approaching the HOS content that is currently difficult to understand: establishing the relationship between relevant results for science construction and discarded or "weird" contents*

The episode is full of weird contents from the current point of view, such as the idea of active principles in matter, attraction and repulsion as hidden forces from Renaissance and God as the cause of gravity.

III.3 *Diachronically approaching different science conceptions in distinguished periods and civilizations*

It is challenging to understand the harmony between science and the neo-Platonic worldview, in the period marked by the break with the magical and alchemical thinking of the Middle Ages and Renaissance. Newtonian science has been generally portrayed as emblematic of this break, capping the use of mathematical abstraction, quantification, and experimentation in the investigation of natural phenomena. Little recognition is seen from experimentation that already occurred in Classical Antiquity and throughout the Middle Ages (Martins 2017) and the broader contributions of Renaissance humanism to modern science, in addition to the Renaissance abstract and theoretical mathematics related to natural phenomena and mystical and religious issues (Debus 1978).

III.4 *Presenting contemporary thinkers working with the same methodological assumptions can help the criticism of prejudice and anachronism*

The effort to show Newton's contemporaries, such as Robert Boyle and other Cambridge neo-Platonists, involved with the same conceptions, intends to problematize a simplistic, presentist, and stereotyped interpretation of these thinkers.

III.5 *Presenting examples of theories overcome in different contexts allows naive ideas to be criticized, such as the possible conception that current science can solve all problems*

The mechanistic models seemed to meet "more rational" criteria and they would be devoid of hidden forces and other Renaissance "superstitions." But they failed to explain the gravitational phenomenon, giving room for the alchemical and neo-Platonic ideas "to explain" falling bodies. These aspects were banned from the Newtonian doctrine in the next century, but documents from Newton himself show his contribution. It is important to emphasize that current theories propose explanations for the cause of gravity. The predictive and descriptive success of these models is relevant, but they did not "prove" the cause of gravity.

III.6 *Preparing teachers to identify and problematize whig historical narratives and anachronistic manifestations.*

The subject turns to teacher preparation, discussing anachronisms and stereotypes, offering freedom to each trainer to deepen the aspects they deem appropriate.

IV. Considering the construction of texts and activities

IV.1. *Assessing when it is possible to overcome or circumvent the lack of prerequisites in mathematical, physical, historical, and epistemological knowledge*

Physics and mathematics content are not expected to be problematic in teacher training. The historical and epistemological content is also a goal of the proposal.

IV.2. *Combining different didactical strategies and resources can compensate for the lack of knowledge in certain physical and mathematical content*

We suggest the second year of the licentiate, seeking greater contact of the students with physics, mathematics, and other intrinsic content to teacher training. The reading of texts, the debate in simulated jury, expositive, and dialogic moments are proposed, supported by slides and written production of a short text. Including exercises and open problems facing the mathematization of concepts is suggested.

IV.3. *Defining the level of depth and discursive formation of the epistemological content*

Epistemological issues (that emerge from the episode itself (Forato et al. 2017)) are approached implicitly and explicitly as they were established in the pedagogical goals (initial parameters). It is not intended to define and to memorize philosophical content, but to understand the processes in the context of the development of this episode of science. Four of the secondary sources were suggested because they have been written for teachers (Assis 2006; Forato 2006b; Martins 2006; Braga et al. 2008), and one of the papers is aimed at science dissemination (Forato 2006a).

small excerpts from primary sources (Newton [1713] 1952; 1733) are proposed after discussing the content in class. There are still other suggestions for deepening.

IV.4. *Defending a new idea conflicting with the ones prevailing in the cultural repertoire of students requires the use of strategies able to create discomfort that allows the questioning of pre-established ideas*

The dispute between the conceptions of action at a distance and by contact can engage students with the assumptions of the period to contextualize the idea of divine activity in matter. The challenge of building this approach can cause this mobilization of learning discomfort.

IV.5. *Allowing students to experience aspects of the debates between rival theories favors the understanding of NOS aspects:*

Debates between different models and concepts of nature mobilizes aspects currently considered as unscientific, exemplifying that each period and branch of science has its own aspects of NOS.

IV.6. *Choosing topics that arouse the curiosity of the intended age group*

Alchemical or religious aspects in the Newtonian thought have awakened, in general, the interest of students. The "myth of the apple" is usually resorted to, but here it seeks to promote a critical reflection.

IV.7. *Pondering on the number and depth of the texts*

The appropriate amount of texts is assumed for the duration of the teacher training.

IV.8. *Keeping in mind the different social roles of academic and school knowledge*

The future teacher does not have to be a historian of science to understand the episode and to implement it in their teaching practice. A summary of the episode and references for those who want to deepen their knowledge are suggested.

IV.9. *The timeline with commercial movies can aid in the dilemma length x depth*

Select a commercial movie showing the cultural context in England in the period.

IV.10. *Questioning every objectified message on science in different didactical activities and different historical episodes:*

The proposal involves a single historical episode and the pretended epistemological aspects were considered in different didactical activities: texts, lectures, debate, writing, production.

The analysis that delimited the historical approach and the didactic proposal also included the educational perspectives presented below.

18.3.2 Elaborating the Didactic Proposal

The analysis for the delimitation of the historical episode was also supported by prospects of didactics of science and a conception of education, considering the inseparability between the science view, epistemic and non-epistemic aspects of its development, its teaching and learning (Forato 2009). Committed to this consistency, this delimitation considers that transposing the HOS expertise to classrooms requires mediating the historiographical preciousness to meet the other social function, the training of physics and science teachers. Moreover, historical narratives carry the influence of the historian's personal values and interests in choosing the object and the interpretation of documents, making it essential to prepare the teacher to put historical interpretations in perspective, as they intrinsically promote a particular view of science (Forato et al. 2017).

There are no foolproof recipes for educational processes aimed at any subject, but it is considered that experimental methodologies involving HOS contributes to teachers' learning about NOS and to promote their autonomy in further actions in Basic School (Adúriz-Bravo and Bejarano 2015). HOS also provides a valuable way to understand the construction of science as a collective process, in which controversies, debates, and discarded ideas also contribute to its development (Matthews 2012). It is believed that it is important that students discuss case studies and problems that led to the proposition of concepts and explanations (Allchin 2011) and that they experience historical controversies in activities such as debates, simulated jury, games, performing a play, (Forato 2009), in which they promote explicit reflections on NOS (Rudge and Howe 2009). Such activities provide spaces of interpersonal interaction in educational activities aimed at a reflective transformation (Yáñez and Maturana 2009). This perspective resonates with a critical education that promotes dialogue and awareness, as an educational praxis for the world transformation (Freire 1996). Therefore, for the construction of didactical activities a social perspective of knowledge construction was adopted, guided in a critical and reflective practice (Carvalho and Gil-Perez 2001).

Pozo and Crespo (2009) consider that learning a new concept requires recognition of its meaning by the student, and they defend a relationship with previous knowledge being established. Thus, this proposal starts in the review of the concepts of forces by contact and at a distance, seeking to review content already known, to create a safe "pavement" from which new concepts are presented and developed. In each class, new texts and ideas are being added, deepening discussions.

18.4 The Didactic Proposal

This didactic proposal (Chart 18.1) seeks to set up as an example, offering suggestions for the linking of contents and activities, allowing teacher trainers to select, delete, and deepen what they deem appropriate. Support texts are suggested to

Day	*Content*	*Activities / Text*
Pre-class	Action by contact and action at a distance.	1. Review of Assis (2006).
1	Context of the seventeenth century. How to explain the falling bodies?	2. Overview in slides. 3. Reading and questions about Forato (2006a).
Extra Class	Context of the seventeenth century. Myth of the apple.	4. Movie about the period and reading of Martins (2006).
2	Universal Gravitation. Newtonian alchemical conceptions and God as the cause of gravity.	5. Review of the theory. Reading of Forato (2006b).
Extra Class	Huygens' criticism for Newtonian Gravitation.	6. Elaborating questions about Martins (1989).
3	Action by contact and at a distance.	7. Discussion on students' questions. 8. Group work from Martins (1998); Newton (1713). (Presentation of the proposal for the debate).
Extra Class	What is the cause of gravity?	9. Preparation for the debate: Forato (2008); excerpts of Newton (1733).
4	Gravity: Action by contact *versus* action at a distance. Role of bible prophecies in the Newtonian project.	10. Debate between groups. 11. Systematization. 12. Was God excluded from Newtonian mechanics?
Extra Class	Who banned metaphysics from the Newtonian doctrine?	13. Reading of Braga et al. (2008).
5	Newton: God as the cause of gravity and biblical prophecies to "prove" His existence.	14. Slide presentation. Overview of the episode.
Extra Class	Influence of personal beliefs in Newtonian Gravitation.	15. Development of narrative or story for Basic School, or dissemination. (Evaluation).

Chart 18.1 Summary of the classes

maintain the historiographical approach. This suggestion was planned for four or five physical meetings, with extra class readings in the context of a Licentiate in Physics or Sciences, adaptable to other related courses.

Suggested content:

1. Action at a distance and by contact from a historical perspective
2. Falling bodies and gravity in the context of the seventeenth century
3. Law of universal gravitation
4. Neo-Platonism in Newton
5. Science as part of culture.

The teacher in training is expected to:

1. Understand and explain, in historical perspective, the concepts of gravity and interaction forces in matter; law of universal gravitation

2. Understand and perform the debate "action at a distance *versus* action by contact" in the seventeenth century
3. Understand the development of science as part of the culture of each period, influenced by metascientific aspects and personal values of the thinkers
4. Enhance critical reading of texts, the ability to develop arguments, and the ability to work in a group.

References for the teacher in training:

Assis (2006), Braga et al. (2008), Forato (2006a, b). Martins (1989, 1998, 2006), and Newton (1952, 1733).

References for deepening:

Alfonso-Goldfarb (2000), Westfall (1987), Dobbs (1991), Forato (2008), Mcguire and Rattansi (1966), Martins (1993, 2017), and Rattansi (1988).

18.4.1 Development of the Activities

1. Extra class: Request reading of Assis (2006), who discusses different historical models to make the interaction between bodies explicit. The teacher could propose guiding questions, for example: "*When the author presents Faraday's understanding on the electrical and magnetic interactions, do they seem absurd or are they in accordance with the scientific assumptions of his time? Does the author offer elements in the text for us to deduce these conclusions?*" A review of the text is suggested for evaluation.
2. The teacher may start the class by identifying the students' conceptions, such as: "*Do you believe that faith or religious orientation issues influenced the work of thinkers throughout the history of science?*", "*Do you know any historical episode in which the personal beliefs of a thinker or group of thinkers have influenced the construction of models considered as scientific?*" Next, present an overview of the cultural context of the seventeenth century and the episode. The inclusion of poets, painters, musicians and their works, historical and political facts in different parts of the world, some of the natural philosophers, science inventions and discoveries is suggested. Resuming the concepts of interaction in matter by contact and at a distance and to introduce different historical explanations for gravity, electric, and magnetic phenomena is also suggested; introduce the ideas of Descartes (action by contact) and interaction at a distance; comment on the intellectual environment in Cambridge and contemporaries who defended a neo-Platonic world view; comment on mechanistic models of the period, devoid of hidden forces of the Renaissance, but that they had limitations in explaining the gravitational phenomenon; highlight different contemporary interpretations aimed at criticizing the naive empiricism and show science integrated into the cultural environment.
3. Students in pairs or trios: Reading and preparation of questions about Forato (2006a), which summarizes the main ideas of the proposal and introduces alchemical conceptions in Newton (Dobbs 1991). Discuss the proposed ques-

tions in a plenary. Start an explicit reflection on the differences between Newton's neo-Platonism and classical mechanism.

4. Extra class: Movies on the period, exemplifying the social, cultural, and political context of the period. Reading of Martins (2006).
5. Review of the law of universal gravitation, problematizing the myth of the apple. The relationship between variables allows the idea of forces at a distance to emphasized, because the concept of the gravitational field was only invented in the late nineteenth century.
6. Extra class: Reading and preparation of questions about Martins (1998), highlighting the arguments of René Descartes against the conception of action at a distance.
7. In small groups, or plenary discussion, ask students to discuss the questions they prepared. Guide them so that they also emphasize phenomena and limitation in different models to explain the falling bodies.
8. Guide and organize groups for a fictitious debate in a simulated jury to occur in the next class: "*We are in the seventeenth century and there is no consensus about the best explanation for the phenomenon of falling bodies. There are mechanical models of action by contact and action at a distance. A certain scientific society proposes a public debate between possible representatives of both conceptions, and a jury composed by renowned natural philosophers of the period will evaluate the arguments and choose the best-justified conception. The room will be divided into three groups. One group to defend each conception and a third group to compose the jury. The teacher or a student may play the role of a judge or mediator. Depending on the time available, the teacher can set 15 minutes for each group to present their defense, and 15 minutes for each replica. If the time allows, it is possible to propose a rejoinder. The jury will also have 15 minutes to ask questions to both groups, seeking to better understand each conception. Both groups will leave the room while the jury discusses and elaborates the verdict (15 minutes). When the discussion is finished, the groups come back to hear the verdict. All the arguments of the three groups must consider ONLY the perspective of the seventeenth century. Further theories, ideas and concepts cannot be used for any of the three groups. The jury will decide on the best theory taking into account only the arguments of both groups and not further knowledge of the history of science.*"

   One can imagine other characters in the context, for example, witnesses who can present data to support or refute hypotheses. Depending on the context, this activity may have a few students participating more actively in the debate. Teachers can add other bibliographic references they deem relevant.
9. Preparation for the debate: groups meet in an extra class, requiring an interval between these classes. Reading of Forato (2008).
10. Debate between groups.
11. Systematization, reviewing concepts and complementing what is needed. Highlighting aspects of the intended NOS is suggested.
12. At the end of the class, propose the question: "*Why such Newton's ideas are not included in textbooks and are not discussed? How was metaphysics removed from the Newtonian doctrine?*"

13. Extra class: Reading of Braga et al. (2008). Enlightenment context that led to the withdrawal of metaphysical influences from encyclopedias throughout the eighteenth century and from science manuals in the early nineteenth century. Reflect on the heritage that prevailed in physics teaching and the science view generally encouraged in many textbooks.
14. Class for overview of the episode, emphasizing the complex relationship between Newton's science and other culture fields of the seventeenth century. It is important to emphasize that there was predominance, in a way, of a mechanistic conception of the Universe, for many philosophers of the European continent, such as Descartes, for example. In England, on the other hand, a group of thinkers adhered to the neo-Platonic view, accepting the action of forces at a distance, which inscribed implicitly or explicitly, the existence of active principles in matter. For Newton, such actions were due to the presence and action of God in the natural world, He being the cause of gravity. He used biblical prophecies as proof of God's existence, as one of his "weapons" in the fight against atheism. To fight the stereotype of Newton as the last of the wizards, or someone who has gone mad in old age, turning to the study of religion, it can be emphasized that Newton and many contemporaries defended ideas relevant to their cultural context. It is important that the trainer clarifies that the confluence between religion and science in this episode is not intended to promote relativism, nor devalue science, but it seeks to problematize myths, dogmatism, prejudice, and disrespect with different worldviews.
15. As an evaluation activity, requesting the students to write a small text is suggested, in the form of a story or free narrative on the subject: "*Isaac Newton, forces at a distance and the cause of gravity.*" In case of a licentiate in physics, the text should be designed for a high school context and it may be followed by a didactic proposal. For related courses, an essay on science dissemination may be considered.

In addition to working content in and about sciences, this didactic proposal is aimed at providing resources for the teacher to deal with conflicting issues between science and religion that occur in classrooms (Leal 2014). It is considered essential to prepare Basic School teachers to position themselves critically and respectfully in front of these discussions, regardless of their personal beliefs.

Religious conceptions or metaphysical ideas are part of the culture and they can be understood as social and historical knowledge. Reflective approaches about such content can promote the respect for diversity, as a way of approaching human rights in classrooms (Oliveira and Queiroz 2015). Linking to other subjects and texts would be possible, but it depends on the intention of the trainer, who will guide goals and discussions.

18.5 Final Considerations

This research is aimed at contributing to the contextualized teaching of physics concepts, expanding epistemological discussions in classrooms, for example, exemplifying how personal values of a thinker or group of thinkers, also have an impact on science. The historical episode discusses Isaac Newton (known as a model of rationality) harmonizing science, God, and alchemy in a neo-Platonic world view, aimed at complexifying the idea of the eternal conflict between science and religion, predominant in a common sense.

Regardless of the students' religious beliefs, it is considered essential that they appropriate themselves of the scientific culture of their time and understand aspects of how such knowledge was built. When we provide the experience of this content in which the students' religious or atheist options are respected, they can acquire elements to respect the beliefs and positions of their future students, always seeking a mediator and inclusive attitude toward different points of view.

The teaching strategies sought to encourage critical engagement of students, but only by analyzing the implementation results of the proposal will it be possible to have evidence for its potential around the established pedagogical goals. Thus, as an implication, we intend to present this proposal to teachers of different universities, to evaluate it in different classroom contexts.

Acknowledgements I thank CNPq and FAPESP for supporting various previous stages of this research.

References

Abd-El-Khalick, F. (2012). Nature of science in science education: Toward a coherent framework for synergistic research and development. In B. J. Fraser, K. G. Tobin, & C. J. McRobbie (Eds.), *Second international handbook of science education* (pp. 1041–1060). Dordrecht: Springer.

Acevedo-Díaz, J. A., & García-Carmona, A. (2016). Algo antiguo, algo nuevo, algo prestado: Tendencias sobre la naturaleza de la ciencia en la educación científica. *Revista Eureka sobre Enseñanza y Divulgación de las Ciencias, 13*(1), 3–19.

Adúriz-Bravo, A., & Bejarano, N. (2015). Finding criteria to support decisions on which philosophy of science is more useful for science teachers. Oral presentation. In *ESERA 2015 – 11th Conference of the European Science Education Research Association*. University of Helsinki, Finland, 31 August – 4 September.

Alfonso-Goldfarb, A. M. (2000). Uma suposta contradição na ciência inglesa do século XVII: Divulgação x sigilo. *Discurso, 31*, 347–363.

Allchin, D. (2011). Evaluating knowledge of the nature of (whole) science. *Science Education, 95*(3), 518–542.

Assis, A. K. T. (2006). Interações na física: Ação à distância *versus* ação por contato. In C. C. Silva (Ed.), *Estudos de História e Filosofia das Ciências: Subsídios para Aplicação no Ensino Silva* (pp. 87–102). São Paulo: Editora Livraria da Física.

Bagdonas, A., & Silva, C. C. (2015). Enhancing teachers' awareness about relations between science and religion. *Science & Education, 24*(9–10), 1173–1199.

Braga, M., Guerra, A., & Reis, J. C. (2008). O papel dos livros didáticos franceses do século XIX na construção de uma concepção dogmático-instrumental do ensino de física. *Caderno Brasileiro de Ensino de Física, 25*(3), 507–522.

Carvalho, A. M., & Gil-Pérez, D. (2001). *Formação de professores de ciências: Tendências e inovações*. São Paulo: Cortez.

Copenhaver, B. (1980). Jewish theologies of space in the scientific revolution: H. More, J. Raphson, I. Newton and their predecessors. *Annals of Science, 37*, 489–548.

Dagher, Z. R., & Erduran, S. (2016). Reconceptualizing the nature of science for science education: Why does it matter? *Science & Education, 25*, 147–164.

Forato, T. C. M. (2006a). Os "poderes ocultos" da matéria e a gravitação universal. *Scientific American Brasil. História: Os grandes erros da ciência, 6*, 38–43.

Forato, T. C. M. (2006b). Isaac Newton, as profecias bíblicas e a existência de Deus. In C. C. Silva (Ed.), *Estudos de História e Filosofia das Ciências: Subsídios para Aplicação no Ensino* (pp. 191–206). São Paulo: Editora Livraria da Física.

Forato, T. C. M. (2008). A filosofia mística e a doutrina newtoniana: uma discussão historiográfica. *Alexandria Revista de Educação em Ciência e Tecnologia, 1*(3), 29–53.

Forato, T. C. M. (2009). *A natureza da ciência como saber escolar: um estudo de caso a partir da história da luz* (2 vols.). Tese de Doutorado, Faculdade de Educação da Universidade de São Paulo, São Paulo, Brasil.

Forato, T. C. M., Bagdonas, A., & Testoni, L. (2017). Episódios históricos e natureza das ciências na formação de professores. *Enseñanza de las Ciências, digital, v. extra*, 3511–3516.

Forato, T. C. M., Pietrocola, M., & Martins, R. A. (2011). Historiografia e natureza da ciência na sala de aula. *Caderno Brasileiro de Ensino de Física, 28*(1), 27–59.

Forato, T. C. M., Martins, R. A., & Pietrocola, M. (2012). History and nature of science in high school: Building up parameters to guide educational materials and strategies. *Science & Education, 21*(5), 657–682.

Debus, A. (1978). *Man and nature in the Renaissance*. Cambridge: Cambridge University Press.

Dobbs, B. J. T. (1991). *The Janus faces of genius: The role of alchemy in Newton's thought*. Cambridge: Cambridge University Press.

Freire, P. (1996). *Pedagogia da autonomia. 42 reimp*. São Paulo: Paz e Terra.

Gil-Pérez, D., Montoro, I. F., Alis, J. C., Cachapuz, A., & Praia, J. (2001). Para uma imagem não deformada do trabalho científico. *Ciência & Educação, 7*(2), 125–153.

Höttecke, D., & Silva, C. C. (2011). Why implementing history and philosophy in school science education is a challenge: An analysis of obstacles. *Science & Education, 20*, 293–316.

Leal, K. P. (2014). *A influência da religião na ciência e da ciência na religião: O ponto de vista discente e a construção de elementos para um juízo de valor mais fundamentado*. Trabalho de Conclusão de Curso, Centro Federal de Educação Tecnológica Celso Suckow da Fonseca, Petrópolis, RJ, Brasil.

Lederman, N. G. (2007). Nature of science: Past, present, and future. In S. K. Abell & N. G. Lederman (Eds.), *Handbook of research on science education* (pp. 831–880). Mahwah: Lawrence Erlbaum Associates.

Manuel, F. (1974). *The religion of Isaac Newton*. New York: Oxford University Press.

Martins, R. A. (1989). Huygens e a gravitação newtoniana. *Cadernos de História e Filosofia da Ciência [Série 2], 1*(2), 151–184.

Martins, R. A. (1993). Huygens reaction to the Newton's gravitational theory. In J. Field & F. A. J. L. James (Eds.), *Renaissance and revolution: Humanists, scholars, craftsmen and natural philosophers in early modern Europe* (pp. 203–213). Cambridge: Cambridge University Press.

Martins, R. A. (1998). Descartes e a impossibilidade de ações à distância. In S. Fuks (Ed.), *Descartes 400 anos: Um legado científico e filosófico* (pp. 79–126). Rio de Janeiro: Relume Dumará.

Martins, R. A. (1999). O que é a ciência vista da epistemologia? *Caderno de Metodologia e Técnica de Pesquisa, 9*, 5–20.

Martins, R. A. (2017). O estudo experimental sobre o magnetismo na Idade Média, com uma tradução da carta sobre o magneto de Petrus Peregrinus. *Revista Brasileira de Ensino de Física, 39*(1), e1601. https://doi.org/10.1590/1806-9126-RBEF-2016-0181.

Martins, R. A. (2006). A maçã de Newton: história, lendas e tolices. In C. C. Silva (Ed.), *Estudos de História e Filosofia das Ciências: Subsídios para Aplicação no Ensino Silva* (pp. 167–189). São Paulo: Editora Livraria da Física.

Matthews, M. (2012). Changing the focus: From nature of science to features of science. In M. S. Khine (Ed.), *Advances in nature of science research* (pp. 3–26). Dordrecht: Springer.

Mcguire, J. E., & Rattansi, P. M. (1966). Newton and the 'pipes of pan'. *Notes and Records of Royal Society, 21*, 108–126.

Newton, I. ([1713] 1952). *Mathematical principles of natural philosophy. Optics* (Trans: Andrew, M. [2a.ed]. Chicago: Encyclopaedia Britannica (Great Books of Western World, v. 34).

Newton, I. (1733). *Observations upon the prophecies of Daniel, and the Apocalypse of St. John*. London: J. Darby and T. Browne.

Oliveira, R. D. V. L., & Queiroz, G. R. P. C. (2015). *Olhares sobre a (in)diferença: formar-se professor de ciências a partir de uma perspectiva de educação em direitos humanos*. São Paulo: Editora Livraria da Física.

Pozo, J. I., & Crespo, M. Á. G. (2009). *A aprendizagem e o ensino de ciências: do conhecimento cotidiano ao conhecimento científico*. Porto Alegre: Artmed.

Rattansi, P. M. (1988). Newton and the wisdom of the ancients. In J. Flauvel, R. Flood, M. Shortland, & R. Wilson (Eds.), *Let Newton Be!* (pp. 185–210). New York: Oxford University Press.

Rudge, D. W., & Howe, E. (2009). An explicit and reflective approach to the use of history to promote understanding of the nature of science. *Science & Education, 18*, 561–580.

Westfall, R. (1987). Newton and the Christianity. In F. T. Birtel (Ed.), *Religion, science, and public policy* (pp. 79–94). New York: Crossroads.

Yáñez, X. D., & Maturana, H. (2009). Hacia una era post moderna en las comunidades educativas. *Revista Iberoamericana de Educación, 49*, 135–161.

Chapter 19
The Enlightenment *Paideia*: The French Origins of Modern Science Teaching

Marco Braga, Andreia Guerra, and José Claudio Reis

19.1 Introduction

When we take part in a culture, we usually do not think about it too deeply because everything seems natural for us. We believe that all things have always been this way thus far and will always be this way.

Schools have a culture composed of practices, content, and rituals. All actions are guided by philosophical principles embedded in the organization of classrooms, such as how classes are conducted and how choices of curriculum are considered important. Science teaching culture, in the same way, works by assessing and determining what constitutes a good teacher or what the right way to teach is (Hottecke and Silva 2011). The model of a good science teacher considered by traditional education is someone who teaches the truth about science; hence, there is no space for questions and discussions. In this way, the didactic culture of science teaching could be an obstacle to implementing pedagogical practices in an historical–philosophical approach. The historical–philosophical approach, in general, brings to science classes questions and discussions about science and its development throughout history. Based on these considerations, it is possible to argue that it is important to reflect on science teaching culture. The problem is that the principles that guide the science teaching culture are hidden and not apparent in each small action or in material elements of a classroom and school. Everything appears natural for both teachers and students.

However, it is important to understand that this science teaching culture has an origin. Throughout history, all societies have designed forms of education with

M. Braga (✉) · A. Guerra
Federal Center for Technological Education (CEFET), Rio de Janeiro, RJ, Brazil
e-mail: marcobraga@namelab.education

J. C. Reis
State University of Rio de Janeiro, Rio de Janeiro, RJ, Brazil

M. E. B. Prestes, C. C. Silva (eds.), *Teaching Science with Context*, Science: Philosophy, History and Education,
https://doi.org/10.1007/978-3-319-74036-2_19

inherent philosophical principles, beliefs, and assumptions (Matthews 2009a). Therefore, we are heirs of concepts about science teaching created in the past that have often been reproduced for centuries.

It is a task for philosophers of education to examine and question what these principles are and where they are manifested today.

In the case of science teaching, there are questions that should be posed: what are the principles guiding the project of current science teaching? When were they created and for whom? What are the human and social concepts that have guided this teaching? And the last question: How has the project carried on reproducing science teaching practices?

This is not a simple task. But some clues can be followed that involve history, philosophy of science, and philosophy of education (Schulz 2014).

Modern science established its hegemony in the eighteenth century, at the same time as the Enlightenment emerged as a cultural movement, spreading new ideas about human nature and society across Europe (Clark et al. 1999). Sometime later, Enlightenment principles were exported to other regions of the globe. These ideas had a great influence both on education and on the construction of scientific knowledge. Therefore, all paths converged at this time.

The Enlightenment involved constructing a new project of civilization opposing the past, considered by eighteenth century men as the Dark Ages in Europe. The entire period after the fall of the Roman Empire until the dawn of the named scientific revolution of the sixteenth and seventeenth centuries was considered a historical period based solely on religious grounds. Many historians (Benoit and Micheau 1995; Rossi 2001) have denounced this view as reductionist. For example, this historical period occurred during the height of Islamic culture presence in Europe, which developed numerous contributions to science, astronomy, and mathematics, partly in the Iberian Peninsula (Benoit and Micheau 1995). Even after the Christian reconquest of the territory, the new ideas continued to spread to other European regions with the creation of universities.

The Enlightenment philosophers believed that Europe was experiencing a special moment, different from the medieval times. Therefore, for them it was necessary to create an opposition between these two historical moments, presenting modern times as an illuminated era and the past as darkness. In this way, the Enlightenment philosophers called the past the Dark Ages, associating that age with metaphysical themes.

The Enlightenment was a cultural movement and its construction involved establishing new social parameters. The role of human nature in society was at the center of this project. There was a newfound confidence that humans could choose their own fate and create an ideal society in their own image. Therefore, there was a great deal of discussion and debate as to what vision of society and man should be pursued. In this respect, education became the most important issue of this project. The Enlightenment philosophers drew inspiration from the Greek concept of *paideia* (Jaeger 1965).

19.2 From Greek Paideia to Christian Paideia

In the 1920s and 1930s, the German philosopher Werner Jaeger (1888–1961) extensively researched the Greek concept of "*paideia*" (Jaeger 1965). He was living at a crucial moment in Germany where he believed that values were decaying; hence, his comprehensive study aimed at rescuing the values of Greek civilization. Jaeger can help us to clarify the more important phases of transformation of the concept of *paideia*. This concept could help to understand the birth of science education in the eighteenth century.

The Ancient Greeks developed a concept of human excellence exemplified by the *paideia* (παιδεία). *Paideia* is a word without translation in most languages, yet is translated as "formation," "civilization," and "tradition" in some. This word should not be understood solely in the modern sense of development. The meaning of *paideia* gathers many connotations, but is more than simply the acquisition of knowledge, skills, and competencies. It refers to higher level of human existence. *Paideia* refers to the process of upbringing, enculturation, and education that results in the formation of a new being.

Paideia refers to forming a being in the rough, like the work of a sculptor carving a rough stone to form an image that will be the most beautiful possible.

Another Greek word that can further illuminate this concept is *arete* (ἀρετή). Many scholars translate this word as "virtue." In this sense, the virtuous man is the one who realizes his own destiny. For example, in war, the hero reflects the virtue of heroism by not being intimidated in the face of danger. In Homeric times (850 BC), stories such as the *Iliad* and the *Odyssey* display various dimensions of *arete* for all society, expressing a powerful concrete form of *paideia*. Early on, *arete* was considered solely a concern and trait of the aristocracy; however, through the poetry of Hesiod (650 BC), *arete* changed its face, and the hard work of farmers came to be considered a virtue as well. The hero was not only the aristocratic warrior fighting in battles, but also the ordinary working man toiling day by day.

At this time, education was not considered a means of developing virtues. Vocational training, passed from father to son, taught only necessary work skills, and was not a force for modeling the spirit. In Athens, with the rise of the Sophists, as teachers of free citizens, education became the most important tool for cultivating virtues because knowledge gained an importance that it did not have previously.

For Plato, this change would allow *paideia* to be raised to new heights. *Arete* is linked with knowledge, but it is not determined solely by the quantity of knowledge acquired. Only with Socrates did *paideia* develop to its most mature state, as philosophy became the core of this final concept. Philosophy is the means of reflecting on concepts to determine their veracity. A philosopher is someone who questions all possibilities and decides the best way to go forward. This act is the virtue that generates all others.

Plato argued that all leaders should be trained as philosophers because they are the only people who have *arete*, and therefore, the skills to govern.

The concept of *paideia* was shown to be central to understanding all philosophy of education in subsequent years. Jaeger (1961) published a study reviewing the difficulties that the first Christians had in presenting their message to the Greco-Roman world. As they strove to create a new concept of Paideia to implement this project, the symbiosis of Christianity with ancient Greek philosophy became an important step in this direction (Hooykaas 1988).

As it was necessary to adapt Platonism and Aristotelianism to Christianity, the union of faith and reason was the basis of this new model. Virtue came to be considered as the process of drawing from one's inner light to know the Truth (Augustine of Hippo 1953) and dedicating one's life to acquiring a place in the Heaven.

For the European peasants, the life of saints became the model they emulated. The stories of the saints were told orally aiming to show the virtues that every good Christian should have. Many of these stories were modified to highlight these virtues. The first biographies of saints were very different from the stories told for the people years later. The legends focused on human elements in the lives of saints to highlight their Christian virtues, which represented them as semi-gods. The process worked the same way for the ancient Greek concept of *paideia* with the stories of Homer and Hesiod.

19.3 The Enlightenment *Paideia*

The Enlightenment was a cultural movement that joined science, philosophy, literature, painting, politics, and other expressions of human culture (Cassirer 1968; Clark et al. 1999). In philosophical studies, it is not possible to find a doctrine that could characterize it. The Enlightenment comprised various intertwining threads forming a broad fabric. The most important element uniting these threads was the rejection of metaphysics. This rejection was gradually expressed throughout the eighteenth century. The first step involved expurgating inherited elements from the religious past, although the early enlightenment thinkers did not deny God. They denied the so-called revelation, the basis of the three most important monotheisms. For these thinkers, known as Deists, there was a God who created the world, but this God never talked directly with humans about this project.

Three philosophers made important contributions to science in early Enlightenment: Descartes, Newton, and Leibniz. Although there were differences among the thought of these three philosophers and their studies cannot be considered strictly part of the Enlightenment, many elements of their studies were valued by early Enlightenment thinkers (Clark et al. 1999).

Some theories that were accepted in the eighteenth century had a deist basis. For example, the construction of the momentum conservation principle was derived from this vision of nature: when creating the world, God gave movement to all elements of nature. Based on this principle, from this time forward, transformations have happened, but all movement was conserved. Both Descartes and Leibniz were adherents to this view. Leibniz proposed that God put all his intentions in the world when creating it (Leibniz 1979b). As a result, this world is not a random version, but the best possible world that could exist. This vision led to the concept of "laws of nature." Leibniz proposed that God had imbued some matter with a vital principle, which was the distinguishing feature between living beings and inanimate objects (Leibniz 1979a). This principle was not well accepted in Enlightenment as it was considered metaphysical, because there are no means of experimentally proving its existence. Descartes' materialism was considered more acceptable as it proposed that matter only existed in bodies and movement, even though the question of the existence of life was not fully explained.

From the second half of the eighteenth century onward, many deist philosophers such as La Mettrie and Buffon began to deny this view and to change their position to a full materialist view. The most important challenge for these thinkers was to explain the creation of life. At this time, for example, theories of spontaneous generation emerged (Hankins 1998).

The materialist philosophers denied all religious and alchemical texts of Newton, trying to protect the myth that Enlightenment philosophers, such as Voltaire, had transformed Newton. The rational side of Newton embodied all the features that the enlightenment movement expected from the new man dedicated to science.

In the *Mathematical principles of natural philosophy*, Newton draws on four principles to guide the new science (Newton 1962). These principles were named rules of reasoning in philosophy.

Rule 1. We are to admit no more causes of natural things, than such as are both true and sufficient to explain their appearances (Newton 1962, p. 398).

Rule 2. Therefore, to the same natural effects we must, as far as possible, assign the same causes (Newton 1962, p. 398).

Rule 3. The qualities of bodies, which admit neither intension nor remission of degrees, and which are found to belong to all bodies within reach of our experiments, are to be esteemed the universal qualities of all bodies whatsoever (Newton 1962, p. 398).

Rule 4. In experimental philosophy we are to look upon propositions collected by general induction from phenomena as accurately or very nearly true, notwithstanding any contrary hypotheses that may be imagined, till such time as other phenomena occur, by which they may either be made more accurate, or liable to exceptions (Newton 1962, p. 400).

These principles call for the construction of a program of research and the fourth principle stresses that it should deny any metaphysical elements in the construction of scientific knowledge. The Enlightenment movement embraced this program and elected Newton as the model of the new man of science (Hankins 1998).

Despite the fact that Newton had proposed this new program, he did not always practice what he preached. Many propositions of his theory of universal gravitation had metaphysical elements (Burtt 2003). Many natural philosophers in the eighteenth century considered the most important element of Theory of Universal gravitation, known as action at a distance, metaphysical, because it was not possible to experimentally prove its existence. During the eighteenth century, most of Newton's critics denounced this element as bearing a resemblance to the alchemical concept of affinity (Fanning 2009).

Actually, many historians considered that Newton had been influenced a great deal by alchemy and the Christian world view as he had read several works by alchemists and wrote more texts about religion than about science. One of the most important debates in the early eighteenth century was the correspondence between Leibniz and Clark about many elements of Newton's theory. In one of these letters, Leibniz denounced the concept of "Sensorium Dei" as a prison for God in three dimensions (Leibniz 1979b). Clark, friend and defender of Newton's ideas, used arguments to reject these criticisms (Leibniz 1979b). This debate revealed metaphysical elements considered by the Enlightenment movement to be without merit, knowledge from past eras that had to be eliminated. These metaphysical elements would have tarnished Newton's image as a role model for the new man of science. Nevertheless, this problem was not exclusive to Newton as all of the founders of modern science also had metaphysical elements in their work. To consolidate the Enlightenment movement, it was necessary to purge and erase all religious and metaphysical influences from the texts of the early founders of science.

The process of cleansing the discourse of its metaphysical elements was linked with the education of new generations. The first step focused on the education of new natural philosophers and the second was linked with the popularization of this new world view.

How should new generations of natural philosophers be educated? Generally, young philosophers learn the fundamentals of their area by reading the writings of their predecessors. They received this heritage from the past and proposed new ideas by maintaining a constant dialogue with these writings. However, in the process of constructing a new science, a different approach was needed, as it was necessary to erase the metaphysical elements from the writings of the founders of science. An uncritical reading of the founders could allow the new generation to adopt and believe in metaphysical elements. This fact generated a wave of treatises and books by scholars exposing only the empirical core of the theories, free of metaphysical elements. It was not necessary to discuss the process of constructing theory, only to present the results. The new focus was on the "laws of nature" (Braga 1999).

In many cases, it was necessary to create a new language. In 1789, in the *Traité Elementaire de Chimie* or *Elements of Chemistry* (1990), Lavoisier proposed a new nomenclature, differing from that used by the alchemists (Hankins 1998). After some years, the newly trained chemists did not understand the alchemical manuals, and soon they were forgotten entirely.

At the end of the eighteenth century, upcoming generations of chemists did not read the original writings of the founders of the new science; instead, they studied their science in textbooks referred to as treatises, which summarized the original works of the founders.

For the general population, many books were written to disseminate the new ideas of science. In France, Voltaire played a very important role in the process. Besides being a philosopher, he also wrote novels. Voltaire studied Newton's work in a period of exile in London, and when he returned to Paris in 1738, he wrote a highly influential book called *The elements of Sir Isaac Newton's philosophy* (Voltaire 1991). His intention was to propagate Newton's thought in France (Casini 1995). Voltaire also wrote *Candide, or optimism* in 1758, a novel in which he ridicules the metaphysical idea of Leibniz, that this is the best of all possible worlds.

Another important innovation in spreading this new worldview was the introduction and circulation of encyclopedias. The name "encyclopedia" was based on the idea of creating a circle of *paideia* (έγκυκλοπαιδεία) or a circle in which all people could receive education. All encyclopedia entries were aimed at being accessible to all people. However, the intention was to introduce all knowledge from an Enlightenment perspective. *The Chambers cyclopaedia: a universal dictionary of arts and sciences*, published in 1728, was one of the first encyclopedias aimed at distributing scientific knowledge from an Enlightenment perspective. Jean D'Alembert and Denis Diderot organized and published the *Enciclopédie*, a French version issued throughout the entire eighteenth century.

Both in the field of training of new natural philosophers and for promulgating this new world view, the Enlightenment project grew strong. By the late eighteenth century, much of the European population had been exposed to the foundations of the new science. New schools of thought began to formulate ideas in philosophical terms, giving rise to positivism from Auguste Comte to the Vienna Circle. However, it was still necessary to deepen the project. The case of France was the most representative because it was born of a Revolution that sought to transform the entire society through new parameters. In this process, new institutions were created towards the formation of a new human being.

19.4 Science As Practical Knowledge

From the end of the Middle Ages onward, technology became profoundly important in Europe. In Antiquity, manual labor was denigrated by most societies as the activity of slaves. With the advent of the Modern Age, this view changed in many societies. Technical work began to be seen as a producer of wealth and an essential tool for transforming society. Modern science acquired many characteristics of this work, such as experimentation. It became fundamentally a practical knowledge because natural philosophy joined with practical work to create technology. More than simply knowing nature, it was necessary to change nature. In the early years of capitalism, scientific knowledge became a valuable tool for gaining profit.

The eighteenth century European society witnessed great transformations in this sense. The First Industrial Revolution relied on mechanics who knew little of scientific theories. Only nearly one century later, the theoretical foundations of thermodynamics provided an explanation of how thermal machines worked. The empirical knowledge was independent from this theory. When the electrical motors began to be introduced in the industry in the mid-nineteenth century, this logic was changed. The electrical motors were a direct consequence of the development of a theory and their transformation into a commercial device was very quick. For this change to occur, a transformation was required to train men to work with these new theories and techniques. The workers who knew practical mechanics became engineers when the theory was included in their training.

Many changes and initiatives were necessary to facilitate this transformation in many countries of Europe. In France, the movement was based on government support; yet, it was not until the French Revolution that this support flourished. The founding of the *École Polytechnique de Paris*, in 1793, was the beginning. There were many engineering schools that provided practical training to students (Braga 1999). Many mathematicians and physicists who passionately took part in the revolution believed in the role of theory in this training. They proposed combining mathematical teaching with practical training.

In contrast, British scientists tried to integrate new science ideas with the everyday work world. The Royal Society played an important role in this process, supporting the studies of free thinkers such as John Theophilus Desaguilier or John Banks. They taught the foundations of Newtonian mechanics to workers, entrepreneurs and some people curious about the new principles of mechanics and published a related textbook (Soares 2009). From this action, Newtonism began spreading throughout England during the Industrial Revolution in the cities and factories. This movement was not a government initiative. It was a movement based on the actions of free citizens who supported the Royal Society. Thus, initially, there was no government policy supporting education.

The French Revolution represented a unique historical moment in education because times of radical change and upheaval were far more conducive to experimenting with new ideas. The institutions and practices of the old order were shaken. Many revolutionaries were natural philosophers and this was a crucial factor supporting them to establish a new educational system with a cutting-edge science education. The new science teaching developed in higher education was transposed to elementary education with modifications. Despite the content being simplified and adapted to the lower grades, the basic principles of the philosophy of education were the same (Braga et al. 2011). The understanding of this philosophy is the understanding of the roots of our own current science teaching.

19.5 Enlightenment *Paideia* in Practice

In this work, we use the term "French roots" because when applying the idea of an Enlightenment *paideia* to the concrete practice of teaching, each of the countries that were at the forefront of the movement defined their own model. In this study, we take only one of these paths, the French project. This is because of its replication in several peripheral nations, such as Brazil, that were heavily based on French culture. Other roots of great importance such as the British, for example, are not addressed in this study. Yet, it should be noted that these models are not composed of independent paths, as there were dialogues and certain characteristics common to them all.

The French way is referred here as the *Polytechnique* model because it was in the *École Polytechnique* where this concept of teaching was first developed. *École Polytechnique* was a technological university founded in Paris in 1793 by natural philosophers who had taken part in the French Revolution. Engineering training was provided by specialized institutions such as the *École des Mines* and *École de Ponts et Chaussés* that had practical curricula. The goal of the *École Polytechnique* was to present the foundations of mathematics, physics, and chemistry applied to practical problems to future engineers (Miquel 1994). The French natural philosophers wanted to provide theoretical knowledge for a profession that historically worked only with practical knowledge. This was the first step in creating an institution that taught the core of modern technology.

Over the years following its founding, the *École Polytechnique* became an institution aimed at providing more than scientific knowledge for future engineers. Modern science brought a new view in which nature was considered to be like a machine. Nature was no longer viewed as enchanted or influenced by magical forces as in medieval times (Hankins 1998). A machine model of the world allowed parameters to be calculated and future events to be determined. Over time, society also began to be compared with the workings of a machine. The increasing rationality imposed on social relations required leaders who understood this rationality. Therefore, engineers with scientific training came to be seen as model leaders. In the Greek *paideia*, the philosopher was a model leader. In the new Enlightenment *paideia*, the leaders would be the engineers. At the turn of the nineteenth century, the *École Polytechnique* began changing from a university focused on engineering education to an institution focused on training leaders (Belhoste et al. 1995; Adler 1999). In this new kind of specialized university, a new model of education developed based on turning out social leaders and a specific kind of science teaching (Miquel 1994). The *Polytechnique* model spread to many countries during the nineteenth century.

The Enlightenment *paideia* manifested in a concrete form in the *Polytechnique* model (Braga 1999). During the nineteenth century, other specialized universities were gaining importance and formed a group known as the *Grandes Écoles*. Nevertheless, the *École Polytechnique* was still very important because, in its first decades of existence, new ways of science teaching were tested. At the *École Polytechnique*, an educational philosophy was created for science education. This philosophy was exported to other education grade levels along with appropriate modifications of contents (Braga 1999).

We argue that this new philosophy of science education is particular to the French education system, but during the latter decades, this model was popular in many countries, not only in France. Although there are characteristics unique to each country, such as England and Germany, several features of this model have been absorbed. Today, worldwide, a student would have no trouble recognizing the style of a science class in any country.

19.6 Characteristics of a New Philosophy of Science Education

The new philosophy of science teaching developed during the first decades of the nineteenth century in the *École Polytechnique* had certain critical features, which are considered in detail below.

19.6.1 Dogmatic Characteristic

One the most important sources of information elaborating on the philosophy of science teaching from the *Polytechnique* model are the textbooks issued in the first decades of the nineteenth century. Two textbooks were chosen for the study because of the number of editions they had throughout the nineteenth century (Duhamel 1853; Lamé 1840). These textbooks were used in several countries that have adopted the polytechnic model, such as Brazil (Braga et al. 2011). These textbooks present minor differences in structure, but, essentially, shared the same philosophical principles, beliefs, and assumptions.

The first interesting characteristic of these principles was the way in which they presented science content. The textbooks followed the same pattern of the eighteenth century treatises (Braga et al. 2011). The authors of these textbooks transcribed revised versions of treatises, which had removed metaphysical references in early scientists' studies.

It is possible to find in all textbooks the presentation of laws and theories as being the result of immutable pure knowledge. These laws and theories were presented as discoveries of a single person and not as results of collective work. These discoveries were not constructions of the human mind, as according to this perspective, the person who discovers a law reveals a truth that exists in pure form in nature (Braga 1999).

Therefore, from this perspective, science is not considered a construction. Constructions imply that metaphysical elements are involved in the process. The presentation of science in these textbooks does not consider this perception.

These features anticipated the ideas proposed by Hans Reinchenbach (1938) by nearly a hundred years, that is to separate the context of discovery from the context of justification, believing that the former could still contain metaphysical elements but not the latter. To guarantee validity, the presentation of a theory should be made only through the context of justification.

Auguste Comte (1988) also believed that science teaching could be implemented in two ways. The first was the historical way and the second was the dogmatic way. He recognized the importance of the historical way, but this way is fraught with controversy. He believed that studying the construction of scientific knowledge would be a waste of time. The dogmatic way would be more efficient.

Both Comte and Reinchenbach rejected the process (the discovery context for one and the historical way for another) because it would be necessary to discuss metaphysical elements in science.

19.6.2 Instrumental Characteristic

The fact that the model was developed in an engineering school marked the second characteristic. The goal of an engineer's training is not to understand nature, but rather to apply scientific knowledge to practical problems. This fact explains the emphasis on presenting laws that could be applied quickly to solve practical problems. The engineers did not need to know the process of constructing theories.

In this respect, the three laws of Newton were considered more important than the law of gravity. Despite these four laws being linked, the application of three classical laws in problems linked to everyday work is more direct; hence, they were more valuable.

Dogmatic and instrumental characteristics marked the *Polytechnique* model, which was in harmony with the objectives of education for a new kind of engineer, a professional who should be less of a craftsman and more of a technological master, mixing practical and theoretical knowledge. The model was also in harmony with the project of the education for a new kind of leader: a pragmatic person who when facing problems of society should use technical rationality.

For Plato, the new leader should be a philosopher, a man who solves the problems of the *polis* by reason. In the Enlightenment *paideia*, technical rationality supplants reasoning, as society is considered similar to a machine. The engineer is the new leader, supplanting the philosopher.

19.7 Final Remarks

When science teaching began to gain importance in elementary schools, the *Polytechnique* was fundamental to structuring a new educational model. Despite the content being simplified, the educational philosophy embedded in it was preserved. The rejection of metaphysics resulted in the rejection of epistemology as well. To avoid metaphysical references, science teaching had to abandon discussions of the process of constructing theories in favor of solely presenting its products (Matthews 2009b).

This fact had dire consequences for education. When students were faced with the products without considering the process, they understood theories and laws as discoveries of an immutable truth in nature. For them, these laws and theories exist purely in nature, and scientists have the power to remove the veil that covers the truth of their nature. This perception empowers scientists.

The French historian of science Pierre Thuillier provided an important insight in an interview:

> Science is seen in society as an absolute instance, exactly how God is seen by the Church. As said the priests who burned heretics in the Inquisition: "I do not have a choice, it is what God wants", so our technocrats, when making decisions, say they are not responsible, but science is. (Thuillier 1989, p. 20)

When students study science as a construction, they can follow the epistemological process, in which there are controversies and discussions about experiments or phenomena. This fact reinforces the human character of scientific activity. This is what Thomas Kuhn had in mind when he noted that textbooks provide an idea of science similar to that provided by tourist brochures about the culture of a country (Kuhn 1996).

The Enlightenment *paideia* proposed a concept of nature, science, education, society, truth, and humanity. In the twenty-first century, these concepts could not be more different. Society changed, and the ideals of the Enlightenment were called into question throughout history. However, the Enlightenment *paideia* remains part of the heritage of science education nowadays. Contemporary educational practices and culture of science teaching have certain characteristics that originated in the Enlightenment *paideia*. Studying the characteristics of this heritage is essential to question the educational culture of science disciplines, and thus to discuss the role of history and philosophy in science teaching. These analyses indicate the importance of philosophy in discussing science education. History, philosophy, and, in particular, the philosophy of science and the philosophy of science education should not be ignored by teachers and researchers (Schulz 2014). It is important that they consider the study of these four disciplines to understand their day-to-day practice.

References

Adler, K. (1999). French engineers become professionals; or, how meritocracy made knowledge objective. In W. Clark, J. Golinski, & S. Schaffer (Eds.), *The science in enlightened Europe* (pp. 94–12). Chicago: The University of Chicago Press.

Augustine of Hippo. (1953). *Of true religion in Augustine: earlier writings*. Philadelphia: The Westminster Press.

Belhoste, B., Dalmedico, A. D., Pestre, D., & Picon, A. (1995). *La France des X: Deux siècles d'histoire*. Paris: Economica/EDF.

Benoit, P., & Micheau, F. (1995). Arab intermediary? In M. Serres (Ed.), *A history of scientific thought: Elements of a history of science* (pp. 160–190). Cambridge: Blackwell Publisher.

Braga, M. (1999). *A nova Paidéia: A ciência como princípio educativo na formação do homo-industrialis*. Tese de Doutorado, Universidade Federal do Rio de Janeiro, Rio de Janeiro, RJ, Brasil.

Braga, M., Guerra, A., Reis, J. C. (2011). The legacy of first positivistic textbooks to science education. In *11th International IHPST: Science & Culture, (Tessaloniki), Epikentro*, Athens, Greece.

Burtt, E. (2003). *The metaphysical foundations of modern science*. New York: Dover.

Casini, P. (1995). *Newton e a consciência europeia*. São Paulo: Editora UNESP.

Cassirer, E. (1968). *The philosophy of enlightenment*. Princeton: Princeton University Press.
Clark, W., Golinski, J., & Schaffer, S. (1999). Introduction. In W. Clark, J. Golinski, & S. Schaffer (Eds.), *The sciences in enlightened Europe* (pp. 3–31). Chicago: The University of Chicago Press.
Comte, A. (1988). *Introduction to positive philosophy*. Cambridge: Hackett Publishing Company.
Duhamel, J. M. C. (1853). *Cours de mechanique de L'École Polytechnique*. Paris: Bechelier.
Fanning, P. A. (2009). *Isaac Newton and the transmutation of alchemy: An alternative view of the scientific revolution*. Berkeley: North Atlantic Books.
Hankins, T. (1998). *Science and the enlightenment*. Cambridge: Cambridge University Press.
Hooykaas, R. (1988). *A religião e o desenvolvimento da ciência moderna*. Brasília: Editora da UnB.
Hottecke, D., & Silva, C. C. (2011). Why implementing history and philosophy in school science education is a challenge: An analysis of obstacles. *Science & Education, 20*, 293–316.
Jaeger, W. (1961). *Early Christianity and the Greek Paideia*. Cambridge: Belknap Press of Harvard University Press.
Jaeger, W. (1965). *Paideia: The ideal of Greek culture* (Vol. 3). New York: Oxford University Press.
Kuhn, T. (1996). *The structure of scientific revolutions*. Chicago: University of Chicago Press.
Lamé, G. (1840). *Cour de physique de L'École Polytechnique*. Paris: Bachelier.
Lavoisier, A. (1990). *Elements of chemistry*. Chicago: The University of Chicago.
Leibniz, G. (1979a). Discurso da metafísica. In *Leibniz* (pp. 75–110). São Paulo: Abril Cultural.
Leibniz, G. (1979b). Correspondência com Clark. In *Leibniz*, 403–468. São Paulo: Abril Cultural.
Matthews, M. (2009a). Science and worldviews in the classroom: Joseph Priestley and the photosynthesis in science. In M. Matthews (Ed.), *Worldviews and education* (pp. 271–302). Dordrecht: Springer.
Matthews, M. (2009b). Teaching the philosophical and worldview components of science in science. *Science & Education, 18*, 697–728.
Miquel, P. (1994). *Les polytechniciens*. Paris: Plo.
Newton, I. (1962). *Sir Isaac Newton's mathematical principles of natural philosophy and his system of the world, vol 2: The system of the world.* 1st ed. based in translation by Andrew Motte in 1st ed 1729. Berkeley: University of California Press.
Reichenbach, H. (1938). *Experience and prediction: An analyzes of the foundations and the structure of knowledge*. Chicago: University of Chicago Press.
Rossi, P. (2001). *O nascimento da ciência moderna na Europa*. Bauru: EDUSC.
Schulz, R. (2014). Philosophy of education and science education: A vital but underdeveloped relationship. In M. Matthews (Ed.), *International handbook of research in history, philosophy and science teaching* (pp. 1259–1316). Dordrecht: Springer.
Soares, L. C. (2009). John Theophilus Desaguilier: um newtoniano entre a patronagem e o mercado. *Revista Brasileira de História da Ciência, 2*, 82–95.
Thuillier, P. (1989). O contexto cultural da ciência. *Ciência Hoje, 9*, 18–23.
Voltaire. (1991). *The elements of Sir Isaac Newton's philosophy (1738)*. Birmingham: Gryphon Editions.

Chapter 20
Design and Implementation of a Lesson Plan for High School Students: A Case Study with Oersted's Experiment

Ana Paula Bispo da Silva, José Antonio Ferreira Pinto, and Éwerton Jéferson Barbosa Ferreira

20.1 Introduction

Although the inclusion of the history of science (HOS) in science education has been considered to potentially improve students' knowledge of science for decades, many challenges to this approach lie ahead in teaching practice. Remarkably, in Brazil, where schools do not have an ideal physical structure and many students have social problems, teachers face several obstacles to implementing the teaching of HOS in their classes. Many of these obstacles, such as the difficulty of finding HOS teaching material and using instructional design principles for HOS lessons, are similar to those found internationally (Höttecke and Silva 2011; Höttecke and Henke 2015).

In this work, it is intended to address this lack of historical studies that could better inform science teaching by considering an important historical episode accompanied by hands-on activities. For this purpose, we chose the Oersted experiment with the compass of 1820 and prepared a lesson plan that would enable students to explore it from three points of view: the historical, the epistemological, and the experimental. The methodological aspects considered were historiographical research based on primary and secondary sources, followed by educational research grounded in contextualized and inquiry-based laboratory teaching (Rosa and Rosa

A. P. B. da Silva (✉)
Department of Physics, State University of Paraiba, Campina Grande, PB, Brazil

J. A. F. Pinto
State Elementary and Secondary School of the State of Paraiba, Department of Physics, State University of Paraiba, Campina Grande, PB, Brazil

É. J. B. Ferreira
State University of Paraiba, Campina Grande, PB, Brazil

M. E. B. Prestes, C. C. Silva (eds.), *Teaching Science with Context*, Science: Philosophy, History and Education,
https://doi.org/10.1007/978-3-319-74036-2_20

2012; McComas 2005). One of the authors of this paper (Pinto) is one of the teachers who planned and implemented the lesson plan. We also considered the possibility of exploring the historical episode from an epistemological point of view. Thus, the historical material prepared promotes the discussion of aspects of the nature of science, such as the fact of observations to be theory-laden. Another aspect that this historical episode allows students to explore is that scientific reasoning cannot be compelling without appealing to social, moral, spiritual, and cultural resources, contexts, beliefs, and values.

Our hypothesis was that the combination of historical context, experimental activities, and an inquiry-based laboratory approach might enable the students to develop a comprehension of scientific practice and science as a cultural activity, and to learn electromagnetism.

20.2 Historical Episode: First Steps in the Study of Electromagnetism

During the nineteenth century, the relationship between electricity and magnetism was a main topic of study among natural philosophers. After Volta's invention of the pile, more experiments were designed and conducted to improve comprehension of the nature of electricity. Different conceptions of electricity existed simultaneously: they involved one fluid, as among Benjamin Franklin (1706–1790) and his followers; or two fluids, as among Charles Du Fay (1698–1739) and others. Magnetism was also an object of study. Some philosophers ascribed magnetic phenomena to a fluid, but the resemblance to electricity did not hold because the north and south poles could not be examined independently (Heilbron 1979, pp. 431–448, Darrigol 2000, pp. 1–5). Around the 1820s, there was no consensus on the nature of electricity or magnetism, because both phenomena could be explained by different theories and could reach equal results.

However, reports showing that lightning provoked problems with compasses during storms suggested that some connection might exist between electricity and magnetism (Darrigol 2000, p. 3). In 1813, Hans Christian Oersted (1777–1851), a Danish natural philosopher, professor in the University of Copenhagen, predicted a relationship between electricity and magnetism. He based this relationship on his conceptions of Nature and the existence of a unique Force, which ordered all physical phenomena (Williams 2008). However, the contemporary academicians did not consider his speculations. By the end of 1819, during his lectures, Oersted observed that the needle of a compass seemed to move when an electrical current passed through a near coil. He made several observations, trying to understand the nature of the phenomena. Indeed, at the time of this observation, the effect of electricity on compasses was expected. However, there was a relationship with the symmetry of the effect, because the linear movement of the electrical current seemed to provoke a circular movement in the needle. This conclusion was not intuitive. The simplest explanation involved linear symmetry, as maintained by philosophers who supported

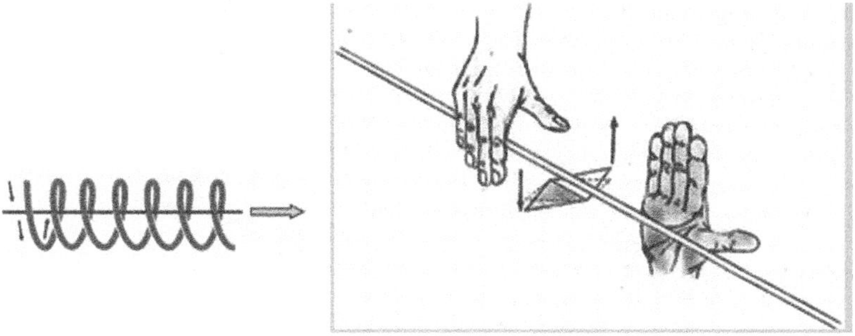

(a) (b)

Fig. 20.1 Oersted's interpretation. (**a**) Electrical conflict makes a loop and pushes the needle. (**b**) The drawing shows how the right hand can be used to explain the movement of the electrical conflict following Oersted's interpretation

Newton's concept of force. Even knowing that his ideas were controversial, Oersted continued his experiments and reported them in the 1820 Latin publication *Experimenta circa effectum conflictus eletrici in acum magneticam* (Experiments on the electric conflict on the magnetic needle). Oersted's explanations were not accepted immediately, with most resistance coming from those who supported Newton's concept of force (Darrigol 2000, p. 5).

In his manuscript, Oersted described the effect of the needle's position with regard to the wire, as reported by Cavicchi:

> During the summer break, Oersted experimented and made out patterns from the needle's behavior. What the needle did depended on its position around the wire, and on the direction of electricity going through that wire (determined by which end of the wire was connected to the battery's zinc end). In one experiment, the wire was above the needle and both wire and needle were aligned with earth's north-south line. When that wire was connected to the battery, the needle reoriented parallel to the east–west perpendicular direction. If, instead, the needle was above the wire and the same battery connection made, the needle oriented in the opposite sense. When needle and wire were in the same horizontal plane, the needle tipped up (on one wire side) or down (on the other). The needle's orientations reversed when the battery connection was inverted. (Cavicchi 2003, p. 158)

Oersted interpreted this phenomenon as reflecting electricity in the space around the wire and pushing the needle (Fig. 20.1):

> All effects presented here, related to the North Pole, are easily comprehended if one supposes that the negative electrical force or electrical matter courses along a right-bending spiral line, and pushes the North Pole, but does not act on the South [Pole]. (Oersted in Martins 1986b, pp. 121–122)

Although Oersted interpreted his results as the reflection of an electrical conflict turning around the wire, the concept of a magnetic field induced by the electrical current can also be represented by a loop around the wire. This movement is usually represented using the right hand, as illustrated in Fig. 20.2: the right thumb indicates the direction of the electrical current and the other fingers, turning around the wire, represent the region where the magnetic field is acting.

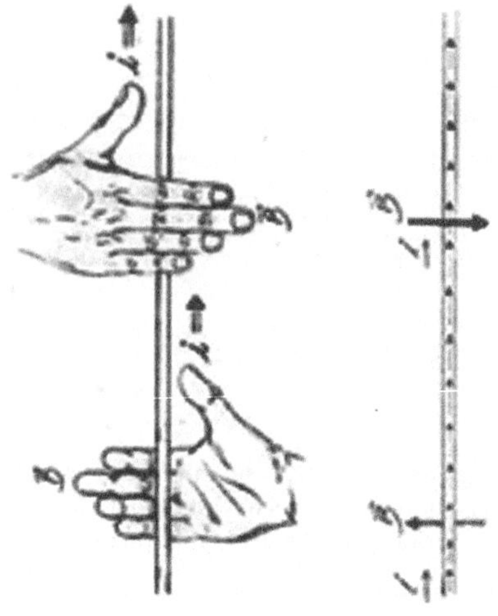

Fig. 20.2 Use of the right hand to represent the magnetic field

Oersted's interpretation was consistent with the movement of the needle and with his theoretical background; for example, the component of electrical conflict was based on the two-fluid theory of electricity. Oersted was a follower of *Naturphilosophie*, a philosophical current that supported Schelling's ideas of Nature. Nature, as Schelling advocated, was formed by a unique force, which was always divided into opposite forces (Caneva 1997). Thus, followers of *Naturphilosophie* expected that relationships among all forces, such as electricity, magnetism, heat, and chemical forces, existed. That thought made it easy for Oersted to accept the existence of a relationship between the electrical current and the magnetic effect on the compass. Many articles have described Oersted's results and the influence of *Naturphilosophie* (Stauffer 1957; Gower 1973; Cunningham and Jardine 1990; Caneva 1997; Brain et al. 2007), in addition to the contribution of this historical episode to science education (Seroglou et al. 1998; Tu 2000; Binnie 2001; Cavicchi 2003; Kipnis 2005).

20.3 Theoretical Framework to Design the Lesson Plan: Personal Constructivism, Inquiry-Based Approach, and Historical Experiments

Personal constructivism assumes that laboratory activities are more than cookbook-like activities. A laboratory activity must have some impact on learners, allowing them to acquire practical and intellectual skills beyond simply following rules (Koponen and Mäntylä 2006; McComas 2005). The introduction of HOS content into a laboratory activity enables emphasis on the process of producing knowledge in different historical periods (Golin 2002). Thus, a laboratory activity incorporating personal constructivism and an HOS approach at the conceptual level could fulfill the goals of this research.

We adopted a model of laboratory activity proposed by Rosa and Rosa (2012) that follows the guidelines of Golin (2002), McComas (2005), and Koponen and Mäntylä (2006). The model comprises three stages: pre-experimental, experimental, and post-experimental. In the pre-experimental stage, the teacher provides only the problem and reminds the students of some basic concepts required for the activity. They are stimulated to design their investigations to solve the problem, with no idea of the "correct" answer. During the experimental stage, students are asked to carry out their investigations and to write down their reasons for choosing the procedures they use. They also take notes on their "ways of thinking," their hypotheses, and possible problems and solutions. This stage can encompass more than one class period, depending on the study problem. In the post-experimental stage, students share their results with classmates and the group tries to reach consensus on the interpretation of the experimental results. This research focused on the first and third stages because they are moments of discussion during which students can delve deeply into experimental and theoretical concepts, interact with other groups, and improve their argumentative skills.

We believe that the experimental activity based on these stages promotes critical analysis of the physical phenomenon and avoids the cookbook-like approach to laboratory activities. Students are stimulated to create a scientific context inside the classroom and to reproduce some of the steps of scientific research (McComas 2005). The teacher plays a fundamental role in achieving pedagogical efficiency. It is the teacher's responsibility to evaluate the students' work without being intrusive; she/he must guide the students with indirect, thought-provoking questions without providing final answers or imposing rules.

The lesson plan presented here explores the idea that the conclusion regarding circular movement around the wire is not intuitive and, consequently, that Oersted's conclusions are not the results of experimentation alone. The first (historiographical) step of the research conducted during the development of the lesson plan was to prepare suitable material for the exploration of the historical episode in a high-school educational setting. For this purpose, we performed a bibliographic search for primary and secondary sources describing the context of Oersted's experiments. Based on the information contained in these sources, we prepared two one-page texts to use in the lesson plan: text 1 is a biography of Oersted, based on the *Dictionary of Scientific Biographies* (Williams 2008); text 2 describes the compass experiment, the assumptions of *Naturphilosophie*, and Oersted's interpretations.[1] Text 2 was based on information from the bibliographic sources and the Portuguese translation of Oersted's manuscript (Martins 1986a, b). In writing these texts, we considered the modern historiography of science, focusing on the scientific and cultural contexts of Oersted's work, his philosophical influences, and the difficulties that he faced in gaining acceptance of his ideas among contemporaneous scientists.

As Golin (2002) argued, Oersted's experiment is essential, as it represents the birth of a new branch of physics. Thus, our lesson plan is based on an experimental activity related to Oersted's work and to electromagnetism. We also built an

[1] Due to space considerations, the texts are not included here.

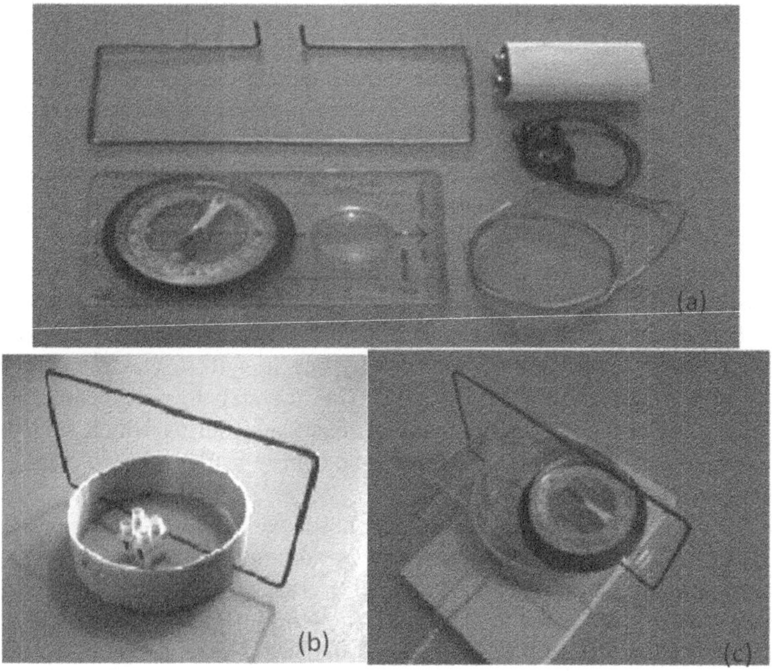

Fig. 20.3 Experimental kit used to reproduce Oersted's observations. (**a**) Materials: a coil, a compass, wires, and a battery. (**b**) Framework for the coil. (**c**) Set model

experimental kit with which Oersted's observations could be reproduced (Fig. 20.3) and based the lesson plan on a challenge laboratory (McComas 2005).

The second step in this research was to develop a methodology for the elaboration and implementation of the lesson plan. Our main goal was to develop a methodology that incorporated scientific ideals for physics teaching and learning (based on personal constructivism). Accordingly, we determined that the lesson should improve students' understanding of the role of experiments in science. We also expected to provide the students with argumentative skills regarding scientific concepts and physical phenomena. Consequently, the lesson plan should improve students' knowledge of electromagnetic phenomena and electromagnetic theory.

We also determined that the methodology should include HOS as part of the approach. Considering the pedagogical knowledge of the teacher[2] and the goals of the lesson plan outlined above, we chose to emphasize HOS at the conceptual level by exploring the context of discovery (Seker 2012). From this perspective, the historical episode frames the problem to be investigated (what is the relationship between the electrical current and the magnetic force that results in the observed effects on the compass?) and part of the solution (can a plausible explanation be

[2] We are referring to one of the authors (Pinto).

Table 20.1 Lesson plan summary

Class meeting	Lessons	Topics
I	1. Inquiry-based approach to electromagnetism	Short initial discussion on electromagnetism and its fundamental principles, present in objects such as electric guitars, blenders
	2. Experimental activity and report writing	Short explanation of and guidelines for kit materials (battery, wire, compass)
		Students divided into groups, conduct experiment, record their observations
II	3. Sharing of observations and discussion of results	Students present their observations and assumptions during the experiment
	4. Reading Oersted biography (text 1) and debating the "discovery" of electromagnetism (text 2)	Guided dialogue among groups, searching for points in common
		Short presentation of Oersted's biography
		Discussion of text 2
III	5. Review of experimental activity and writing of final report	New experiment and recording of results
	6. Evaluation	Discussion of first and second reports
		Students write final report showing points in common, changed procedures, and concepts related to the experiment and the historical episode (texts)

reached based only on observations?). We designed the experimental activity in the lesson plan as a "real" problem, with students designing and carrying out their own investigations as the nineteenth-century scientist did (Koponen and Mäntylä 2006).

Our lesson plan (Table 20.1) lasted of 45- to 50-min lessons, to fit the standard high-school class period in Brazil. It covers three class meetings, each with one presenting two lesson components and incorporating the three-stage model (Rosa and Rosa 2012). The division of the plan into three stages and three class meetings is coincidental; one stage can occupy more than one class period, and an intermediate class period can be inserted to complete the experiment. This theoretical approach (Rosa and Rosa 2012) presumes the possibility of taking up previous discussions to review, reinforce, or correct material from previous stages. Thus, some class meetings can involve more than one stage.

20.3.1 Pedagogical Guidelines for the Lesson Plan

20.3.1.1 Class Meeting I

In the first class meeting, the teacher should present equipment that functions based on electromagnetism. The teacher must provoke students into asking how this equipment works, but without declaring the relationship between electricity and magnetism. For instance, she/he could present the materials in the experimental kit. The battery produces an electrical current when its poles are connected to a conductor, and the compass is sensitive to magnetic changes. Based on these explanations, the teacher can remind students of previously learned concepts (charge, electrical current) and properties (attraction and repulsion). She/he can ask about similarities and differences between magnetic and electrical materials, prompting the students with information to make them think about the subject.

For the experimental activity, the students can be divided into groups, depending on their number and the physical structure of the laboratory. We recommend groups of four students. Each group must use this lesson to design and carry out the experiment, taking notes of all assumptions and drawing the equipment and set-up. Students must also record procedures and results.

We suggest that the teacher collect the students' notes at the end of the class to avoid any modification before the next class. It is not desirable for students to search for information on the phenomenon before the next class and discover the "right" answers.

20.3.1.2 Class Meeting II

This class meeting must include students' sharing of their experimental results. The teacher must stimulate them to speak, and to try to reach consensus among groups. The students should engage in debate until the teacher leads them to conclude that the experiment alone is not adequate to reach an answer about the relationship between electricity and magnetism.

This class meeting also includes reading the texts about the historical episode. We suggest that the teacher ask students to read text 1 first, then, text 2. Those texts focus on Oersted's difficulties in understanding the phenomenon and the philosophical influence of *Naturphilosophie* on his work. We expect these readings to help the students to accept the lack of a single unique answer to a phenomenon, and to understand that observations are susceptible to external influences, such as philosophical currents. The teacher must link the text content with the previous experience of the experiment and stimulate the students to read more about the historical episode.

20.3.2 Class Meeting III

In this class meeting, students conduct a new experiment, following the theoretical discussion based on the texts about the historical episode and the information shared during class meeting II. The main idea of this class is to provide the students with information that will enable them to perceive changes in their methodologies and explanations of the phenomenon, once they are oriented by theory. Students must write a new report about their observations, which will likely differ from the first report because of the theoretical background knowledge gained.

The second part of this class meeting involves evaluation. The teacher should again provoke debate and ask students to compare their two reports, analyzing their observations and assumptions related to the experimental activity. We expect students to be able to recognize and explain the phenomenon, and to identify how their interpretations changed, by the end of this class meeting. We also expect students to understand that the interpretation of the experimental results related to the "right-hand rule" is based on a complex conclusion about the interaction between experiment and theory.

20.4 Implementation of the Lesson Plan

The teacher, who was also the researcher and one of the authors (Pinto), was responsible for implementing the lesson plan, recording (audio/video) and transcribing the classes, and collecting the students' reports. The lesson plan was implemented in a Brazilian state school, with two groups of 12th grade students, in November 2015. Students are from a low social class, they have no good social conditions at home and, usually, they have to work to help their families financially. Thirty students who participated in this study were divided into groups of three to four students and taken to the same school laboratory at the same time. During this research, the school had a special organizational structure that kept students there all day. During extra class periods, students could look for teachers to discuss and resolve their doubts.

The first lesson began with a brief talk about the importance of electricity and how it is part of our routine. The teacher reminded students of previously learned physical concepts related to electricity, such as electrical properties, charge interaction, and technologies that use it. Students were divided into nine groups of either three or four students each. The groups had to identify what they already knew about electricity using parts of broken equipment. They analyzed the electrical motors of a blender, a printer, and other home appliances, and identified magnetic material inside all these motors.

We expected students to ask why magnetic material is necessary for equipment that functions with electricity. However, only one student showed interest in this question, and he interpreted the magnetic material as part of the structure of the equipment, classifying it as just one more piece, like the screws.

After that, the teacher introduced the magnet and its characteristics. He described the properties of magnetism, with emphasis on the similarities and differences between magnetism and electricity. The teacher asked the groups to manipulate some magnets to understand how they interact with each other. Students had to answer two questions: "If electricity and magnetism are so similar, are they related to each other?" and "If so, what is the nature of this relationship?" Students' written answers were collected and the whole discussion was recorded.

The teaching strategy used prompted the students to answer the first question affirmatively, and then speculate about this answer when considering the second question. During the activity, the teacher noticed that some spoken answers revealed conceptual misunderstandings, showing that students did not yet have sufficient abstract concepts to comprehend the phenomena. One student answered: "*magnetism and electricity] must have some relation because both magnetic sides are full of electric charges; on one side there is one kind of charge, on the other side there is another kind of charge, and then both sides interact.*"[3]

Spoken and written answers also showed that students thought that, despite their similarities, electricity and magnetism do not influence each other. Some students concluded that the only relationship between electricity and magnetism was that "*similar things happen in the interaction between magnet and magnet, and charge and charge,*" but that each phenomenon occurs in its own field.

However, some students answered that electricity and magnetism were related to each other if they were properly physically bonded. Students who held this conception argued that "*there should be a way to plug magnets into wires or a battery to make electricity stronger.*"

The groups were asked to use these three conceptions as their hypotheses for the following experimental activity. Each group received a box with the experimental kit (Fig. 20.3), and the teacher explained how each part of the kit worked. However, he gave no information about how to associate or connect the components. The students were then asked to start the experimental activity and test their hypotheses. During the experimental activity, the students asked about the absence of a magnet in the kit, which led them to add one more hypothesis: "*there must be a means to make a magnet and the compass to detect it,*" because they knew that only a magnetic field could interfere with the compass.

Most experimental set-ups of the battery and coil were similar; the students plugged the battery into the coil using the wires. The problem was what to do with the compass to find a magnet. Initially, some students attempted to plug the compass into the wires, but this proved to be impossible. At the end of the activity, two groups presented set-ups that generated similar results, but were based on different assumptions. The first group argued that the compass should be in contact with the coil, and that a large contact area would result in a stronger effect on the compass needle. The second group argued that contact between the compass and coil was not necessary to produce an effect because the compass was covered in an insulating material, but they agreed that "the needle moved more strongly" when the compass was near it. For this group, the effect on the needle was related only to the battery and how it was plugged into the coil. Consequently, they argued that a larger battery would have a stronger effect on the needle. The other groups chose one of these ideas, without adding any hypothesis or suggestions. All the students recorded their procedures and handed them in to the teacher as reports. They also took photographs and drew schematics of the set-up.

[3] Transcription of the students' words. The teacher recorded the class and transcribed the talks.

At the second class meeting, the teacher reviewed the experimental hypotheses and the students shared their results, beginning a debate with their classmates. The teacher noticed from the talk among groups of students that some of them had looked for the experiment on the internet and tried to find the "correct way" to conduct it. All groups agreed that the phenomenon happened because the coil worked as a magnet when electric current flowed, and that this depended on energy from the battery, and that it was necessary to know more about the experiment to claim that the conclusions they drew were correct.

This second class meeting showed that the students were used to finding correct and final answers either in their textbooks or on the internet. For them, consulting the textbook and solving the experimental activity problem was enough. At that moment, the teacher's position was essential. He had to interfere, provoking a debate on wrong or distorted information presented in books and on websites and giving examples. The teacher then asked if the students would like to know about the scientist and the original work related to the experimental activity. He indicated that the historical episode about the relationship between electricity and magnetism would be introduced in the next class period.

Each group received the two texts. After reading them, the students discussed the information they contained, the association between the original experiment and the classroom activity, and the degree of agreement between the results obtained by Oersted and those obtained in the classroom. The students also had to look for information about Oersted and his experiment in the textbook. They found only superficial information in the textbook, and stated that the texts were more interesting.

By the end of this class meeting, the students concluded that their experiments were similar to Oersted's, but that their assumptions were related only to the set-up of the model and their focus was on the compass and coil. When the teacher asked if they knew how to describe the interaction between the compass and coil, they reported that they had not bothered with such questions.

Some students argued that the texts dealt with these questions, but that they were irrelevant. The phenomenon was more important because it was something that they could observe. Therefore, the teacher asked them to repeat the same experiment in the next class meeting, and to try to show the interaction between the compass and the electric current.

In the next class meeting, the teacher asked the students to repeat the experiment, demonstrating how the magnetic field acted on the compass. The students asked to change the battery because they supposed that a larger battery would have a stronger effect on the needle. Seven out of the nine resulting reports showed a circular field around the coil. These results came from tests in which the students varied the position of the compass in relationship to the coil and observed its reaction. The other two reports presented some assumptions, but no explanation of how conclusions had been drawn. After the groups presented conclusions involving the circular motion, the teacher presented the "right-hand rule" and discussed its application in the experiment.

After this last experiment and report, the teacher returned each group's first report, asking the students to analyze the production of the first reports and to discuss changes found in the second reports, relating the discussion to the historical texts. Each group was asked to write an essay about the experimental activity, the texts, and the debates that had taken place during the lessons. At the end of the class, the students were asked to share the group's opinions and to debate until they reached a single conclusion. This request, which was not in the lesson plan, helped to clarify doubts, solve conceptual difficulties, and stimulate students' self-confidence regarding their actions related to the experiment.

20.5 Conclusions

We conclude that the experimental activity, presented in association with material about a historical episode, engaged students during all the class meetings. They took a critical attitude during the experiment and in their reports, arguing in defense of their points of view. They seemed to be more self-confident when they found out that a single, unique answer explaining the phenomenon was not required. Thus, they concluded, by themselves, that the experiment was not the only source of knowledge; that the possession of theoretical knowledge before making observations can guide the interpretation of the results. By the end of these lessons, they understood that the "right-hand rule" is only a model used to describe the relationship between electricity and magnetism, which can have different meanings. At the same time, the development of the experimental activity over multiple class periods and its association with the texts about Oersted and his context helped the students to understand that scientific work is complex and challenging.

Students' reports and essays, classes' recordings, and teachers' observations during the classes provided rich material for analyses. Triangulation of these materials revealed that the inquiry-based method stimulated students' participation in all steps of the experimental activity. The students' engagement was corroborated by the approval of their paper to be presented at a Latin American Conference of Young Scientists in December 2015, designed to stimulate high-school students to follow scientific careers.

Acknowledgements The authors acknowledge the financial support of the National Council of Scientific and Technological Development (no. 474924/2012-2).

References

Binnie, A. (2001). Using the history of electricity and magnetism to enhance teaching. *Science & Education, 10*(4), 379–389.

Brain, R. M., Cohen, R. S., & Knudsen, O. (Eds.). (2007). *Hans Christian Ørsted and the romantic legacy in science: Ideas, disciplines, practices*. Dordrecht: Springer.

Caneva, K. L. (1997). Physics and naturphilosophie: A reconnaissance. *History of Science, 35*(1), 35–106.

Cavicchi, E. (2003). Experiences with the magnetism of conducting loops: Historical instruments, experimental replications, and productive confusions. *American Journal of Physics, 71*(2), 156–167.
Cunningham, A., & Jardine, N. (Eds.). (1990). *Romanticism and the sciences*. Cambridge: Cambridge University.
Darrigol, O. (2000). *Electrodynamics from Ampère to Einstein*. Oxford: Oxford University Press.
Golin, G. (2002). Introducing fundamental physical experiments to students. *Science & Education, 11*(5), 487–495.
Gower, B. (1973). Speculation in physics: The history and practice of naturphilosophie. *Studies in History and Philosophy of Science, 3*(4), 301–356.
Heilbron, J. L. (1979). *Electricity in the 17th and 18th centuries: A study of early modern physics*. California: University of California Press.
Höttecke, D., & Henke, A. (2015). Physics teachers' challenges in using history and philosophy of science in teaching. *Science & Education, 24*(4), 349–385.
Höttecke, D., & Silva, C. C. (2011). Why implementing history and philosophy in school science education is a challenge: An analysis of obstacles. *Science & Education, 20*(3–4), 293–316.
Kipnis, N. (2005). Chance in science: The discovery of electromagnetism by HC Oersted. *Science & Education, 14*(1), 1–28.
Koponen, I. T., & Mäntylä, T. (2006). Generative role of experiments in physics and in teaching physics: A suggestion for epistemological reconstruction. *Science & Education, 15*(1), 31–54.
Martins, R. A. (1986a). Oersted e a descoberta do eletromagnetismo. *Cadernos de História e Filosofia da Ciência, 10*, 89–114.
Martins, R. A. (1986b). Experiências sobre o efeito do conflito elétrico sobre a agulha magnética. *Cadernos de História e Filosofia da Ciência, 10*, 115–122.
McComas, W. (2005). Laboratory instruction in the service of science teaching and learning: Reinventing and reinvigorating the laboratory experience. *The Science Teacher, 72*(7), 24–29.
Rosa, C. T. W., & Rosa, Á. B. (2012). Aulas experimentais na perspectiva construtivista: Proposta de organização do roteiro para aulas de física. *Física na Escola, 13*(1), 4–9.
Seker, H. (2012). The instructional model for using history of science. *Educational Sciences: Theory and Practice, 12*(2), 1152–1158.
Seroglou, F., Koumaras, P., & Tselfes, V. (1998). History of science and instructional design: The case of electromagnetism. *Science & Education, 7*(3), 261–280.
Stauffer, R. C. (1957). Speculation and experiment in the background of Oersted's discovery of electromagnetism. *Isis, 48*(1), 33–50.
Tu, M. B. (2000). Electromagnetic induction rediscovered using original texts. *Science & Education, 9*(4), 375–387.
Williams, L. P. (2008). *Oersted, Hans Christian. Complete dictionary of scientific biography*. http://www.encyclopedia.com/doc/1G2-2830903215.html. Accessed 27 May 2016.

Chapter 21
Investigating the Didactic Use of Primary Sources on the History of Vacuum and Atmospheric Pressure

Juliana M. Hidalgo F. Drummond, Wesley Costa de Oliveira, and Deyzianne Santos Fonseca

21.1 Introduction

Considerations regarding history and philosophy of science (HPS) in education include philosophical–historical approaches in the teaching and learning of scientific concepts. Therefore, "the understanding of physical knowledge is intrinsically related to the understanding of the problems this knowledge attempts to solve" (Matthews 1994, 50).

Furthermore, the intersection between the students' previous conceptions and philosophical–historical approaches has been indicated for decades (Saltiel and Viennot 1985; Mellado and Carracedo 1993; Villatorre et al. 2008). Empiric research demonstrates that students tend to present deeply rooted ideas regarding physical phenomena, which can be different from what the school intends to teach:

> However, this does not mean they are totally incorrect and should be cast aside in the teaching/learning process. They are, in fact, the starting point of this process. (Longhini and Nardi 2009, p. 9)

Given the different contexts and considering the students' initial explanations of physical phenomena, in addition to the views supported by researchers from the past, we consider, for example:

> [...] the similarities [...] indicated are sufficiently strong to make the knowledge of history of science an important ally in this work. Observing historical examples [...] students [may] [...] realize that there have always been discussions and alternatives in history and that some have had ideas similar to their own, but which were later replaced [...]. (Martins 2006, p. xxvi)

When explaining phenomena related to atmospheric pressure, students from different levels tend toward "plenist" explanations that share similarities with the

J. M. H. F. Drummond (✉) · W. C. de Oliveira · D. S. Fonseca
Postgraduate Program in Teaching of Natural Sciences and Mathematics, Federal University of Rio Grande do Norte, Natal, RN, Brazil

M. E. B. Prestes, C. C. Silva (eds.), *Teaching Science with Context*, Science: Philosophy, History and Education,
https://doi.org/10.1007/978-3-319-74036-2_21

concept historically known as "*horror vacui*"[1] (Solaz-Portolès 2008; Longhini and Nardi 2009). Current scientific explanations oppose these conceptions, and yet, students are expected to accept a "scientific truth" (Martins 2006).

School books normally present knowledge without any reference to the problems which originated this very knowledge:

> [...] they emphasize the results reached by science – accepted theories and concepts and techniques of analysis used – but they do not usually present any other aspect of science. How were theories and concepts developed? How do scientists work? Which ideas were accepted in the past, but are no longer so? (Martins 2006, p. xxi)

In Brazilian school books, the concept of atmospheric pressure is presented as a "finished truth" disregarding the students' initial views and the historical context in which the concept was created. The experiment of Evangelista Torricelli (1608–1647) is de-contextualized, although it occurred exactly in the realm of the discussions concerning vacuum, which could inspire the problematization of alternative conceptions. An interesting didactic perspective could be a diachronic discussion of historical sources dealing with daily phenomena, such as the use of straws to sip liquids. Students could compare past conceptions with their own explanations. Regarding the didactic use of primary sources from the history of science (HOS), several possibilities and potential uses have been identified by current literature.[2]

Historical documents may be used in interventions that favor the building of historical knowledge in an autonomous way. Regarding the acknowledgement of differences and similarities, the interpretation of primary sources allows the retrieval of past explanations. Therefore, careful and diachronic comparisons among scientific conceptions from different times may be established.[3] Past conceptions of certain phenomena may be compared *by the students to their own explanations*, making them feel as part of the process of building knowledge.

The didactic use of primary sources in *investigative dialogical activities* is also connected to certain questionings and postures that have been inserted in the educational context for some time:

> How and why was a certain idea accepted? How and why were these alternatives abandoned? When we investigate the past, we can attempt to answer these questions. (Martins 1989, p. 9)

The diachronic investigative activity promotes a dialogue between past and present and reflects some of the most significant goals related to HOS in the educational context. However, despite the potential uses indicated here, it is quite difficult to find the use of dialogical activities of interpretative nature that conceive primary

[1] The concept that nature avoided the formation of a vacuum was well accepted in Antiquity and in the Middle Ages. This was contested during the Scientific Revolution through new explanations based on the idea that air had weight and exerted pressure.

[2] About the didactic use of primary sources, see: Bueno and Pacca 2009; Briccia and Carvalho 2011; Boss 2011; Dion and Loures 2013; Batista et al. 2015; Silva and Guerra 2015.

[3] A diachronic interpretation of sources in the classroom is based in the synchronization between current historiographical knowledge and their didactic use (Forato et al. 2011; Batista et al. 2015).

sources not as mere illustrations of content, but as registers of processes lived by the characters of HOS that are, therefore, marked by their intentionality.

In light of these considerations, we have proposed a didactic sequence from an historical–philosophical perspective with the intention of contributing to the teaching and learning process of the concept of atmospheric pressure and the discussion of the nature of science in basic education. In this *dialogical* intervention, students are engaged in an "investigative" process that includes the historical construction of the concept of atmospheric pressure through *the dialogical interpretation of excerpts from historical sources.*

This didactic sequence was applied in December 2015 in a public school in the city of Natal, located in the state of Rio Grande do Norte, in the Northeast of Brazil. We used 2 days of mini-courses with a total of 8 h. During the application, we conducted empirical research to investigate the contribution of this proposal in relation to the goals set. Exploratory instruments were used before and after the intervention. Investigative instruments were composed of two parts: the first regarding physical phenomena and the second the theme of the nature of science. Audio-visual registers from the collective discussions were made and the students also registered their reflections individually in material that contained the transcriptions of primary sources. Owing to the restricted number of pages, this work presents only the results regarding the analysis of the answers on physical knowledge that composed the pre- and post-test.

21.2 The Didactic Sequence

The suggested didactic sequence is aimed at contributing to the understanding of the concept of atmospheric pressure highlighting the *historical aspects* of its construction. It privileges the notion that physical knowledge is related to the investigation of the problems.

Moment 1 During Moment 1 – Part 1 we suggest an attempt at an initial collective construction of explanations for an every-day event: the sipping of liquid through a straw. The questions raised become suggestions for a reflection mediated by the teacher.

1. Why does the liquid go up the straw?
2. If there were a hole in the straw, would the liquid go up? Why?

Throughout the intervention it is our intention that the students themselves, through the mediated interpretation of specific excerpts of primary sources from HOS, identify in the science of the past possible explanations that are similar to those proposed originally by the groups, whether they are close to "*horror vacui*" or to possibilities that refer to atmospheric pressure.

During Moment 1 – Part 2, with the mediator's support, students are encouraged to compared their initial explanations with those that emerged from the interpretation of the primary source in which the same phenomenon is explained by Jean Buridan (1295–1363) – a French thinker from the fourteenth century – in the work *Questiones super octo physieorum libros Aristotelis* (Questions commentary on the eight books of Aristotle's Physics) as due to "*horror vacui*":

> When we place the extremity of a [reed] straw in our mouth and dip the other extremity into wine [...] we suck the air from the straw and the wine is drawn up, even though it is heavy. This occurs because it is always necessary for a body to come immediately after the air that is sucked up to avoid the formation of a vacuum [...]. (Buridan apud Martins 1989, p. 19)

Moment 2 The purpose of Moment 2 is to contribute new elements to the understanding of what atmospheric pressure is by recovering aspects of the historical construction of this concept. We suggest carrying out the experiment proposed by Aristotle (384–322 BC), the Greek philosopher (fourth century BC.), in his work *De Caelo* (On the heavens) commenting on the shared reading:

> [...] air is heavy [...]. Evidence of this is that a full balloon weighs more than an empty one. (Aristotle apud Martins 1989, p. 4)

In the sequence, students are invited to interpret the primary sources that suggest an alternative to *horror vacui* and focus on atmospheric pressure. Students come close to current scientific explanations in a collective reflection on the researcher Isaac Beeckman's register from the sixteenth century, in his work *Mathematico-physicarum meditationem* (Mathematical–physical meditations):

> [...] air is heavy [...] it pressures us from all sides in a uniform way. [...]. Things precipitate into an empty space with great force because of the tremendous height of the air above. (Beeckman apud Martins 1989, p. 25)

Moment 3 Moment 3 of the didactic sequence is dedicated to the problematization of the content present in physics textbooks usually used in Brazilian schools. Normally, the concept of atmospheric pressure is presented in a final form with a brief reference that does not do justice to its historical development. Torricelli supposedly discovered and measured atmospheric pressure in his most famous experiment. Some questions are suggested for collective reflection:

1. Was it really Torricelli's intention to measure atmospheric pressure?
2. Is it really possible to see that the suspension of mercury is due to something external (pressure) in Torricelli's experiment? Does the experiment show this result? Did Torricelli encounter difficulties?
3. What did they think of *horror vacui* at the time?

These questions are supposed to instigate imagination and raise suspicion that there are elements missing from the description of Torricelli's experiment in the books.

Moment 4 In Moment 4, we suggest using a fictitious interview with Torricelli in which "questions" are formulated by the character of a journalist to elucidate

Moment 1 Part 1	Initial attempt: explain the rising of liquid through straw.
Moment 1 Part 2	Dialogical activity – Medieval primary source (Jean Buridan - "*horror vacuis*").
Moment 2	Historical experiment (Aristotle) Dialogical activity – Scientific Revolution primary source (Isaac Beeckman: air is "heavy" and "pressures us from all sides").
Moment 3	Problematization from school textbooks
Moment 4	Dialogical activity – Fictitious interview with Evangelist Torricelli. Reflection about unanswered aspects from previous moments. Return to initial question: can the rising of liquid be (re)explained based on atmospheric pressure?

Chart 21.1 The didactic sequence

Torricelli's *intentionality* in "answers" that come from original texts written by the researcher.[4]

Throughout the interview, many interesting aspects for the reflection mediated by the teacher appear. Torricelli's motivation came from his involvement in the discussion about the possibility of vacuum. He had difficulties regarding the certainty that the upper part of the tube was empty and he made an effort to argue this point. He stated that the mercury column was sustained because of the air's pressure over the surface of the liquid. However, he was cautious about this statement, giving the impression that he was partaking in a dissident interpretation and that *horror vacui* was traditional and influential.

Elements present in historical texts may contribute to the understanding that the concept of atmospheric pressure developed in a context of controversies regarding a vacuum in close relation to the phenomena that intrigued thinkers from the past (such as the rising of liquid through a straw, as seen by students).

To conclude the proposal, we suggest returning to the initial question: based on the fact that air has weight and exerts pressure (theoretical assumption), how could the *rising of liquid through a* straw be explained when we sip juice/soda?

Considering similar initiatives regarding other concepts (Bueno and Pacca 2009), we propose, therefore, a didactic sequence (Chart 21.1) in which students are engaged in the "investigative" process allowing them to have: the perception of similarities (and differences) between their own initial explanations and those expressed by thinkers from the past in historical documents, the understanding of elements related to the process of historical construction of the concept of atmospheric pressure, the perception of the inexistence or weakness of this factor (as a theoretical supposition) in their initial explanations, and the possibility of reconsidering in light of this new element. A possible contribution from the understanding of the physical concept of atmospheric pressure can be delineated inasmuch as this

[4] Interview composed by authors of a recently published work (Batista et al. 2015) from a letter written by Torricelli in 1644 addressed to Michelangelo Ricci (1619–1682) (Magie 1969, pp. 70–73).

concept could be used by students in reinterpreting the physical phenomenon initially considered.

21.3 Results from Empirical Research

Thirteen high-school students (identified as A_n, where n varies between 1 and 13) took part in the research conducted during the application of the didactic sequence in a Brazilian public school. They answered a pre-test and a post-test on physical knowledge individually. The tests were composed of three open-ended questions each. Their answers were considered through content analysis criteria (Bardin 2004).

21.3.1 *Pre-test*

The first two questions in the pre-test (Appendix 1) were analyzed together considering the similarity between the phenomena and the possibility of a better understanding of the students' views (Tables 21.1 and 21.2).

The first question was based on a problem situation suggested by Longhini and Nardi (2009, pp. 13–14). It referred to a bottle containing water with artificial coloring turned upside-down into a receptacle containing the same liquid. The prompt asked students to explain why the water in the bottle did not drain completely.

The second question was based on an educational project available on the Ministry of Education's (MEC) website (Monteiro and Gomes 2003). The prompt informed that a heated nail was used to make small holes in the bottom of a plastic bottle, which was then placed standing into a bowl with water. The bottle was filled with water, capped, and removed from the bowl, and was held over it until it stopped dripping. Students were asked why the water remained in the bottle.

Table 21.1 Question 1: Pre-test

Categories	Students
Because of external atmospheric pressure	A9, A12
Because outside pressure is greater than inside the bottle	A4
Because inside pressure is greater than outside the bottle	A5
Because water in the bottle suffers pressure	A10
Because of the interaction between the water in the bottle and in the receptacle	A2, A3, A7, A11, A13
Because air cannot enter the bottle	A6
Because of the vacuum at the bottom of the bottle. There is nothing to replace the water	A8
Because the vacuum created in the bottom of the bottle prevents water from draining	A1

Table 21.2 Question 2: Pre-test

Categories	Students
Because atmospheric pressure outside the bottle is greater than inside	A9, A4, A12
Because inside pressure is greater than outside the bottle (the cap prevents air from escaping: it contracts/sucks air in)	A5, A7
Because the force of the pressure exerted by the cap keeps the water in	A11
Because the wind creates pressure that doesn't let water drain	A2
Because of the vacuum formed at the top	A8
Because the closed bottle stops the circulation of air	A3
Because of the vacuum at the top (it stops atmospheric pressure from acting and making it fall)	A1
Other (water in the bowl, temperature, closing of holes)	A10, A6, A13

In the first question, 38.5% of students attributed the observed effect to an *interaction between the water in the bottle and the water in the receptacle* – this was the most recurrent answer. In these answers, there was *no reference to the pressure exerted by air*. Some comments saying that the water in the receptacle "creates" (A_2) pressure that holds the water in the bottle or that "the water does not drain because the mouth of the bottle is still in contact with the water" (A_{13}) were also registered.

The second question could have destabilized this kind of answer, as water is sustained in the bottle without the presence of another receptacle. The group of students presented various explanations. A_{13} considered simply that cooling the temperature eliminated the holes. The others discussed the effect, considering that the holes remained there.

We could observe statements such as: a certain "force of the pressure" exerted by the cap keeps the water in (A_{11}); the pressure inside is greater than outside the bottle, "like when we breathe; holding the air in" (A_7). In both cases, it is possible that the students were considering the "pressure" that they mentioned as a "pull" inside the bottle that kept the liquid from going down.

For student A_2, the situation proposed in question 2 was "similar to the first, but instead of water creating this pressure, here it was the wind that did it." In referring to the "wind," the student probably had "air" in mind, which would approach an explanation influenced by a current scientific notion, even if still misunderstood or miswritten by the student (highlighting the students' difficulties in written expression). A similar interpretation may be given for A_3's answer: "because the bottle is closed, the air doesn't flow in the bottle and the water stays stable and stops falling."

Returning to question 1, according to two students, A_8 and A_1, a *vacuum created in the bottom of the bottle was responsible for the phenomenon in question.* Table 21.1 presents two different categories for these students' answers which, examined carefully, represent distinct understandings.

According to A_1, "atmospheric pressure inside the bottle becomes smaller, creates a vacuum, and stops the water from draining." This student's focus was exclusively on the inside of the bottle. From the lack of greater clarification, we can even speculate whether the student was referring to a "pull" exerted by the vacuum. However, this hypothesis of *an action of vacuum on water* loses force with the answer to question 2. The effect observed was once again attributed by A_1 to the vacuum at the top of the bottle with the following clarification: "the vacuum created stops the atmospheric pressure from acting on the water so it can drain". Thus, it is possible that for this student, the vacuum represented an impediment for water to be pushed down.

For student A_8, there is one single explanation for the effects: "Because of the vacuum created in the bottom of the bottle, that is, there is nothing to replace the water" (question 1). In the second answer, the student did not prolong the answer, apparently assuming that it had already been given in the first question. There is apparently the conception of a tendency to limit the formation of a vacuum.

A suggestively plenist conception was expressed by A_6, in question 1, stating that the liquid did not fall because air could not enter. This student's answer to the second question proved to be confusing though, not demonstrating the same kind of conception.

In question 1, student A_5 attributed the effect to greater pressure *inside* the bottle. We can speculate whether the student was confused in the writing registering the exact opposite to what was meant. Everything points to fact that this was not the case. The student's answer to question 2 again states that pressure was greater inside the bottle. According to the student, this added pressure was due to the cap, which did not allow the air escape. The student did not offer a complete explanation for the sustained liquid in either case, but signaled that the effect was related to a certain volume of air confined in the top of both receptacles. Although we cannot say with certainty, it is possible that the student attributed this supposed increase in internal pressure to a "pulling" or "sucking" of the liquid.

Some explanations with little detail were noted in question 1. These could approach or pull away from a currently valid scientific explanation, depending on the understanding of aspects that were not made clear by their authors. Student A_{10}, for example, did not explain to what pressure he was referring or how it acted. Considering the variation of answers obtained to this question, it is possible that the student was referring to air pressure, to the pressure of the liquid in the bottle over the liquid in the receptacle, or even to an internal "pull" associated with "pressure." The student's answer to question 2 was confusing and did not contribute to clarifying this point.

In the first question, 23.1% (A_4, A_9, A_{12}) of participants presented answers according to a current scientific viewpoint. Only one student, A_9, gave a complete explanation of how the action of external atmospheric pressure retained the liquid: "When the receptacle is empty, there is draining because internal pressure suffers influence from external pressure. But when the receptacle is filled with water and its water gets into contact with water in the bottle, the external (atmospheric) pressure holds the liquid in the bottle instead of draining it."[5]

Another student, A_{12}, attempted an explanation in this direction by stating that the pressure outside – atmospheric – was greater than inside the bottle. The student made no explicit reference to what this pressure might be, but the same student's answer to question 2 in the pre-test indicates an understanding of what atmospheric pressure is.

In the second question, the same three students (A_4, A_9, A_{12}) were the ones who, among the group of 13 students, explicitly attributed the phenomenon to the fact that the pressure outside (atmospheric) was greater than inside the bottle. Therefore, in both questions, about 23.1% of students came close to a current scientific viewpoint.

The third question in the pre-test was adapted from empirical research conducted by Solaz-Portolés (2008, p. 86). The prompted asked students to explain why a syringe, whose tip was dipped into liquid, filled up when its plunger was raised (results in Table 21.3).

Only student A_9 made an explicit reference to atmospheric pressure, expressing an understanding that the liquid was *pushed* in and thus approaching a current scientific explanation: "When you raise the plunger, it creates a negative pressure inside the syringe and because the environment tends to balance itself, the atmospheric pressure above the liquid forces it into the syringe."

Student A_{12} attempted an answer based on the difference of pressure by stating that it was smaller inside the syringe. However, the student did not specify the role of atmospheric pressure and explained that the water was *pulled* in: "The force of pulling the plunger up pulls the water into the syringe that has room (making more

Table 21.3 Question 3: Pre-test

Categories	Students
Because the pressure created in the action pulls the water	A8
The syringe helps in the pressure: the air in the syringe makes the liquid escape	A10
Because the plunger pulls the air and exerts pressure on what is outside, pulling it into the syringe	A7
The plunger works as a "puller": the air is caught at the top, making the entry of the liquid easier	A5
Because the force of the syringe makes the water enter	A6
Because of the plunger sucking the water	A2, A3, A11
Because it is necessary to fill the empty space in the syringe/because the water is sucked and goes up/ because it is possible as the air in the environment occupies its place in the bowl, avoiding formation of a vacuum	A1, A4, A13
Because there is a negative atmospheric pressure that pushes the liquid	A9
Because there is a difference in pressure (smaller inside, when the plunger is pulled) and the water is pulled in	A12

[5] Teacher recorded the class and transcribed the dialogue.

room) [...] because the pressure is smaller, and it makes the water go up (taking up that space)."

Three students demonstrated plenist conceptions. A_4 and A_{13} explained the effect in a similar way. By itself, water would tend to occupy the empty space and this would explain its rising. As the plunger was pulled, *empty space would appear and soon be filled* by water. They made no reference to the exterior of the syringe or to the action of atmospheric pressure.

For A_1, water was sucked in by the syringe. An ingenious explanation of plenist contours was spontaneously registered to support the impossibility of water rising in case the receptacle was closed: "Because the receptacle with water is under influence of atmospheric pressure and so the water sucked through the syringe is soon replaced by air. If the receptacle were closed, this wouldn't happen because there would be a vacuum."

Atmospheric pressure, in this case, exerts influence over the receptacle with water. However, the student did not indicate that water would be pushed by air. Instead, the syringe would pull the water, which *could* go up *because it was replaced* by air in the receptacle.

The effect mentioned in question 3 was explained in different ways by other students. For A_2, $A_{3,}$ and A_{11}, there was suction of water through the plunger. Something similar was possible indicated by A_6, for whom the *force of the syringe* is what made the water enter. These students, therefore, attributed the effect to a direct action of the plunger/syringe over the liquid. For A_4, the plunger worked as a "puller," the air would be stuck on the top, and would *make the entry of the liquid easier*. The "pull" would be exerted over the air and this would make free room for the water to rise. None of the five students *used the term pressure* in their responses.

Three other students used it, but especially in what seemed to be a reference to the *action of pulling* liquid, and not pushing it. For A8, *pressure* created in the action *pulled the* water. For A_{10}, "the syringe helped in the pressure" and the *air in the syringe made* the liquid go in. "Pressure," in this case, could be an allusion to a "pull" exerted by the air contained in the syringe. Something similar to a pull or suction also seems to have been in mind in A_7's answer "the plunger pulls the air and makes pressure on what is outside of the receptacle into the syringe."

21.3.2 Post-test

The first question in the post-test (Appendix 2) was based on a problem situation suggested by Saad (2005, p. 28). It required an explanation for the workings of a water fountain for birds in cages. The second question was adapted from research conducted by Solaz-Portolés (2008, p. 85). It asked students to attempt to explain how part of the mercury remained in the inverted tube in Torricelli's experiment. These first two questions were analyzed together (Tables 21.4 and 21.5).

For question 1, we obtained very significant results. Seven out of thirteen participants, or 53.8%, gave explanations close to current scientific views: atmospheric

Table 21.4 Question 1: Post-test

Categories	Students
Water is contained by atmospheric pressure/pressured by air	A6, A9, A10, A8, A13, A11, A7
Pressure stops it from overflowing	A2, A3
It does not overflow because of the vacuum on the top of the water fountain	A1
Thin water fountain	A12
Because pressure outside is greater than inside (and expresses a plenist conception)	A5
Confusing answer (uses the expression "atmospheric pressure")	A4

Table 21.5 Question 2: Post-test

Categories	Students
Explained by a group of factors. Internal: Pressure related to the vacuum in the tube and the interaction of liquids in the receptacle. External (additional): atmospheric pressure	A6
Because of air pressure/weight of the air column: The descent stops when there is balance	A2, A3, A7, A8, A9, A11
Because pressure inside the tube is greater and pushes the mercury until there is balance with pressure outside the tube	A4
Does not answer what was asked	A13
Because of the balance between the mercury in the tube and in the receptacle (focus on liquids)	A12
Does not answer what was asked and registers a plenist conception	A5
It does not leak because of the formation of a vacuum	A1
No answer	A10

pressure or air pressure was responsible for preventing water from overflowing. Some answers highlighted the balance of the column of liquid through air pressure: "Because water is light, it is easily contained by atmospheric pressure that keeps it from overflowing" (A_9).

Two students, A_2 and A_3, stated that the "pressure" stopped the overflowing of water in the fountain. In the following question, they referred explicitly to "atmospheric pressure" or to the "weight of the column of air" as being responsible for sustaining the mercury. Considering that these elements contribute to a more detailed understanding of the scientific views, the percentage of answers close to current scientific explanations comes to 69.2% in the first question.

As the second question dealt with Torricelli's experiment, which had been discussed throughout the mini-course, it would seem reasonable to expect students to answer it more easily. But curiously enough, not all students who answered the first question correctly did so for the second one.

In fact, six students repeated explanations that were close to current scientific views. They stated that air pressure or the weight of the column of air influenced the situation. The descent stopped when there was a balance with the heavy mercury.

Of the other three students who had correctly answered the previous case, one left the question blank (A_{10}) and another made general comments that did not apply as an answer to the question (A_{13}). Student A_6 attributed the situation to a group of three factors: first, a certain pressure related to the vacuum in the tube: "because of the pressure, actually at the top there is a small vacuum in the tube." The notion of a "pull" exerted by the vacuum is not explicit in the student's answer, although this is common in the repertoire of alternative conceptions. The second factor was the interaction between the liquid in the tube and the liquid in the receptacle: "pressure from the two receptacles makes the one that is heavier go down." This was the only factor that was mentioned by another student (A_{12}) regarding Torricelli's experiment. The third factor regarding the answer given by A_6 on Torricelli's experiment, put atmospheric pressure explicitly as an *additional influence*: "it is also influenced by an external factor that is atmospheric pressure." However, this same student's answer regarding the water fountain did not present any other factor beyond atmospheric pressure. It is quite curious that on that occasion, the student explicitly related the situation to Torricelli's experiment: "Because of atmospheric pressure, just like in Torricelli's experiment, but with water instead of mercury [...]".

In the first question, A_4 used the expression "atmospheric pressure" in an answer formulated in not such an obvious way. When the water is drunk, the atmospheric pressure inside the water fountain pushed the water outward. In the second question, the student attempted an answer with certain difficulty: Pressure inside the tube was *greater*, pushing the mercury to the tray until there was a balance with the pressure outside the tube. The student was possibly trying to use recurring terms from the mini-course.

Student A_{12} simply attributed the lack of overflow to the thickness of the water fountain. In the same question, A_5 presented a confusing answer. Using fragments from what was discussed in the mini-course, the student mentioned that the pressure outside was greater than the internal space of the water fountain. Without clarifying why, the student concluded that in this situation the space inside the water fountain was "full of air that avoided emptiness." The student did not answer what was asked in the second question. He registered "*horror vacui*," discussed during the mini-course, giving the impression of having incorporated a form of plenist conception: "The space above the mercury is filled by air because nature has horror to vacuum. And also to complete the space of emptiness."

The reference to emptiness can also be seen in the answers given by A_1. In the first question, the student explained that because the water fountain was closed, an empty space formed and "the water didn't overflow because of the vacuum". The same explanation was given for the second question: the liquid did not descend "because of the formation of a vacuum." It is possible that, for A_1, emptiness or a vacuum exerted some kind of "pull."

The third question in the post-test was based on the problem situation suggested by Longhini and Nardi (2009, p. 15). It required a reflection concerning the opening of packages closed by a vacuum, such as cans of food that usually contain a type of seal in their lids (see results in Table 21.6).

Table 21.6 Question 3: Post-test

Categories	Students
Because opening the seal releases internal pressure	A2, A3, A5
Because by opening the seal, exterior air pressure makes the air inside the package escape more easily	A7
Because in the emptied package, internal pressure is greater than exterior and the opening of the seal creates a balance	A12
Because of the vacuum inside the package	Al
Because when the seal is opened, air enters, there is a balance between internal–external pressure/there is no longer pressure on the cap	A11, A13, A8, A4, A9
Because when the seal is opened, air enters and opening becomes easier	A6, A10

Despite the expression "vacuum-packed," for 30.8% of the students the sealed receptacle contained air initially. The explanations proposed by three students are quite similar: the breaking of the seal released the internal pressure (A_5) or the air inside (A_2, A_7) that exerts pressure (A_3), making opening easier (A_2, A_3) or allowing the opening (A_5). These answers were grouped into the first category of analysis. Student A_7 stated that the exit of internal air was made easier by the exterior pressure of the air – without however explaining the answer.

For the other participants, the sealed receptacle was initially evacuated. A_6 and A_{10} gave answers with little detail. They indicated that when the seal was broken air went in (A_{10}) or "the vacuum allowed the air to go in" (A_6) and made the opening easier. In the absence of an explanation as to why this happened, their answers were classified into a specific category.

In another category, we find five students – 38.5% – who gave answers in which the term "pressure" or its balance were present, approaching them to an updated scientific view of the phenomenon. A_9's answer offers us a more complete explanation: "Because atmospheric pressure is so strong forcing the lid in the vacuum, that it is greater than the force applied to open it. But when we remove the seal we cause the entry of external pressure and this quickly increases the internal pressure so that the internal pressure is equal to the external one in a balance. This decreases the pressure that was exerted on the lid before."

The answers of the other students included in the same category contain no clarification regarding the term "pressure." However, there are indications, from the previous answers, that they share the view supported by A_9.

Student $A_{11,}$ for example, stated that "when you remove the seal, the great pressure that existed in the receptacle brakes as the air comes in." The student was not clear as to what was the nature of this pressure, but his previous answers suggest an understanding of what atmospheric pressure is. A similar conclusion can be reached regarding the answers of A_4, A_8, and A_{13}. These students' references to pressure were followed by terms such as "atmospheric" or "air" in the other items of the post-test. In addition, none of them indicated a belief that "pressure" was a "pull exerted by emptiness."

This last observation is significant as these students could, on occasion, conceive the existence of an internal suction caused by emptiness and that would be undone by the entrance of air with the breaking of the seal. This is the conception that may have been supported by A_{12}, to whom initially, internal pressure in the evacuated receptacle was *greater* than the exterior pressure. The same possibility can be speculated regarding the laconic answer given by A_1: "Because of the existence of a vacuum inside the package." In both cases, we cannot dismiss the possibility that they were thinking of a type of suction exerted by emptiness. These students' answers to the previous questions do not allow us to eliminate these conceptions.

21.4 Final Considerations

The participants demonstrated great interest throughout the intervention. In their initial reflections in the pre-test, overall, they made little reference to the concept of atmospheric pressure, even though they had already been formally in contact with the content in the classroom. In this first exploratory instrument, there was a range of explanations based on alternative conceptions similar to those found in the literature. An example was the indication that the liquid ascends to occupy the void left by air. Significant attention was given spontaneously by students to the interior of the objects mentioned in the questions. In general, they attributed less importance to the exterior. The internal sucking was related immediately and sufficiently to the rising of the liquid in objects such as the syringe. In some situations, the term "pressure" was used as indicative of suction, configuring an alternative conception in which the act of sucking is responsible for the creation of a pressure (Solaz-Portolès 2008; Longhini and Nardi 2009).

The didactic sequence seems to have contributed to an approximation to the concept of atmospheric pressure. A significant number of the students expressed adequate answers in the post-test. Satisfactory references to atmospheric pressure appear in a significant number of reflections made by students after the intervention.

Appendix 1: Pre-test

1. We put some water containing artificial coloring for better visualization in a bottle and in a receptacle. The bottle was turned upside-down into the receptacle containing the same liquid. The water in the bottle did not drain completely (Fig. 21.1). Why did it happen?

____

____

____

____

Fig. 21.1 Apparatus for question 1

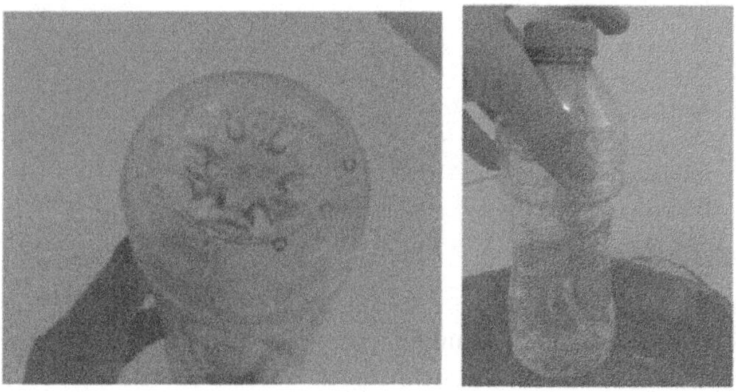

Fig. 21.2 Apparatus for question 2

2. A heated nail was used to make small holes in the bottom of a plastic bottle, which was then placed standing in a bowl of water. The bottle was filled with water, capped, and removed from the bowl, and was held over it until it stopped dripping. Even so, the water remained in the bottle (Fig. 21.2). How was this possible?

____

____

____

____

3. The picture represents a syringe, whose tip was dipped into liquid (Fig. 21.3). Why did the syringe fill up when its plunger was raised?

____

____

____

____

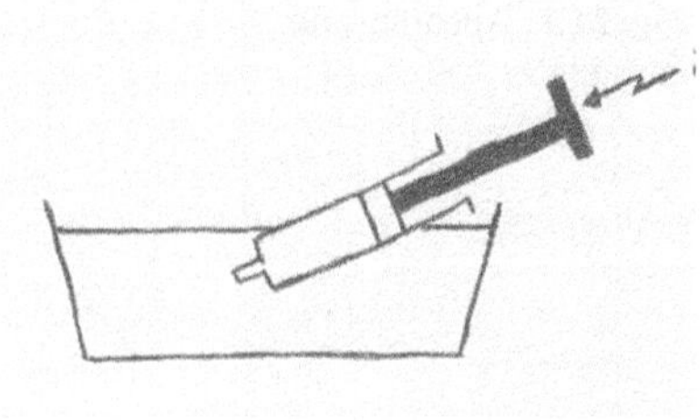

Fig. 21.3 Apparatus for question 3

Fig. 21.4 Apparatus for question 1

Appendix 2: Post-test

1. Many people install water fountains to attract birds to their houses (Fig. 21.4). How do water fountains for birds in cages work?

2. We filled a glass tube with mercury and closed its end with one hand. Then, we put the glass tube upside down in a receptacle containing mercury. Part of the mercury descended to the receptacle, whereas another portion of the liquid remained in the inverted tube. How could you explain that situation? (Fig. 21.5)

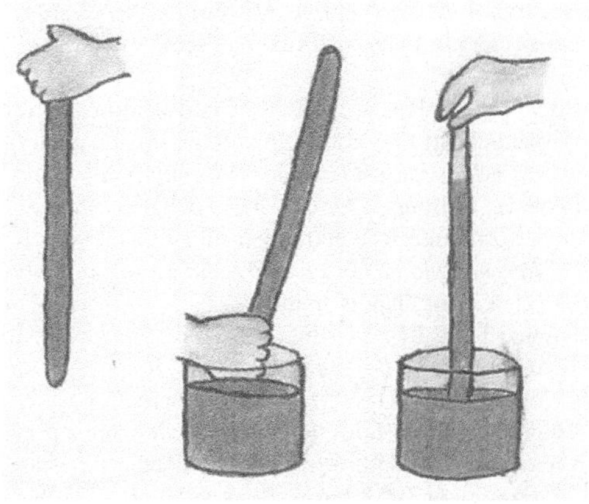

Fig. 21.5 Apparatus for question 2

Fig. 21.6 Apparatus for question 3

3. Vacuum-sealed packages are common in supermarkets. Cans of food usually contain a type of seal in their lids (Fig. 21.6). Why can these packages only be opened when the seal is removed?

____

____

____

____

References

Bardin, L. (2004). *Análise de conteúdo*. Lisboa: Edições 70.

Batista, G., Drummond, J., & Freitas, D. (2015). Fontes primárias no ensino de física: considerações e exemplos de propostas. *Caderno Brasileiro de Ensino de Física, 32*, 663–702. https://doi.org/10.5007/2175-7941.2015v32n3p663.

Boss, S. (2011). Traduções de fonte primária: Algumas dificuldades quanto à leitura e o entendimento. In *Anais do VIII Encontro Nacional de Pesquisa em Educação em Ciências*. Campinas: ABRAPEC.

Briccia, V., & Carvalho, A. M. (2011). Visões sobre a Natureza da Ciência construídas a partir do uso de um texto histórico na escola média. *Revista Electrónica de Enseñanza de las ciências, 10*, 1–22.

Bueno, M., & Pacca, J. (2009). Combinando a leitura de originais da ciência com outras atividades para construir o conhecimento na sala de aula. In *Anais do VII Encontro Nacional de Pesquisa em Educação em Ciências*. Florianópolis, SC: ABRAPEC.

Dion, S., & Loures, M. (2013). O debate realismo/antirrealismo em situações de ensino de física, à luz da interface entre história e filosofia da ciência. In C. C. Silva & M. E. B. Prestes (Eds.), *Aprendendo ciência e sobre sua natureza: Abordagens históricas e filosóficas* (pp. 193–208). São Carlos: Tipographia.

Forato, T. C. M., Martins, R. A., & Pietrocola, M. (2011). Historiografia e natureza da ciência na sala de aula. *Caderno Brasileiro de Ensino de Física, 28*, 27–59. https://doi.org/10.5007/2175-7941.2011v28n1p27.

Longhini, M., & Nardi, R. (2009). Como age a pressão atmosférica? Algumas situações-problema tendo como base a história da ciência e pesquisas na área. *Caderno Brasileiro de Ensino de Física, 26*, 7–23. https://doi.org/10.5007/2175-7941.2009v26n1p7.

Magie, W. (1969). *A source book in Physics*. Cambridge: Harvard University Press.

Martins, R. A. (1989). O vácuo e a pressão atmosférica, da Antiguidade a Pascal. *Caderno de História e Filosofia da Ciência, 2*, 9–48.

Martins, R. A. (2006). Introdução: A história das ciências e seus usos na educação. In C. C. Silva (Ed.), *Estudos de história e filosofia das ciências: subsídios para aplicação no ensino* (pp. xxi–xxxiv). São Paulo: Livraria da Física.

Matthews, M. (1994). *Science teaching: The role of history and philosophy of science*. New York: Routledge.

Mellado, V., & Carracedo, D. (1993). Contribuciones de la filosofia de la ciencia a la didáctica de las ciencias. *Enseñanza de las ciencias, 11*, 331–339.

Monteiro, M., & Gomes, C. (2003). Garrafa furada. In *Banco Internacional de Objetos Educacionais*, Ministério da Educação, Brazil. http://objetoseducacionais2.mec.gov.br/bitstream/handle/mec/11306/garrafa%20furada.pdf?sequence=1. Accessed 20 July 2016.

Saad, F. (2005). *Explorando os fenômenos da pressão do ar e dos líquidos através de experimentos simples*. São Paulo: Editora Preliminar.

Saltiel, E., & Viennot, L. (1985). Qué aprendemos de las semejanzas entre las ideas historicas y el razonamiento espontáneo de los estudiantes? *Enseñanza de las Ciencias, 3*, 137–144.

Silva, A. P., & Guerra, A. (2015). *História da ciência e ensino: Fontes primárias e propostas para sala de aula*. São Paulo: Editora Livraria da Física.

Solaz-Portolès, J. (2008). *Concepciones de los estudiantes e historia de la ciencia: El caso del concepto de vacío*. http://www.eumed.net/libros/2008c/451/. Accessed 7 Mai 2015.

Villatorre, A., Hilga, I., & Tychanowicz, S. (2008). *Didática e avaliação em física*. Curitiba: Editora Ibpex.

Chapter 22
The Status of the Lines of Force in Michael Faraday's Thought: History and Philosophy of Science in the Classroom

Sonia Maria Dion

22.1 Introduction

The propriety of including history and philosophy of science in the teaching of physics is a subject still under debate in the literature. This paper shares Michael Matthews' comprehensive view about scientific education, by postulating the need to help students, in addition to learning science, "to learn *about* science" (Matthews 1994 p. 81), which implies to incorporate the study of questions about "its changing methods, ..., its methods of proof, its interrelationships with the rest of culture and so forth" (Mathews 1994, p. 81).

Besides these methodological questions, there are others about the meaning and conditions related to the use of certain terms from the scientific discourse, such as law, model, verification, falsification, cause, truth, etc. "To learn *about* science" also involves making room for questions of an ontological nature, such as those that point to the basic entities of the world and lead us to wonder whether certain theoretical terms in physics in fact have empirical correspondence.

Furthermore, if discussion about the nature and limits of science is to be included in the classes of physics there arises, almost naturally, a role for history too. However, one should recognize that the articulation between history and philosophy of science in the classroom is not a trivial question. The results from the meta-analysis of history and philosophy of science in physics teaching case studies presented by Teixeira et al. (2012), for example, give an indication of this difficulty, as, out of 11 works chosen for the study, only five propose the simultaneous use of history and philosophy in the classroom.

S. M. Dion (✉)
Department of Philosophy, São Judas University, São Paulo, Brazil
e-mail: prof.sdion@usjt.br

M. E. B. Prestes, C. C. Silva (eds.), *Teaching Science with Context*,
Science: Philosophy, History and Education,
https://doi.org/10.1007/978-3-319-74036-2_22

A relevant aspect involved in this articulation, that adds to the debate concerning opposition to the inclusion of history in physics education[1] is the decision about the way to approach history, as the use of primary sources by teachers wishing to introduce it into science classrooms implies a risk of their improper, biased, less than rigorous appropriation. On the other hand, as the primary source in the history of science (HOS) can be a report of an invention or discovery, there emerges from that material a way to have access to the complex interaction between science and philosophy of science from the perspective of the theory of knowledge and/or ontology of the *authors* of science themselves.

This chapter involves in the interface among science, history, and philosophy in a way that tries to take these considerations into account. To the question posed by Dagher and Erduran, "what aspects of nature of science should be taught and learned" (Dagher and Erduran 2016, p. 147), it suggests as a workable answer the discussion about the status of theoretical terms in physics, among which the nature of the lines of force, as to whether or not they have an empirical meaning, could be seen as a sub-case.[2]

However, to avoid having a discussion *about* philosophy rather than a philosophical discussion, in this chapter, the incorporation of a primary source that allows the discussion to stem from some specific content from physics is proposed as a pedagogical strategy. The content proposed here is the change in the ontological status of the concept of lines of force in the thought of Michael Faraday.[3]

In the second section, the historical case is further examined to display the context underlying the ideas of Faraday and the reasons that justify them. We show that, starting from a conception that was merely instrumental, the scientist evolved to a realist point of view about the lines of force.

Students' views collected from the literature (Pocoví and Finley 2002; Pocoví 2007) indicate the attribution of ontological properties to the lines of force. These views are placed by the authors in the category of "matter based" according to the theory of conceptual change by Chi et al. (1994).[4] Ideas of Faraday also fall into the same category (Pocoví and Finley 2002, pp. 465, 471). In the third section, we propose an alternative interpretation of these results in which the materiality given

[1] For arguments regarding this opposition see Klein (1972), Whitaker (1979), and Russell (1984). For an analysis of the obstacles of implementing history and philosophy of science into classrooms see Höttecke and Silva (2011).

[2] Although lines of force do not have the same function as the theoretical terms, which are immediately present in the equations, we consider them here as a kind of "theoretical term," given their geometrical character, which simply represents an electrostatic or magnetic problem.

[3] It should be kept in mind that texts from primary sources do not always constitute a precise account of a scientific process, either because such an account may have been systematized by the author himself or herself, and emphasizes the results, or because he or she avoided publishing partial failures. We do not consider this to be the case with Michael Faraday.

[4] From this perspective, categorization is built according to the predicates assigned to the concepts. Certain concepts belong to the category "matter based" if, when the attributes associated with this category are predicated with it, the result is a meaningful sentence, either true or false.

to the lines of force is understood as a type of realism that, in the case of the students, will be justified in Bachelard's terms.

In the fourth section, we explore the historical case considered here to introduce a philosophical question in the context of the teaching of physics, once realism is taken as the link for the dialogue between reader and text. At the same time, we draw attention to the care required in this kind of appropriation, given the differences between the historical and the pedagogical contexts. We suggest the search for the reasons underlying the concepts as a way of reading effectively when using primary sources in HOS from the perspective of some type of communication between the ideas of students and scientists.

22.2 Michael Faraday and the Lines of Force: The Change of Ontological Status and Its Foundations

Michael Faraday introduces the idea of lines of force at the very beginning of his studies on electromagnetic induction, reported in his masterpiece *Experimental researches in electricity*.[5] Throughout this work and as his theory of matter changes, a transformation in the ontological status of this concept clearly emerges: interpreting them at first as a mere descriptive tool, a sort of chart, in the final years of his career, he assigns them physical reality and the power to interact with matter itself.

He first mentions lines of force in the context of his experimental research on the induction of electric currents, particularly when seeking an explanation for "Arago's magnetic phenomena."[6] The reference here is to "lines of magnetic forces," a name given to that "which would be depicted by iron fillings; or those to which a very small magnetic needle would form a tangent" (Faraday [1831] 1952 par. 114, n. 1, p. 281). In this case, their function coincides with that found in basic school textbooks nowadays, namely, to allow the visualization of a magnetic situation.[7]

In this early period, Faraday did not dedicate himself to thinking about the ontology of the lines, but employed them simply as a way of expressing himself through visual models, a feature extensively used in his work. His theoretical research focused on the nature of electric forces and was closely associated with a view about matter as consisting of particles capable of interaction with each other. Except in the

[5] Published in three volumes between 1839 and 1855, they consist of 29 series (Faraday [1839–1855] 1952).

[6] The essence of this problem lay in the difference in the effects coming from the existence or not of the relative movement between a copper plate and a magnetic needle suspended close to it: although rotation of one determined the movement of the other, the absence of relative movement between them did not produce any effect at all, neither attraction nor repulsion.

[7] Although today, from a conceptual point of view, we would not affirm that a "small magnetic needle would form a tangent" to the lines at different points, but rather understand that the tangent to a line of force gives the direction of the magnetic field.

case of universal gravitation, however, his theory denied action at a distance between particles, rather postulating an action chain, from particle to particle.

His experiments with electrolysis provided him with an experimental basis to support his conception of induction as action between contiguous particles:

> [...] the whole effect in the electrolyte appeared to be an action of the particles thrown into a peculiar or polarized state. I was led to suspect that common induction itself was in all cases an *action of contiguous particles*, and that electrical action at a distance ... never occurred except through the influence of the intervening matter. (Faraday [1837] 1952, par. 1164, p. 441)

In the action at a distance model, a particle is capable of acting on another one without interference from the intervening ones; thus, action takes place according to a straight line drawn between their locations. His view on "inductive action," on the other hand, proposes action between nearby particles. In other words, action moves through intermediate particles; therefore, the straight line is no longer required, and transmission following curved lines can now be admissible. In addition, curvature will become the characteristic Faraday sought to serve as an argument for contiguous action:

> If in straight lines only, though not perhaps decisive, it would be against my view; but if in curved lines also, that would be a natural result of the action of contiguous particles [...]. (Faraday [1837] 1952, par. 1166, p. 442)

This understanding about the propagation of electric action was not standard at the time. Faraday devoted the eleventh series of the "Researches" to provide responses, based on experiments, to possible objections from contemporary supporters of action at a distance. One of the experimental resources he created for this purpose was the "inductive apparatus." This device basically consisted of a set of two concentric conductive spheres of different diameters with the space between them filled by some nonconductive substance that would be put to test in the presence of electric tension.

From these and other similar experiments, Faraday concluded that "the courses of inductive action" were "disturbed [...] from their rectilineal form" ([1837] 1952, par. 1224, p. 452), a result interpreted as evidence for his theory:

> All this appears to me to prove that the whole action is one of contiguous particles, related to each other, not merely in the lines which they may be conceived to form through the dielectric [...] but in other lateral directions also". (Faraday [1837] 1952, par. 1231, p. 453)

In this model of particulate matter and electric action transmitted from particle to particle, lines of force are dissociated from matter: "I use the term *line of inductive force* merely as a temporary conventional mode of expressing the direction of the power in cases of induction [...]" (Faraday [1837] 1952, par. 1231, p. 453).This is a diagram, a visual model, which the scientist uses because of its representative power.

In the mid-1840s, Faraday starts to reformulate his theory of matter, which will lead to the change in the status of the lines of force.

Two articles published in 1844[8] mark this transitional period in his conception of matter, and the ideas presented there represent a step halfway toward the transfer of the action to the medium, in the form of lines of force.[9] These are articles of a speculative character, in which the scientist turns to extra-empirical criteria, such as simplicity and economy, and to an aesthetic principle that drives him to choose what he believes to be a "more beautiful" view about the constitution of the bodies (Faraday [1844] 1952, p. 854).

This is the view of atoms as "centers of force."[10] Beginning with the premise that "all our perception and knowledge of the atom [...] is limited to ideas of its powers" ([1844] 1952, p. 853), Faraday denies that the atom can be an impenetrable piece of matter, provided with an "atmosphere of force" around itself; his atom comes to be a mere mathematical point, from which forces expand. Thus being,

> [...] matter is everywhere present, and there is no intervening space unoccupied by it [...] that which is truly the matter of one atom touches the matter of its neighbours. (Faraday [1844] 1952, p. 854)

This model expresses a dynamic conception of nature, in which matter is indistinguishable from its powers. This notion may have resulted from the cross combination of extra-empirical influences, such as the tradition represented by Joseph Priestley (1733–1804) (Heimann 1971; Heimann and McGuire 1971), the belief in the fundamental identity of the forces of nature (Williams 1965), or even the conception of nature from his theology (Williams 1965; Levere 1968).

Priestley's vision exemplifies a dynamic conception of matter prevalent in Great Britain in the first half of the nineteenth century and that dated back to Gottfried Wilhelm Leibniz (1646–1716) and his choice of the concept of force as an explicative principle. For Priestley, the essential characteristics of matter are due to powers associated with it – "take away attraction and repulsion, and matter vanishes" (Priestley apud Heimann and McGuire 1971, p. 273).

Belief in the fundamental unity of nature also permeated the physical sciences in this period. This belief appears clearly when Faraday affirms that the forces of nature are "different manifestations of one fundamental power" (Faraday [1850] 1952, par. 2702, p. 670). Such conviction leads to the priority of force over matter.

Finally, although Faraday prized the explanatory potential of the dynamic conception of the atom as such, theology may have influenced his choice of this model, as his understanding that "God worked through powers and through forces" (Levere 1968, p. 101) would justify the existence and priority of the powers of the centers of force, "because it made the world more truly continuous with the Divine activity" (Levere 1968, p. 101).

[8] "A speculation touching electrical conduction and the nature of matter" ([1837] 1952, pp. 850–855), and "Matter" (published in Levere 1968).

[9] Although decisive for the evolution of the concept of lines of force, this period is not taken into consideration in the research of Pocoví and Finley (2002).

[10] In spite of the fact that Faraday explicitly ascribes this model to Boscovich (Ruggiero Giuseppe Boscovich, 1711–1787), several authors challenge this source, which triggered a historiographical problem that is beyond the scope of this chapter and, therefore, is not discussed here.

In the dynamic view, the concept of contiguity becomes irrelevant for the transmission of action, because matter is contiguous, it makes no difference whether particles are close or distant. In 1846, in the paper "Thoughts on ray-vibrations," the lines of force are associated with the centers of force. Faraday, however, did not turn to the investigation of the intimate nature of the lines of force until around July 1851:

> The time has arrived, when the idea conveyed by the phrase [lines of magnetic force] should be stated very clearly, and should also be carefully examined, that it may be ascertained how far it may be truly applied in representing magnetic conditions and phenomena [...]. (Faraday [1851] 1952, par. 3070, p. 758)[11]

He initially recalls the properties of the lines of force as a symbolic construct that facilitates thinking about the electric and magnetic powers of matter. In a subsequent passage, however, after revisiting some experimental results, he introduces the possibility of its physical existence:

> [...] but that wherever it may seem to represent the idea of the *physical mode* of transmission of force, it expresses in that respect the opinion to which I incline at present. (Faraday [1851] 1952, par. 3175, p. 778)

In his paper "On the physical character of the lines of force," published in *Philosophical Magazine*, in 1852, Faraday devotes himself to asserting the reality of the lines of force and its independence from matter. He argues for their physical existence based on the distinction between the properties of radiation, on the one hand, and gravity, on the other:

> When we turn to the radiation phenomena, then we obtain the highest proof, that though nothing ponderable passes, yet the lines of force have a physical existence independent, in a manner, of the body radiating, or the body receiving the rays. They may be turned aside in their course, and then deviate from a straight into a bent or curved line. (Faraday apud Fisher 2001, par. 3247, p. 572)[12]

The curvature of the action seems to be the decisive argument in favor of the reality of the lines. When considering the case of a magnetic bar placed in a free space, he affirms:

> It appears to me, that the outer forces at the poles can only have relation to each other by *curved* lines of force through the surrounding space; and I cannot conceive curved lines of force without the conditions of a physical existence in that intermediate space. If they exist,

[11] For Williams, Faraday's presentation and defense of his theoretical conceptions in this particular period owes much to the fact that his long career of experimental discoveries was finished in around 1850. This does not mean that the scientist abandoned the experimental approach, "but his experiments were now overtly the ammunition with which he supported theoretical positions taken up publicly and in print. His purpose was nothing less than to supply a general view of the modes of action of force" (Williams 1965, p. 445).

[12] The association between lines of force and the phenomenon of radiation is not accepted any longer; according to present understanding, light rays are a geometric resource that gives us the direction of the propagation of the wave fronts.

> it is not by a succession of particles [...] but by the condition of space free from such material particles.[13] (Faraday apud Fisher 2001, par. 3258, p. 572)

Previously, the lines represented phenomena associated with the particles of matter, but his final view postulates its independence of matter – lines of force are capable of existing and acting by themselves, they are real structures, and have a place and the power to interact with matter.

As seen above, the change in the ontological status of the lines takes place while Faraday reformulates his theory of matter. Their final character comes out when action is transferred to the medium, in terms of continuous forces filling space. This is an interpretation built on experimental results to which, however, a dynamic view of the world is added, and in which force has priority over matter. As a consequence, lines of force become the primary entities in the explanation of electric and magnetic phenomena.

It can be said, therefore, that, starting from a conception that we would classify as instrumentalist, Faraday ends up holding a realist position about the status of the lines of force.

22.3 Student's Ideas about Lines of Force and Their "Naive" Realism

One identifies in the students' thinking at least two ways of interpreting the lines of force: one compatible with its exclusively geometric nature, describing the electric configuration of a problem, and another that assigns materiality to them and responsibility for the transmission of the electric action (Pocoví and Finley 2002).

Predicates such as "describe," "are depicted," "are geometric lines," and "do not intervene in the electric action," when applied to the lines, are regarded as indicating the first kind of understanding (Pocoví 2007, p. 112). On the other hand, expressions such as "transporting charges," "having to be followed by the charges," "transporting forces," and "containing the charge" (Pocoví and Finley 2002, p. 469) indicate the attribution of ontological characteristics to the lines.[14]

---

[13] Faraday's explicit proposal that the lines exist independently of a "succession of particles" leads us to disagree with Pocoví and Finley, for whom the fact that the lines represent "a way of filling the space between the intervening charges or poles" "show his deep philosophical commitment to the idea of contiguous action" (2002, p. 472).

[14] We do not find it necessary here to look further at the theory of Chi et al. (1994), already mentioned in note 4 above, and the category of "matter-based" derived from it. For an explanation of its use in this context, see Pocoví (2007, p. 108). As our purpose is not to propose a method of conceptual change, it suffices to take as well grounded the statement that there is attribution of materiality to the lines by the students.

Even after experiencing a differentiated instructional activity,[15] students continued to assign ontological characteristics to the lines; the results of the interviews by Pocoví show that:

> [...] there is no doubt about the strong material commitment of the matter-based students: lines are material entities that can play different roles in the transmission of electrical effects. (Pocoví 2007, p. 129)

This inappropriate ontological categorization is not exclusive of the case of the lines of force. Chi et al., for instance, ascribed the difficulty in learning certain concepts to the fact that there is "incompatibility between the categorical representation that students bring to an instructional context, and the ontological category to which the science concept truly belongs" (Chi et al. 1994, p. 33). The question, formulated by Matthews and illustrative of the way in which students think, "excuse-me, but if no one has ever seen atoms, why are we drawing pictures of them?" (Matthews 1988, p. 168), seems compatible with the trend toward materialization diagnosed in the literature.

What are the reasons behind the attribution of materiality to the lines of force? If we pose the question in these terms, we may not get a satisfactory answer. We propose, instead, to say that students are *realist* about the lines of force, i.e., about the drawings whose function is to help visualize the geometry of a physical problem. If we formulate the question in these terms, the possibility of an answer comes out, and presents itself in Bachelard's terms.

For Bachelard, realism and substantialization correspond with each other. The philosopher understands substantialization as an epistemological obstacle that links the elements describing a phenomenon to their respective substance; it leads to the substantialization of the immediate qualities perceived by our direct intuition (Bachelard [1934] 1967, p. 118). It is not only a way of expressing or translating a phenomenon (ibid., p. 119), but a *conviction* that leads to the explanation of properties through the substance.

This substantialization of qualities or characteristics of a phenomenon follows "the method of immediate realism" (Bachelard [1934] 1967, p. 118). This realism is "naive" (Bachelard [1940] 1966, p. 42), instinctive, and has an affective basis; it constitutes the only "innate philosophy there is" (Bachelard [1934] 1967, p. 149). Such primal realism leads, for example, to the gathering of the forces in the substance "without realising that every force is a relation" (ibid., p. 117). From this perspective, substantialization would be homologous to being realist about the properties of and relationships between phenomena.

The image that Bachelard suggests for the progress of reason is a geometrical axis along which different philosophical positions are placed, and realism stands in the region of origin. This axis is not arbitrary; on the contrary, it "corresponds to the regular [or normal] development of knowledge" (Bachelard [1940] 1966, p. 46).

[15] Povocí's (2007) research used material from the history of science and was aimed at testing its effect on the way in which students learn the concept of lines of force as a geometrical entity, although avoiding the extensive use of exact citations from primary sources "because most of the times they are not easy to understand" (Pocoví 2007, p. 116).

Similarly, substantialism appears as "the first obstacle" to be overcome when developing objective knowledge (ibid., p. 52). We thus have, from these views about the initial form of knowledge, a justification to establish a relation of homology between substantialization and naive realism.

In the case of the lines of force, the materialization involves a concept that is only formal. This materialization occurs at the price of assigning to the lines properties such as "to contain" or "to transport" charges. Lines of force exist by themselves and are capable of influencing the space around them; in Bachelard's terms we see here a substantialization, a kind of "conviction" that expresses itself through an "immediate realism."

Therefore, if we take Bachelard as a frame of reference, the materialist perspective of the lines appears to result from the "regular development of knowledge" and, being so, presents itself somehow as necessary. From this understanding, at least two conclusions arise. The first one, cognitive in character, is the perception that the realism of entities, particularly regarding the lines of force, should be part of our expectations in relation to the initial conceptions of the students. The second, of methodological character, is discussed in the next section.

22.4 Faraday's Texts and the Views from the Students: Realism Concerning Lines of Force and the Required Knowledge of the Context

In Sect. 22.2 above, we saw that Faraday's ideas come close to the understanding of the lines as a tool that helps to visualize an electric or magnetic problem, for example, when he refers to them as "imaginary" or, in the case of magnetism, suggests drawing them from the figures obtained through the iron filings. The ideas of the students when they employ expressions such as "describe," "are depicted," "are geometric lines," etc., would also come close to this instrumental and desirable view about the lines of force.

On the other hand, there is in his texts from the 1850s an explicit realism concerning this concept. His position is revealed in expressions such as "*physical mode* of transmission of force" (Faraday [1851] 1952, par. 3175, p. 778), the statement about the "physical existence independent, in a manner, of the body radiating, or the body receiving the rays" (Faraday apud Fisher 2001, par. 3247, p. 572), and in his final and patent belief in the "physical existence" of the lines (ibid., par. 3258, p. 572).

If we take into account the analysis of the predicates, we find here as well, in principle, a correspondence between the thought of the scientist and the students when they make use of expressions such as "transporting charges," "having to be followed by the charges," "transporting forces"; in both cases, we notice the materiality, substantializing and realism regarding the lines of force.

From a pedagogical point of view, what use could we make of this parallelism? We see here a perspective of communication between the ideas of students and the scientist, at least two possible paths to establishing a dialogue between reader and text.

To begin with, we agree with Kampourakis in his seeing the question of "how to effectively address students' conceptions of science" (Kampourakis 2016, p. 2)[16] as the main challenge to the teaching of the nature of science, and in his conviction that "prior to such teaching, student's conceptions should be uncovered and analysed in detail" (ibid.). We understand that Faraday's texts can be useful in the process of identifying and analyzing the students' conceptions about the lines of force, given the identification of a common ground; namely, the presence of realism.

We saw, in particular, that Faraday's texts pose an objective philosophical problem, in which different views are presented and justified. To ask yourself about and understand the meaning of these views and the reasons that support them is to carry out a philosophical exercise; actually, an exercise enclosed in a virtuous circle, because the philosophical discussion may result in a better understanding of the physical content that initiated it.

On the other hand, the dangers involved in an inadequate appropriation of primary sources of HOS are well known and may result in a superficial or distorted history. No doubt this is a potential problem, as "it is unrealistic to expect students or prospective students to become competent historians [...] or philosophers of science" (Matthews 1988, p. 168). The historical case of Faraday illustrates this danger well. A more detailed analysis of the parallelism mentioned above shows that some restrictions must be put in place when context is taken into consideration from the perspective of either the scientist or the students.

In the case of the students, the realism/materialism was identified by examining the predicates that they associated with the lines of force. In the case of the scientist, though, no such analysis would be able to grasp the extent of his belief or the foundations of his realism.

We saw that the change in the ontological status of the lines takes place while Faraday reformulates his theory of matter. Having initially denied the particle model in which "powers" are added to matter, the scientist goes on to establish the concept of continuous action, in terms of forces filling the space, and only in the end does he turn the lines of force into agents, primary entities in the explanation of electric and magnetic phenomena.[17] His realist position concerning the lines, therefore, is a belief, a way of interpreting experimental results from the perspective of a dynamic vision of the world in which force has priority over matter.[18]

[16]We believe, however, that our "innate philosophy" and its requirement of realism should be added to the conditions that he presents as the foundations of these conceptions, i.e., "images imposed by culture, media, and other social representations of science" (Kampourakis 2016, p. 2).

[17]The fact that, in the case of Faraday, realism stems from a conscious process leads us to disagree with Pocoví and Finley regarding the classification of the scientist's view as "matter-based" in the sense given by Chi et al. (1994) (Pocoví and Finley 2002, p. 465).

[18]In current terms, his thought would resemble the trend of realism that affirms it to be possible to

Here, the parallelism starts to become more diffuse, as, although for Faraday the realist conception of the lines is a gradual and conscious construction, a choice, in the case of the students, it reflects the naive realism of an "innate philosophy." Also, whereas for the students the materiality of the line is their starting point, because realism constitutes their initial way of knowing, for Faraday, it comes *at the end* of the process, as a view built on and justified by the scientist's own terms and change of perspective concerning the relationship between matter and force.

From these considerations, mastery of both the ideas of the students and the thought of the scientist emerges as necessary. As we saw, the study of Faraday's *reasons*, connected with his different views about the structure of matter, and his *justifications*, grounded in experiments, but also on extra-empirical beliefs, imposes limits on the parallelism of their ideas. To take this into account is not only useful, but even necessary, to introduce rigor in the pedagogical appropriation of the text.

Taking these points into account, realism considered as an item of the historical case of the lines of force results in a link for the dialogue between reader and text, and as a workable way of introducing history and philosophy of science into the classroom too.

22.5 Conclusion

This chapter shows that the association of a philosophical question with the teaching of physics acquires meaning and relevance once a text from HOS is considered as a source to identify ontological points of view. The historical case of the lines of force, thus considered, can be taken as a first step to introducing a wider question such as the status of theoretical terms in physics and its place in the current debate between realist and antirealist positions in science.

On the other hand, we saw that the pedagogical appropriation of Faraday's ideas requires that certain limits are drawn. Although a sort of correspondence between the views of the scientist and the students can be established, we saw that their justifications are distinct. We then suggested that limits can be set by identifying the different reasons present in these different contexts, namely, historical and in the classroom. Taking the search for reasons as the guiding principle in the study of a text necessarily leads us to look into the history of the author's thought, problems that occupied him or her, and his or her relationship with contemporary thinkers, and this increases the rigor of the text's interpretation.

---

know reality at both observable and non-observable levels. In other words, true theories may refer to both observable and non-observable things and the relationships between them. Being so, it is assumed that theoretical terms in physics can refer to entities, which, although not directly observed, exist, in fact, in the physical world. For a summary of the varieties of realism, see Niiniluoto (1999). For an example of a view according to which we should be realist in relation to both the observable and the non-observable entities, see Devitt (1997); and for a presentation of epistemological realism illustrated with examples from the history of science, see Matthews (1994, chapter 8).

Although the risk of performing bad historiography in the classroom remains a potential problem, the more detailed textual studies about other content and authors are produced, and the more they are explained, the closer we come to the possibility of an effective and rigorous use of primary historical sources to introduce history and philosophy of science into the teaching of physics.

References

Bachelard, G. ([1934] 1967). *La formation de l'esprit scientifique*. Paris: Librairie Philosophique J. Vrin.

Bachelard, G. ([1940] 1966). *La philosophie du non*. Paris: Les Presses Universitaires de France.

Chi, M. T. H., Slotta, J. D., & de Leeuw, N. (1994). From things to processes: A theory of conceptual change for learning science concepts. *Learning and Instruction, 4*, 27–43.

Dagher, Z. R., & Erduran, S. (2016). Reconceptualizing the nature of science for science education: Why does it matter? *Science Education, 25*, 147–164.

Devitt, M. (1997). *Realism and truth*. Princeton: Princeton University Press.

Faraday, M. ([1839–1855] 1952). Experimental researches in electricity. In *Great books of the western world: Lavoisier, Fourier, Faraday, Robert Maynard Hutchins* (pp. 255–898). Chicago: Encyclopædia Britannica.

Fisher, H. J. (2001). *Faraday's experimental researches in electricity: Guide to a first reading*. Santa Fe: Green Lyon Press.

Heimann, P. M. (1971). Faraday's theories of matter and electricity. *British Journal for the History of Science, 5*, 235–257.

Heimann, P. M., & McGuire, J. E. (1971). Newtonian forces and Lockean powers: Concepts of matter in eighteenth century thought. *Historical Studies in the Physical Sciences, 3*, 233–306.

Höttecke, D., & Silva, C. C. (2011). Why implementing history and philosophy in school science education is a challenge: An analysis of obstacles. *Science Education, 20*, 293–316.

Kampourakis, K. (2016). (The) nature(s) of science(s) and (the) scientific method(s). *Science Education, 2*, 1–2.

Klein, M. J. (1972). Use and abuse of historical teaching in physics. In S. G. Brush & A. L. King (Eds.), *History in the teaching of physics* (pp. 12–18). Hanover: University Press of New England.

Levere, T. H. (1968). Faraday, matter and natural theology: Reflections on an unpublished manuscript. *British Journal for the History of Science, 4*, 95–97.

Matthews, M. R. (1988). A role for history and philosophy in science teaching. *Educational Philosophy and Theory, 20*(2), 67–81.

Matthews, M. R. (1994). *Science teaching: The role of history and philosophy of science*. New York: Routledge.

Niiniluoto, I. (1999). *Critical scientific realism*. Oxford: Oxford University Press.

Pocoví, M. C. (2007). The effects of a history-based instructional material on the students' understanding of field lines. *Journal of Research in Science Teaching, 44*(1), 107–132.

Pocoví, M. C., & Finley, F. (2002). Lines of force: Faraday's and students' views. *Science & Education, 11*, 459–474.

Russell, C. (1984). Whigs and professionals. *Nature, 308*, 777–778.

Teixeira, E. S., Greca, I. M., & Jr, O. F. (2012). The history and philosophy of science in physics teaching: A research synthesis of didactic interventions. *Science Education, 21*, 771–796.

Whitaker, R. J. M. A. B. (1979). History and quasi-history in physics education, Parts I, II. *Physics Education, 14*(108–112), 239–242.

Williams, L. P. (1965). *Michael Faraday*. New York: Basic Books.

Chapter 23
Analysis of Pedagogical Practices Carried Out in Continuing Education Activities for Physics Teachers: Limits and Possibilities

Sandra Regina Teodoro Gatti and Roberto Nardi

23.1 Introduction

Throughout the past few decades, research into science education has highlighted the relevance of the role played by history and philosophy of science (HPS) in the teaching and learning of sciences. At events, congresses, and in specialized journals on science education, there are many papers published dedicating particular space to this subject in Brazil (Martins 2007). However, an analysis made by Menezes in 1980 is still very up-to-date when dealing with the issue of actually embodying the theme in the classroom:

> [...] In the conventional teaching of physics, historical considerations etc. discoveries, has a merely illustrative or anecdotal role rather than being effective part of the educational process. You learn a science that seems to be structured marginally to the social and socio-economic conditions. (Menezes 1980, p. 94)

In this context, it is relevant to question what HPS is proposed? Allchin (1995) discusses how we should *not* teach history of science (HOS), by emphasizing serious issues such as presenting exclusively its products, neglecting the construction process, the difficulty in exposing mistakes, revealing the scientific method as an algorithm that leads to truth, among others.

Our study assumes that HPS should be incorporated into science teaching practices, not only to humanize the curriculum, but also to making explicit personal, ethical, and political interests involved in the historical contexts of the science construction. This implies a concern with teacher training, as the initial training with a license degree to teach cannot be expected to offer a final product, but it should be understood as a first point in continuous and extensive training (Garcia 1999).

S. R. T. Gatti (✉) · R. Nardi
Department of Education, School of Sciences, São Paulo State University, Bauru, SP, Brazil
e-mail: sandragatti@fc.unesp.br; nardi@fc.unesp.br

M. E. B. Prestes, C. C. Silva (eds.), *Teaching Science with Context*,
Science: Philosophy, History and Education,
https://doi.org/10.1007/978-3-319-74036-2_23

23.2 The Research

It is in this scenario of discussions on the importance of HPS in physics teaching and its consequences for teachers' training that our study is set.

From the results of a survey on the profile of sciences, physics, chemistry, biology, and mathematics teachers of a city in the state of São Paulo, Brazil, we proposed a course entitled "The history and philosophy science in the pedagogic practice of physics teachers", a 40-lesson continuing program designed for in-service high school physics teachers.

We followed the development of five in-service teachers who engaged in the course during the first semester of the school year of 2008 and were working with the discipline of physics in high school. We highlight that only two of them had a license to teach physics. That means that they had concluded the license in physics, an undergraduate program designed to train teachers for secondary and high schools in Brazil.

From the reflections occurring during the course, the teachers developed minicourse projects taking into account HPS aspects to be included in the teaching, in addition to considering the students' alternative conceptions. The high school teachers developed the minicourses in real classroom situations, supervised by professors during discussion meetings at the university.

Three years later, we got in touch with the same five in-service teachers, seeking a new partnership, but this time, while supervising internship projects of future physics teachers.

The idea was to invite the teachers to receive pre-service teachers as interns in their lessons and supervise them during activities regarding the above-mentioned course.

One of the five teachers who took part in the training course decided to receive, follow, and supervise the activities of a physics student, a future high school teacher, regarding her internship. The focus was on bringing HPS closer to the teaching, according to arrangements between the supervised practice advisor, the first author of this paper, and the in-service high school teacher who had taken part in the project formerly.

The focus of this step was to understand the reflexes of the continued training course previously taken by the in-service teacher, in terms of:

1. Teacher training: what remains in their discourses?
2. Reflexes in classroom teaching practice.

This study seeks to revisit the continued training actions and discuss their implications for the context of physics education and research through a qualitative approach. The treatment of information obtained was guided by the *content analysis*, according to Bardin (1994).

Table 23.1 summarizes the whole project.

Table 23.1 Research steps

First step
Gathering conceptions
Second step: the training course
1. Theoretical foundations about the teaching and learning processes of physics
2. Philosophy and history of science and physics teaching
3. Recent research on physics education
4. Teacher training and professionalization
5. Development of activities
Third step
Applying the "minicourses" proposals in high school
Running seminars with reports on the participants' experiences
Analyzing the participants' perceptions at the end of the process
Three years later
One of the five teachers who took part in the course (Denise) supervised the academic practice of an undergraduate student (from the physics license). The focus is both on the classroom preparation and on the development of activities involving history and philosophy of science aspects
Seeking marks of the previous training from two points of view:
The teacher's discourse
The teacher's practice

23.2.1 *The Activities of the Course*

After the early stage, which involved the gathering of the participants' conceptions on the scientific knowledge construction (Aikenhead and Ryan 1992), the processes of teaching and learning, and the limits and possibilities in bringing HPS closer to teaching, we proceeded with the development of the training course proper.

The first part was dedicated to the study of a few aspects of philosophy of science, such as distinguishing myth and scientific thinking, in addition to the main scientific conceptions of science: rationalist, empiricist, and constructivist (Chalmers 1993; Chauí 1999).

Subsequently, we discussed papers on the theories of contemporary science philosophers (Ostermann 1996; Silveira 1996), such as Kuhn, Popper, and Lakatos, seeking to work with secondary sources with the objective of debating different models to define how science is elaborated in contrast with the conceptions revealed during the initial collection.

The arguments that justify bringing philosophy of science closer to physics teaching were also discussed (Adúriz-Bravo et al. 2002). Regarding history of science (HOS) in physics education, we used a few texts (Matthews 1994; Monk and Osborne 1997; McComas et al. 1998; Martins 2007). These texts present theoretical grounds of the research defending the incorporation of HPS into education to:

1. Contribute to the humanization of science, revealing personal, ethic, cultural, and political interests

2. Make the lessons more interesting
3. Stimulate the discussion and formation of critical thinking
4. Overcome the mere reproduction of formulations and equations, that often have no meaning for the students
5. Understand that the theory currently accepted had been the object of serious opposition
6. Provide support for the current conceptions to be analyzed critically.

The historical content developed from the text was prepared by the authors seeking to establish a discussion on the construction of the gravitational attraction theme in an attempt to demystify the view of a neutral, cumulative, linear science (Teodoro 2000; Gatti 2005). In this context, our concern was to select the historical content that would allow us to achieve our objectives.

We also debated topics related to studies in the field of science education such as research into *alternative conceptions* and *conceptual change* (Santos 1991), curricula approaching *science, technology, society, and environment – STS* (Santos and Mortimer 2002; Cazelli and Franco, 2001) in addition to *methodological pluralism in science and science teaching* (Laburu et al. 2003).

During the development of the training course, participants were able to select one theme to build up their minicourse proposals.

23.3 The Proposals of the Minicourses, Impressions and Reflections on the Results of High School Classroom Application

For the discussion in this paper, we selected Greice's activity, assessed with the highest degree of engagement, and Denise's, assessed with the lowest degree of engagement. The classification regarding the level of engagement refers to the participant's concern in employing the aspects studied throughout the training course in their activities.

For this purpose, we initially present a synthesis of the participants' lesson plans, including a discussion on the proposal suggested by the teachers and what was actually developed in the classroom, with their justifications and impressions exposed at the seminars for further presentation and reflection on the experience developed.

23.3.1 Greice's Minicourse

As summarized in Table 23.2, Greice's minicourse proposal encompasses the theme "energy," seeking to incorporate the innovations discussed throughout the course by exposing the students' conceptions, using HOS to establish the reflections with the students and discussing the STS issued.

Table 23.2 Synthesis of the planning suggested by teacher Greice

<table>
<tr><td colspan="5">Greice</td></tr>
<tr><td colspan="5">Second year of secondary school</td></tr>
<tr><td colspan="5">Theme: energy</td></tr>
<tr><th>Lessons</th><th>Content</th><th>Activity developed</th><th>Objectives</th><th>Assessment</th></tr>
<tr><td>1 and 2</td><td></td><td>Applying a question to collect the students' conceptions on what energy is</td><td>Assess the students' alternative conceptions on the theme</td><td>Individual answers</td></tr>
<tr><td>3 and 4</td><td>1. Historical evolution of the energy concept</td><td>Reading and discussion of the text prepared by the teacher</td><td>Allow students to recognize physics as a human construction, aspects of its history and relationships with the cultural, social, political, economic, and philosophical contexts</td><td>Students' participation</td></tr>
<tr><td rowspan="2">5 and 6</td><td>2. Sources of renewable and nonrenewable energy</td><td>Research in small groups as a task</td><td rowspan="2">Discuss matters such as the importance of saving energy and environmental impact</td><td rowspan="2">Question: Why should we save energy?</td></tr>
<tr><td>3. STS relationships</td><td>Debate</td></tr>
<tr><td rowspan="2">7 and 8</td><td>Kinetic, potential gravitational, solar, thermal, chemical, nuclear, wind, light, and sound energy</td><td rowspan="2">Dialogue-based expository lesson</td><td rowspan="2">Employ instruments of mathematical calculations to solve problems involving energy conservation</td><td rowspan="2">Solving problems</td></tr>
<tr><td>Conservation of mechanical energy</td></tr>
</table>

About the justification on the theme decision, the teacher reveals that:

> […] I wanted to approach a theme that was closely related to their daily routine. I wanted to lead them to reflect on the increasing necessity to use new energy sources and the environmental impact it has generated. I thought it would be a very good theme to approach the STS issue that we had studied in the course. (Greice' testimony)

Regarding the use of HPS, Greice comments that:

> […] The ground to develop the planning was that text by Matthews (1994) we had studied, where he talks about the advantages of using Science, how to make knowledge more interesting, reveal a view of Science that has been built up. (Greice' testimony)

The activities proposed for lessons 3 and 4, entitled "Historical evolution of the energy theme," were developed by using a text extracted from the book "Conceptual physics" (Hewitt 2002) and adapted by the teacher, who sought to discuss, among other aspects, mathematics as the language of science, scientific methods, and scientific attitude.

The teachers' attempt was to complement the discussion by arguing that the concept of energy still poses a challenge to the scientists, despite having been unable to treat the historical matter more deeply owing to the short time available.

The focus of the remaining activities is the STS relationships.

After the minicourse, the teacher's speech reveals that:

> [...] I tried to show that science, in contrast with what people generally believe, is not something finished. Every single day a scientist is working on a model. [...] I sought to reinforce them that it means that what we studied today on books can be rewritten tomorrow, replaced with a new theory. (Greice' testimony)

It is important to emphasize that the change in the teacher's posture regarding her practice along the development of the research was significant, as, considering her major in mathematics, she reported many difficulties in developing activities that enabled her students' participation, having felt insecure toward their possible doubts about the subjects.

Despite the difficulties, the teacher reveals that the students approved the distinguished and reflexive approach.

The students enjoyed the distinctive activity, but the major difficulty is that they have trouble reading and interpreting texts. In addition, they had an initial resistance derived from the fact that they think reading is a practice associated only with history or Portuguese classes.

About the impact of the experience on their practice as a teacher, she reveals that:

> I used to work Physics through short, summarized definitions and exercises. That would be all. I never spoke of History or Philosophy of Science. The different thing I did was using articles from magazines or newspapers. Everything else was just applied Mathematics. Now, I have realized a new way to understand these issues. (Greice' testimony)

At the end of her description, when questioned about the possible contributions of the experience developed to her training/practice, and whether she intended to continue to use the innovations discussed during the course, Greice reveals that:

> Currently I have been more concerned because I feel more responsible for monitoring the students. That issue of the teacher's autonomy, investigating the practice (...) it also makes us more responsible. Now I try to plan my lesson based on what I can do considering the things we had studied and I have been seeking to analyze the approach with the best results. (...) The course brought many contributions to my training, among others, it consolidated methods that were already part of my classroom practice, enlarged the "universe" of researches diversifying the pedagogic material used in class and promoted discussions on themes related to the students' daily routine clarifying doubts and indicating solutions. (Greice' testimony)

23.3.2 *Denise's Minicourse*

The proposal of this teacher's minicourse approaches the theme "The structure, properties and transformations of matter," as shown in Table 23.3. Denise elaborated on her planning from the historical construction of the theme as a way of

Table 23.3 Synthesis of the planning

Denise				
Second year of high school				
Structures, properties, and transformations of matter				
Lessons	Content	Activity developed	Objectives	Assessment
1 and 2	1. Ancient history: the early days of the chemistry theory on matter	Reading and discussion of the text prepared by the teacher	Arouse the interest of the students in the subject	Students' participation
	2. Empedocles, Aristotle, Leucippus, and Democritus		Show the evolution of the concepts	
3 and 4	3. Alchemy	Reading and discussion of the text prepared by the teacher	Show the evolution of the concepts	Students' participation
	4. The scientific revolution: Paracelsus and Stahl		Discuss the construction of the scientific knowledge	
	5. The phlogistic theory			
	6. Lavoisier and the modern age			
5 and 6	7. Atomic models: Dalton, Thonson, Rutherford, Böhr	Reading and discussion of the curriculum of the state of São Paulo (2008)	Discuss the construction of the scientific knowledge	Students' participation
7 and 8	8. Continuing the previous lesson	Reading and discussion of the curriculum of the state of São Paulo	Discuss the construction of the scientific knowledge	Solving problems

introducing the discussion presented in the new curriculum of the State of São Paulo (2008) regarding the discipline of chemistry.[1]

The lessons planned by the teacher were supposed to be developed within 4 weeks, with the first two lessons dedicated to studying the historical construction of the theme from the texts that she had prepared. The last part of the planning involves the curriculum of the State of São Paulo.

Because of the short time available, in addition to the schedule of activities suggested in the curriculum, the lessons had to be condensed, and ended up lasting 2 weeks.

> I was not able to proceed as had planned since the school strongly demands that all chapters of the textbook are satisfied. […] They say that the teacher can discuss other issues, can go further […] But we hardly have time to do what is proposed. I had little time to cover the historical part because it takes too long, demands reading, you know? [..] I wanted them to see the evolution of Science. (Denise' testimony)

[1] Despite her experience as a high school physics teacher, Denise chose to elaborate her minicourse with a focus on chemistry, her early training.

Denise's speech expresses a certain frustration about the loss of teacher's autonomy with the introduction of the new curriculum of the State of São Paulo. The proposal does not include gathering students' conceptions or the STS relationships. The assessment has only one stage, at the end of the sequence, using a list of exercises.

Despite these issues, the teacher tried to incorporate the themes discussed during the course, such as the construction of scientific knowledge and HOS, revealing an intention to make them part of her teaching practice:

> They enjoyed the historical section that I had elaborated. I sought to emphasize the construction, *the dynamic movement of Science. Despite the short time available, I intend to try to incorporate history whenever possible in my lessons.* (Denise' testimony, with emphasis on the teachers' intention)

At the end of her speech, when questioned about possible contributions of the experience to her training/practice, and whether she intends to continue using the innovations discussed during the course, Denise reveals that:

> It enlarged my awareness on the aspect of introducing finished contents from textbooks and/or pedagogic proposals. There should be greater reflection on how to introduce the content, always seeking greater accuracy of the scientific foundations involved. Nowadays, people in general, especially youngsters and teachers, assimilate a finished knowledge, we acquire and use many technologies without even knowing their functioning principles or the scientific knowledge that these technologies incorporate. A lot of History remains untold, unknown and without understanding! In addition, I have been thinking about my practice so much more by trying to identify ways to facilitate the students' learning. (Denise' testimony)

23.3.3 *Brief Synthesis*

Table 23.4 is a synthesis of the ideas suggested by the teachers regarding the themes discussed during the course that should be incorporated in the proposals. It is important to emphasize that the table analyzes data referring to two teachers who carried out all of the activities.[2]

23.4 The Sequence of the Training Program: Reflections and Reflexes 3 Years Later (Gatti and Nardi 2016)

We carried out the following research steps aiming at investigating, 3 years later, whether the teachers who took part in the training course were able to incorporate those discussions into their teaching practice. Furthermore, we intended to analyze

[2] Teacher Viviane participated in all of the activities of the course during the first semester of the school year. During the second semester, when we intended to follow the elaboration and application of the minicourses, the teacher was not present. Therefore, we do not have data concerning the work that she developed.

Table 23.4 Comparing the minicourses

Level of the proposal engagement	Aspects incorporated into the minicourses of the teachers	Alternative conceptions	History of science	Philosophy of science	CTSA	Activities for students' development and reflection	Assessment
More engaged	Greice	Yes	Yes	Yes	Yes	Yes	Formative
	Denise	No	Yes	Yes	No	No	List of exercises at the end of the sequence
Less engaged							

how that previous training could influence their work as advisors in supervised practice for future physics students. We invited all of the teachers, but only Denise was able to take part in this stage.

Undergraduate student Fernanda followed teacher Denise throughout the second semester of the school year of 2011 in activities of supervised academic practice in addition to participation in study and discussion meetings. Her work was followed by researchers of the university, and was supervised and assessed by the teacher of the school unit.

At the end of the process, Fernanda was supposed to elaborate and apply a pedagogical sequence about a given physics topic previously discussed and arranged with the high school teacher. The paper should necessarily take into account the idea of bringing HPS closer to the teaching.

At this stage, we sought to analyze the potential of a collaborative effort regarding the early training of teachers in addition to investigating the participation of the school unit teacher searching for the marks brought by her previous training in her current supervising actions.

At the end of the activities developed in the school, we interviewed the participants and obtained their impressions on the experience, including their main difficulties.

23.4.1 Fernanda's Impressions

During Fernanda's supervised academic practice in the school, one of the matters she was supposed to observe involved the development of teacher Denise's lessons, as our intention was to investigate whether the aspects of the discussions carried out during the course, in 2008, were present in the teacher's actions.

Fernanda's observations report the teacher's attempt to introduce some of the discussions on HOS into her lessons.

> I realized that she tries to talk about some issues with the students in an effort to introduce this matter of the construction of Science. However, it happened sort of in isolation [...] It seems like she was just giving some hints about it, you know? I talked to her about it and she told me that the classroom daily routine demands too many activities from the textbook, which hampers an enlarged approach. (Fernanda' testimony)

Once again, the curriculum of the State of São Paulo is mentioned, in addition to the pressure involving the completion of the schedule of activities, as aspects that hamper the introduction of actions to take into account matters of a more reflexive nature regarding the scientific knowledge construction.

23.4.2 *Teacher Denise's Impressions: Where Is the Interference of the Educational Research into the University Training Course?*

At the end of the internship, we interviewed Denise. She participated with us in the activities, offering support, suggesting material, and advising the student Fernanda. The fact that this teacher suggests that the student develop the same theme she had worked with during the course is interesting. Her justification still revealed an insecurity to work with the theme, preferring to stay in her comfort zone.

The exchange of experiences becomes evident in her speech, for example, when she reports that she did not know the resources available on the website of the Colorado University (www.phet.colorado.edu) and her desire to understand and make use of them in the classroom.

Regarding the general assessment of the student's work, Denise remarks that:

> The lesson developed based on the subject structure (evolution of atomic models) was well planned, considering the content approached, methodology, and multimedia resource. The approach on the historical aspects was very well conducted regarding the development of different atomic models by many scientists back in their own time as well as the reasons that led them to elaborate different atomic models. The representational aspects of the atomic models to explain the theories on the matter structure was very well explored, making it clear for the students that this is an important aspect of Science. The student had a great posture, interacting very well with the students through questions and answers, demonstrating clarity, objectivity in the process of building up the knowledge approached. (Denise' testimony)

Regarding the introduction of the aspects referring to HPS included in the lessons, the teacher reveals that:

> This aspect was very well discussed, once it was asked what the thoughts on the structure of matter throughout the ages, from ancient Greece to the philosophical speculations on the theory of the four elements evolving into the first atomistic scientific conceptions of the chemical elements (18th century), grounded on empiricism. (Denise' testimony)

Denise's speech reveals the contributions of bringing HOS closer to education, aspects that were discussed during the course in 2008, according to her report below:

> I think it is important to approach both the historical and the philosophical aspects of science teaching since it represents a way to humanize Science, making it clear for the students that it is a human construction throughout time. The scientific knowledge is built up at different ages and historical, cultural contexts, which should be taken into account and not only presented to the students, as something finished. (Denise' testimony)

Regarding the difficulties involved in this approach, the teacher refers to the necessity of covering the content present in the curriculum of the State of São Paulo in addition to the need to prepare the students to take the Brazilian College entrance examination, for example, hampering an approach that would take longer to develop. Furthermore, the teacher also reports her need to seek deeper knowledge on the theme.

In synthesis, the marks of the previous training appear:

In the teacher's speech

- Importance of emphasizing the scientific knowledge construction in education
- Suggestions of reading and references studied on the training course
- Justifications present in the literature for bringing the HPS aspects to education.

In her actions

- Incorporates some aspects specifically
- Reports difficulties in terms of managing the time available along with the necessity for further development.

23.5 Final Remarks

Our study sought to introduce a discussion on the possibility of bringing HPS closer science education in the context of activities to be included in the early training and professional development of in-service teachers (Gatti and Nardi 2016).

The central issue was to assess the impact of a continued training project on the teaching actions of the participants. The analysis of the research trajectory allows us to point out some important aspects and their implications both to science education and to teacher training.

We started our study by revealing the preconceptions of the teacher on the scientific knowledge construction and the processes of teaching and learning, in addition to assessing their opinions on the possibility of introduce HOS into teaching (focus group interview, from which the testimonies were extracted that were reported in Sect. 3).

The results reveal indicators that corroborate research in the field of science education, highlighting the presence of common sense ideas on science and processes of teaching and learning in addition to their influence on teaching actions (Gil Perez 1991; Mellado 1996; Hewson et al. 1999; Levy and Sanmartí Puig 2001, among others).

Regarding the conceptions of the teaching/learning processes, the teacher demonstrated great attachment to the traditional model, based on passive transmission and receipt of knowledge.

Despite recognizing the importance of bringing HOS closer to teaching, the teacher revealed a series of obstacles for it to be put into practice, such as the lack of interest and students' previous knowledge, in addition to low salaries and terrible work conditions for the teachers.

Another significant aspect is that the teacher had not had contact with philosophy of science; therefore, she was not able to analyze its possible contributions to the teaching of physics or chemistry.

These initial data provided an overview that was used to plan the actions during the training course; in this context, we developed our study aimed at building along with the teachers an investigative attitude toward the practical problems.

The training model suggested by a course proved important by contributing to increasing the participants' knowledge, in addition to offering moments of reflection on teachers' professional practice. However, it is important to emphasize that the teachers' training proposal does not refer to a stagnant development, too theoretical and without a concern with practical application.

The training course was just a starting point and the live activities were complemented with practical actions developed in real scenarios of the high school.

It is worth mentioning the following aspects:

1. The course was the first contact of all of the participants with philosophy of science
2. The debates generated reflections on the processes of teaching and learning, as the traditional form of introducing the content as historically accumulated facts was questioned
3. The activities developed were aimed at emphasizing the investigative role of the teacher such as when gathering the students' conceptions to lead to a further group reflection on the results, contributing to the questioning of the traditional view of the processes of teaching/learning
4. The curriculum of the state of São Paulo, by incorporating the results of research in the field of science teaching, does not contribute to the teachers' training if they do not have access to the field discussions
5. Despite the good results obtained and the active participation of the teachers in the process, the possibilities of a permanent approach in the teachers' actions regarding our results are still restricted by unfavorable conditions of the educational reality, such as excessive workload, overcrowded groups, and the pressure to meet the curricular objectives integrally, among others.

The seminars to discuss the results of the development of activities in real high school scenarios enabled an interchange between the teachers' experiences, establishing an open dialogue that unveiled their difficulties.

Although we considered that the activities selected were adequate and well developed, this does not ensure permanent changes in the actions of the participants.

Even though the teachers had pointed out the importance of the elements debated during the course to science teaching and their intention to continue to employ the innovations, at this stage we did not have elements to indicate what the repercussions of the experience would be for the teachers' actions.

For this purpose, 3 years later, we reestablished the discussions, seeking to understand possible influences on the training program.

Denise's speech shows marks of her previous training, carried out at the university. Her supervision actions bring out concepts associated with HPS, including reflections on the advantages of bringing these aspects closer to education, according to the literature (Matthews 1994; Adúriz-Bravo et al. 2002).

We can also observe her concern in suggesting the theme, the same theme she worked with during the course at the university, aimed at facilitating her interaction with the undergraduate student, a future physics teacher. In contrast, it also reveals the teacher's insecurity regarding specific content, which led her to suggest a theme with which she was already familiar.

Her actions reveal attempts to introduce aspects of HPS on several occasions; however, it occurred almost exclusively with comments while introducing a new concept.

In light of the references that support our research encompassing teacher knowledge (Tardif 2000), autonomy (Contreras 1997), and the intellectual, critical role of the teacher (Giroux 1997), we can discuss the results seeking to understand the main difficulties for the consolidation of new practices into the daily routine of schools.

One of the main points that we can highlight is the difficulty with specific content both associated with physics itself and with the historical development; it is essential to understand the process as difficulties with the knowledge of different subjects tends to create insecurity and consequently resistance. In this context, it is important to highlight that the teacher who we followed up to the end of this process did not have an early training in physics.

In addition, the issue of the curriculum of the State of São Paulo was pointed out on several occasions as restrictive for the development of pedagogical sequences that would involve more time and deeper knowledge.

Despite the difficulties involved in the process, the model of collaborative work between the school and the university proved to be an important instrument for both the initial and the continued training of teachers, which required extension, aimed at including more secondary and high school teachers and students as partners and not mere consumers of the results of the research developed in universities.

References

Adúriz-Bravo, A., Izquierdo, M., & Estany, A. (2002). Una propuesta para estructurar la enseñanza de la filosofía de la ciencia para el profesorado de ciencias en formación. *Enseñanza de Las Ciencias, 20*(3), 465–476.

Aikenhead, G. S., & Ryan, A. G. (1992). Students' preconceptions about the epistemology of science. *Science Education, 76*(6), 559–580.

Allchin, D. (1995). How NOT to teach history in science. In F. Finley, D. Allchin, D. Rnees, & S. Fifield (Eds.), *Proceedings of the third international history, philosophy, and science teaching conference* (Vol. 1, pp. 13–22). Minneapolis: University of Minnesota.

Bardin, L. (1994). *Análise de conteúdo*. Lisboa: Edições 70.

Cazelli, S., & Franco, C. (2001). Alfabetismo científico: novos desafios no contexto da globalização. *Ensaio, 3*(1), 1–18.

Chalmers, A. (1993). *O que é ciência afinal?* São Paulo: Brasiliense.

Chauí, M. (1999). *Convite à filosofia*. São Paulo: Ática.

Contreras, J. (1997). *La autonomía del profesorado*. Madrid: Morata.

Garcia, C. M. (1999). *Formação de professores: para uma mudança educativa*. Portugal: Porto Editora.

Gatti, S. R. T. (2005). *Análise de uma ação didática centrada na utilização da história da ciência*. Tese de doutorado, Faculdade de Educação, Universidade Estadual de Campinas, Campinas, SP, Brasil.

Gatti, S. R. T., & Nardi, R. (2016). Reflexões sobre um projeto de formação continuada de professores de física envolvendo a história e afilosofia da ciência: o que permanece nas intenções e o que chega à sala de aula. In S. R. T. Gatti & R. Nardi (Eds.), *A história e a filosofia da ciência no ensino de ciências: A pesquisa e suas contribuições para a prática pedagógica*. São Paulo: Escrituras.

Gil Perez, D. (1991). Que hemos de saber y saber hacer los profesores de ciencias? *Enseñanza de las Ciencias, 9*(1), 69–77.

Giroux, H. A. (1997). *Os professores como intelectuais: rumo a uma pedagogia crítica da aprendizagem*. Porto Alegre: Artes Médicas.

Hewitt, P. (2002). *Física conceitual*. Porto Alegre: Bookman.

Hewson, P. W., Robert Tabachnick, B., Zeichner, K. M., & Lemberger, J. (1999). Educating prospective teachers of biology: Findings, limitations and recommendations. *Science Education, 83*, 373–384.

Laburú, C. E., de Melo Arruda, S., & Nardi, R. (2003). Pluralismo metodológico no ensino de ciências. Ciência e Educação, 9(2), 247–260.

Levy, M. I. C., & Sanmartí Puig, N. (2001). Fundamentos de un modelo de formación permanente del profesorado de ciencias centrado en la reflexión dialógica sobre las concepciones y las prácticas. *Enseñanza de las Ciencias, 19*(2), 269–283.

Martins, A. F. P. (2007). História e filosofia da ciência no ensino: Há muitas pedras nesse caminho. *Caderno Brasileiro de Ensino de Física, 24*(1), 112–131.

Matthews, M. (1994). Historia, filosofia e enseñanza de las ciencias: la aproximación actual. *Enseñanza de Las Ciencias, 12*(2), 255–271.

McComas, W. F., Almazroa, H., & Clough, M. P. (1998). The nature of science in science education: An introduction. *Science & Education, 7*, 511–532.

Mellado, V. (1996). Concepciones y prácticas de aula de profesores de ciencias, en las formación inicial de primaria y secundaria. *Enseñanza de las Ciencias, 14*(3), 289–302.

Menezes, L. C. (1980). Novo (?) método (?) para ensinar (?) Física (?). *Revista Brasileira de Ensino de Física, 2*(2), 89–97. http://www.sbfisica.org.br/rbef/pdf/vol02a19.pdf

Monk, M., & Osborne, J. (1997). Placing the history and philosophy of science on the curriculum: A model for development of pedagogy. *Science Education, 81*(4), 405–424.

Ostermann, F. (1996). A epistemologia de Kuhn. *Caderno Catarinense de Ensino de Física, 13*(3), 184–196.

Santos, M. E. V. M. (1991). *Mudança conceptual na sala de aula: Um desafio pedagógico*. Lisboa: Livros Horizonte.

Santos, W. L. P., & Mortimer, E. F. (2002). Uma análise de pressupostos teóricos da abordagem CTS (Ciência, Tecnologia e Sociedade) no contexto da educação brasileira. *Ensaio, 12*(2), 1–23.

São Paulo, Secretaria da Educação do Estado de São Paulo. (2008). A Proposta Curricular do Estado de São Paulo. In *São Paulo: SEE*.

Silveira, F. L. (1996). A filosofia da ciência de Karl Popper: o racionalismo crítico. *Caderno Catarinense de Ensino de Física, 13*(3), 197–218.

Tardif, M. (2000). *Saberes docentes e formação profissional*. Petrópolis: Vozes.

Teodoro, S. R. (2000). *A história da ciência e as concepções alternativas de estudantes como subsídios para o planejamento de um curso sobre atração gravitacional*. Dissertação de Mestrado em Educação para a Ciência, Universidade Estadual Paulista, Bauru, SP, Brasil.

Part IV
Philosophy of Science

Chapter 24
Philosophy of Science in Science Teacher Education: Meeting Some of the Challenges

Ana C. Couló

24.1 Philosophy of Science in Science Teacher Education

A reasonable familiarity with philosophical content, skills, and attitudes can play an important role for a science teacher (or any teacher) in becoming an educator "as distinct from just coach, instructor or merely the teacher of subject matter" (Matthews 2015, p. 412). G. Biesta (2012) argues for educational wisdom and a focus on the ability to make sound educational judgments in teacher education rather than the exclusive concentration on educational competencies and psychological or social theories of learning that generally seem to rule the field of teacher education. Science teacher education seems to be no different in this respect (Lee et al. 2009). Questions of educational purpose, content, and relationships should have pride of place in teacher education and practice, and here philosophy can make an irreplaceable contribution. Along similar lines, philosophy has an important role to play when teachers or researchers concern themselves with general normative questions of duty, obligation, necessity or desirability, or with a well-educated disposition for critical inquiry. In L. Darling-Hammond's words,

> […] such teachers would be prepared to wrestle deeply with curricular questions and assume leadership based on serious examination of the social and pedagogical implications of different decisions. (Darling-Hammond 2008, p. 1317)

When thinking about philosophy for science teacher education, epistemological and philosophy of science (POS) content stands in the limelight. Philosophy of education also comes to mind, whereas the relationship between science education and philosophy of education has been labelled as "vital but underdeveloped" (Schulz

A. C. Couló (✉)
Instituto de Filosofía & Group of Epistemology, History and Didactics of the Natural Sciences, GEHyD, Facultad de Filosofía y Letras, Universidad de Buenos Aires, Buenos Aires, Argentina

M. E. B. Prestes, C. C. Silva (eds.), *Teaching Science with Context*, Science: Philosophy, History and Education,
https://doi.org/10.1007/978-3-319-74036-2_24

2014). More recently, formal and informal logic, and argumentation theory have been steadily gaining attention (Adúriz-Bravo 2014; Erduran & Jiménez-Aleixandre 2008; Lee et al. 2009). Other philosophical fields can also enrich and sustain fruitful and well-founded teaching practices and research. Ethical and political philosophical topics not only underlie socio-scientific or STSE issues, but an acquaintance with them may promote critical awareness and help teachers to avoid unconsidered or tacit standpoints about epistemic and non-epistemic values that could emerge as a hidden curriculum. Both scientific inquiry and science educational practices are concerned with axiological dilemmas such as the moral or political relevance of research or teaching decisions, responsibility for the consequences of research or teaching choices, or adequate scientist or teacher professional behavior (Couló 2016; Forge 2008; Jones et al. 2010; Kitcher 2001; Lacey 1999; Longino 1990). Metaphysics or philosophy of religion can also help in the recognition, analysis, and understanding of different world views, cultures, and religious viewpoints (Matthews 2009), and in fostering the abilities needed to enable a more rational expression of differences. Even aesthetic expression can be a shared field of reflection. From another point of view, familiarity with philosophical issues may help teachers to recognize, support, and deepen students' spontaneous inquiries about metaphysical, epistemological, ethical or political matters, construing them as potential openings for reflection, rather than disruptive misconceptions. In addition, a working knowledge of philosophical issues and methods may facilitate fruitful interdisciplinary interactions and collaboration between colleagues from different subjects in designing joint school projects.[1]

Nevertheless, philosophy and other foundational studies courses have lately lost their stand in many teacher education curricula (science-related or otherwise). Engagement with social and educational problems seems to have become an area for social scientists rather than philosophers, probably because of philosophy's skeptical turn (Stengel 2002). Sometimes, "philosophical competencies" replace philosophy courses, which usually means that philosophy courses can be eliminated, and its contents "integrated" into other "generally introductory" courses (Hayden 2012). In other cases, more practice-oriented classes are replacing them. To support these decisions, philosophy lessons have frequently been dismissed as being elitist, irrelevant, unnecessarily obscure or merely an old-fashioned ornamental addition to teacher culture with little or no relevance to current classroom practices. Furthermore, philosophers and educators do not have many opportunities for interaction, and, in the worst-case scenario, they do not really seem to care about what the other community is doing.

There are different ways of addressing this situation. One of them is represented by this very publication and other relevant spaces for collaboration and communication among philosophers, scientists and educators concerned with education in philosophy and in science. With regard to science education, this can be illustrated by the International History, Philosophy and Science Teaching Group (IHPST)

[1] *Mutatis mutandis* some of these rationales (and others) can be applied to POS education for scientists (Grüne-Yanoff 2014, pp. 117–123).

community and its biennial Conferences[2]; the Inter-Divisional Teaching Commission (IDTC) of the International Union for History and Philosophy of Science (IUHPS), or journals such as *Science & Education* or *Enseñanza de las Ciencias*.

On the other hand, a possible rationale could be suggested, as sketched above, by referring to the kind of significant issues that philosophy can bring to the development of an educator.[3] However, in this paper, I want to take up the issue from a slightly different point of view. I want to focus on POS teaching; thus, I will not address the connections and divergences between teaching POS and nature of science (NOS) (Adúriz-Bravo 2004, 2011; Couló 2015; Matthews 2012). Furthermore, I will not concentrate so much on the philosophical content to be taught (*what* is taught) but rather on the way in which that content is brought to the classroom (*how* it is taught). In any case, the aims and purposes served by teaching philosophy to science teachers (*why* it is taught) should always be present as a standard against which all possible suggestions ought to be judged.

24.2 Some Approaches to Teaching Philosophy to Pre-service Science Teachers

How should we teach POS to pre-service science teachers? Philosophers usually teach their discipline to philosophers-to-be, sometimes to philosophy teachers, but rarely to scientists or to science teachers. This may have an undesirable outcome: those issues and problems that are part of a standard POS-for-philosophers curriculum may not be adequate for a POS-for-science-teachers program. Likewise, some usual philosophical teaching styles and manners may not be suitable for a different audience.

What do philosophers teach in a standard POS course? By comparing nine POS texts of considerable circulation in the last 17 years, Grüne-Yanoff (2014) identifies 24 different topics, mostly related to epistemic issues. The topics that appear in almost all textbooks are – not surprisingly – explanation, hypothesis testing, induction and confirmation, realism, and scientific progress. Non-epistemic issues are almost completely absent, except for a lone mention of ethics. This leaves out many POS-relevant questions that are part of curricular proposals for science education, whether from a NOS or a features of science[4] perspective (Hodson 2014; Matthews 2012).

However, beyond the issue of content, it would be relevant to focus attention on the alternative modalities of POS teaching. POS courses for nonphilosopher stu-

[2] http://ihpst.net/

[3] For a thorough analysis of this point see Matthews (2015) especially Chap. 12: "Philosophy and Teacher Education."

[4] The idea of "features of science" advanced by Michael Matthews is more congenial to a POS perspective. However, the differences in focus and choice of content still stand.

dents usually retain a traditional style, simplifying some issues, or abstracting "difficult" authors or topics as a sole consideration for a different audience. For instance, they may reduce teaching POS to teaching *a history of* POS (Lécourt 2000, p. 29).[5] Classes sometimes start at classical Greek philosophers (Aristotle) and go on to empiricism–rationalism debates (Descartes and Hume). Usually, an illustration of 20th century POS schools is present: positivism, Popper's version of hypothetico-deductivism (with a visit to Semmelweis' work on childbed fever as portrayed in Hempel 1966) and on to culminate in the so-called new philosophy of science (usually Kuhn's *Structure* version thereof). Emphasis tends to be on philosophers, and sometimes on philosophical schools or traditions. Occasionally, a (slightly) more recent philosopher or "ism" appears at the end of the course (i.e., Lakatos, Laudan, Feyerabend, a sketch of semanticism or the concept of a scientific model) (Adúriz-Bravo 2017, 2011). Philosophers and schools are addressed on their own, as self-contained items, or else compared with a view to showing the differences or disagreements between them (Grüne-Yanoff 2014). A recounting and tracing through discussions and debates from the early to late twentieth century (McLeod 2015) constitutes the standard kind of POS curriculum that may be adequate for philosophy students, but not necessarily for science and science-teacher students.

The relevance or interest of these ideas or philosophers may be (more or less) obvious for a philosophy student, but certainly not for a science teacher. Besides, philosophy teachers on POS courses for nonphilosophers should take into consideration the fact that their audience would probably have had little familiarity or training in the humanities after leaving high school. They cannot presuppose conceptual knowledge or procedural skills in the history of philosophy, formal logic or argumentation theory. Nor can they presume that their students will feel motivated by philosophical issues in general, and by these POS questions in particular. Moreover, this kind of de-contextualized approach may result in students learning "tenets," understood as declarative knowledge, rather than engaging with genuine philosophical questions (Clough 2004) or in their assuming an intellectually *passive* approach to teaching and learning (Davson-Galle 2008).

Philosophy teaching correspondingly has its own predominant style, aiming at exact reading of great philosophical texts, precise conceptual analysis, relevant historical categorization and, on those courses taught by philosophers related to the analytic tradition, discussion of thought experiments (Grüne-Yanoff 2014). These practices can be valuable tools in their proper setting, but, in other contexts, they run the risk of resulting in the mere reproduction of a philosophical canon. At best, this can be thought of as replacing philosophy with a history of ideas. At worst, it may mean eschewing the opportunity for genuine philosophical inquiry, and for developing critical thinking and rationality (Siegel 1989). The risk is especially high if there is no time to deepen reflection on the significance and consequences of each

[5] This conflation between "teaching philosophy" and "teaching the history of philosophy" is sometimes an unconsidered choice, following on from "what has always been done". However, it can also be derived from a deliberate, well-founded approach. We will come back to this point.

standpoint, which is usually the case in POS-for-nonphilosophers classes. Ultimately, this may result in quite un-philosophical habits, such as rote learning of pseudo-philosophical slogans. In the best possible light, although those practices can be relevant and productive in teaching philosophers-to-be, they would be still much less significant in science teacher education. Even in the first case, there is some controversy, and the philosophy teaching community itself has critically examined traditional styles of philosophy teaching. Fortunately, there are several philosophical teaching approaches than can offer richer and more creative experiences for science student-teachers. Therefore, although the interaction between science teaching researchers and philosophers of science is certainly appropriate, further collaboration between the fields of didactics of science (i.e., science teaching) and didactics of philosophy (i.e., philosophy teaching) is also, or perhaps even more, desirable.

24.3 Different Approaches to Teaching Philosophy

English-speaking countries seldom include mandatory philosophy courses as part of secondary, not to mention elementary, education. On the contrary, one or more philosophy courses have long been required in several European countries (Spain, Italy, France, Greece, Poland, to mention just a few) and in most South and Central American curricula, and in French-speaking Canada (Québec). In other places (Germany or Belgium) ethics courses are part of the high school curriculum, sometimes as an option to religion classes. As a result, research and scholarship on philosophy education for nonphilosophers, whether high school, elementary school or graduate education, has produced a valuable corpus of theory and innovation that has been steadily growing in the last few decades. The outcomes of that research can be found in books, conferences and congresses,[6] and academic journals such as *Paideia* (in Spanish), *Diotime* (in French), *Childhood & Philosophy* (in Portuguese, Spanish, and English), or *Zeitschrift für Didaktik der Philosophie und Ethik* (in German). In English, *Teaching Philosophy* started publication in 1975, the American Philosophical Association distributes an on-line *Newsletter on Teaching Philosophy*, and UNESCO has published a comprehensive study on the subject (2007) and a series of regional studies (2009, 2011).

Relevant work is taking place on multiple topics, including inquiry pertaining to teaching and learning to ask philosophical questions and the art of posing suitable problems, which can be an asset to teachers in every field. The same can be said about the skills required to recognize, analyze, and take a stand on the different positions related to controversial issues in educational, epistemic, ethical or political matters. It is widely accepted that first encounters with philosophy require nonphi-

[6] Among others, the *World Congress of Philosophy* includes a section in "Teaching Philosophy". 23rd World Congress http://www.wcp2013.gr/files/items/6/644/programmejuly29.pdf/20th World Congress – Paideia Project https://www.bu.edu/wcp/MainTeac.htm

losophy students to deal, both intellectually and emotionally, with the impossibility of reaching a sole, commonly accepted answer that is the hallmark of philosophy (Rabossi 2008; Rescher 1985). However, they can subsequently find the possibility of engaging in lively yet rigorous dialogue, reasoning, and argumentation more congenial to their interests and abilities (for instance, in science education, see Ratcliffe and Grace 2003; McKim 2010).

Within this framework, philosophy educators have outlined different approaches from which philosophy teaching can be envisioned. Rabossi[7] (1993, 2000) discusses three different approaches to teaching philosophy, namely the dogmatic approach, the eclectic approach and the critical approach. There are three criteria to distinguish between them: first, whether a corpus of philosophical ideas may be considered definitively ascertained, and therefore exclude or minimize other approaches; second, the kind of relationship that each approach establishes with the history of philosophy, and finally, the way in which each approach envisions progress and creativity within philosophical practice. For instance, the *dogmatic* approach assumes that there is genuine philosophical knowledge as long as one specific philosophical system has been established as *the* true one. Every other system or point of view, ancient or contemporary, is regarded as a missed effort, an incomplete sketch, anticipation or attempt at the true one, or a plainly erroneous failure. There is no sense in trying to create new philosophical ideas, the only rational goal is to extend or adjust the existing (true) philosophical corpus. Therefore, sound philosophy teaching requires rigorous transmission of that one true system, and learning philosophy means achieving deep comprehension of its meaning.

Second, the *eclectic* approach considers that philosophical truth cannot be found on any one particular system, but that it is somehow distributed over several different philosophical stances or schools of thought. Therefore, philosophical truth has been developing throughout the history of philosophy, and that is the place where we can find it: all or most philosophical systems are worthy of study, and there are no definitive criteria for demarcation to sort out the merits (or demerits) of philosophical ideas and schools. Philosophical practice is aimed at deepening our understanding of philosophical ideas, and at establishing the different relationships among them (inclusion, inference, contradiction, compatibility, etc.). From a pedagogical point of view, sound philosophy teaching is, above all, teaching the history of philosophy, and within those limits, it implies a careful, rigorous presentation of issues and arguments such as can be found along philosophical history. Correlatively, learning philosophy is learning how to immerse oneself into one or a few philosophical systems, reconstruct their concepts and arguments, and become adept at arguing for its stances.

Finally, the *critical* approach emphasizes philosophical problems rather than philosophical systems or schools of thought. This does not imply rejecting the idea of "philosophical truth," but this term acquires a restricted sense: a philosophical

---

[7]Rabossi was mainly interested in teaching philosophy in philosophy departments at university level. However, these categories can be extended to philosophy teaching at other educational levels.

truth can be recognized by two traits: it is the object of a rationally founded certainty, and it can (and has) survive(d) rational criticism. Practitioners of the critical approach have an "anachronistic" relationship with philosophy's past: philosophical problems and philosophical theses of the past are relevant only inasmuch as they can be addressed from a contemporary way of philosophizing and with regard to contemporary philosophical issues. Therefore, sound philosophy-teaching is mainly aimed at teaching to philosophize, that is, at teaching a theoretical practice. In Kant's much cited words,

> No one at all can call himself a philosopher who cannot philosophize. Philosophizing can be learned, however, only through practice and through one's own use of reason. How should it be possible to learn philosophy anyway? Every philosophical thinker builds his own work, so to speak, on someone else's ruins, but no work has ever come to be that was to be lasting in all its parts. Hence, one cannot learn philosophy, then, just because it is not yet given. But even granted that there were a philosophy actually at hand, no one who learned it would be able to say that he was a philosopher, for subjectively his cognition of it would always be only historical. (Kant [1800] 1992 p. 538)

Therefore, there is no actual philoso*phy* to teach (in the sense of an accepted system), we can only teach (and learn) how to philoso*phize*. However, in philosophy teaching and learning at school and at university, this particular practice presupposes a specific content, in the sense that the issues and questions addressed are usually those that the philosophical community recognizes as philosophical, as a way of avoiding mere common sense debates. Learning philosophy implies, correlatively, the development of the disposition and ability to personally (*subjectively*) pose, reflect on, elaborate, and argue for and against, i.e., *philosophize*, on philosophical issues.

Although Rabossi's view is about philosophy and philosophizing as a profession, Obiols' (2008) is foremost about teaching and learning philosophy as a particular way of philosophizing. He starts by distinguishing two great approaches to philosophy teaching: the *historical* approach and the *problem-based* approach. On the one hand, the historical approach holds that philosophy lives in its history, and therefore teaching and learning the history of philosophy does not mean studying *history*, but studying *philosophy*. The only way to gain access to the philosophical realm is through its historical gates. On the other hand, the problem-based approach points out that philosophy typically addresses a set of specific questions or issues, and that philosophy teaching should pivot around those questions. Although this approach leans toward the Kantian perspective, the former is more at home with a Hegelian viewpoint. These two approaches are predominant in the teaching of philosophy, but with a more restricted scope, it is possible to recognize a third position, the *doctrinarian* approach, close but not identical to the dogmatic approach in Rabossi's view. This approach presents and upholds a sole philosophical standpoint, and builds teaching programs and classes only within the boundaries of that particular standpoint.

At the same time, each one of these approaches can and has been criticized, especially when it grows exclusive and suffers from some sort of hypertrophy. For instance, practitioners of the dogmatic-doctrinarian approach not only select a restricted set of authors and theses (which is inevitable), but that selection systematically excludes rival theories or presents them in a derogatory light. This goes against the very rationales usually offered for teaching philosophy to nonphilosophers. In extreme cases, teaching under a doctrinarian persuasion may slide perilously close to indoctrination. The historical approach can be accused of favoring extension over depth and answers over questions, of hiding problems and weaknesses of the different systems, of lacking a sense of the relationship between philosophical systems and socio-historical events, or, conversely, of becoming a history of ideas (becoming *history*, instead of *philosophy*). On the other hand, the critical/problem-based approach has been accused of a lack of historical awareness, and of an anachronistic comparison of philosophical stances that only superficially refer to the "same" question. Occasionally, the problem-based approach may divert toward an analysis not of philosophical problems, but of philosophical themes or areas (such as ethics, metaphysics, epistemology, philosophy of Science, etc.), which frequently slip into a *history of* ethics or metaphysics or epistemology or POS. As noted above, epistemology, POS (and NOS) courses in science teacher education frequently lapse into this perspective. Teaching POS is understood in terms of a historical sequence of philosophical stances or schools. Sometimes, for lack of a unifying theme, the course is organized around a list of authors (Carnap, Popper, Kuhn, etc.). In these cases, philosophical content becomes doubly decontextualized (Clough 2004). Not only does it abstract philosophical issues about the nature of science away from science content, but it also presents single authors or schools of thought in a historical sequence, without referring to specific philosophical problems, let alone NOS topics or ideas (Adúriz-Bravo 2017). In a similar vein, the rigorous analysis of philosophical texts and arguments that is desirable in good philosophical practice can sometimes become an obsessive dissection of words that in the end misses the actual philosophical issue at stake, and becomes more philological than philosophical in nature.[8]

24.4 Teaching Philosophy from a Critical/Problem-Based Approach

Both philosophy and science have problems and questions, as a verbal form of expressing problems, at their roots (Tiedemann 2012). Thus, problem-based or critical approaches are at the same time didactically richer and epistemically more congruent with their respective content matter. Teaching philosophy to science teacher-students is aimed at their becoming educators rather than instructors

[8] F. Nietzsche once wrote "It has thus come to pass that, in place of a profound interpretation of the eternally recurring problems, a historical – yea, even philological – balancing and questioning has entered into the educational arena: what this or that philosopher has or has not thought; whether this or that essay or dialogue is to be ascribed to him or not; or even whether this particular reading of a classical text is to be preferred to that. It is to neutral preoccupations with philosophy like these that our students in philosophical seminaries are stimulated [...]" (Nietzsche 1910, pp. 129–130).

(Matthews 2014); at helping them develop a richer and better-founded teaching practice, at avoiding the inadvertent transmission of unconsidered or tacit standpoints, at their recognizing and supporting students' spontaneous inquiries, at facilitating fruitful interactions and collaboration between colleagues from different subjects. Given these purposes, it seems desirable to bring an emphasis to the critical/problem-based approach over the historical, and furthermore over the doctrinarian/dogmatic approaches. There is very little, if anything, that philosophers would agree on. At the same time, sound philosophical practice implies that they offer (or at least have) plausible reasons for justifying their particular points of view. Acknowledging that to achieve "rationally founded certainty" statements or theories must submit to rational critical inquiry, argumentative interchange becomes essential. In this way, the problem-based approach seems to be inherent to philosophy itself, or at least to several concepts of philosophy.[9] For instance, this perspective seems to be closely related to the time-honored tradition of Socratic dialogue (Nola and Irzik 2005; Burbules 1993). Emphasis is on reflecting and analyzing problems and also on proposing and assessing arguments, rather than on achieving a definite solution.

In any case, it is important to remember that all three perspectives are ideal types. Theoretical discussion highlights differences: philosophical commitments for each perspective may be incompatible. However, in actual teaching practice, there is no *pure* form of approach, but rather an emphasis on one of them. Both the problem-based and the critical approaches emphasize philosophical activity (philosophizing) over philosophical truths. From a philosophical–pedagogical point of view, this means aiming at critically revising the "obvious" through the posing of relevant, articulated problems, the formulation of alternative responses, their development, and the evaluation of their respective merits. Certainly, this does not mean eschewing the history of philosophy or the analysis of philosophical texts. On the contrary, philosophical problems can only be established with regard to actual and past philosophical communities, controversies, and texts. Choice of problems and analysis of possible solutions should turn to the philosophical canon as the relevant source. Philosophizing cannot take place in a vacuum, and it should not be reduced to mere common sense debates. Nevertheless, philosophy teachers who are sensitive to their students' interests and necessities should be able to select and organize philosophical content around a set of key questions that are relevant to their specific educational aims. Problem-based, critical approaches do not need students to be fluent in philosophy history. By addressing specific cases rather than general, abstract principles or overall philosophical schools POS courses can become relevant to science teachers. Schools and principles should be there, certainly, but instead of being centre-stage, they can be in the background until they become necessary to address the particular problem that has come to preoccupy students. POS teachers have the

[9] There is no unique, universally accepted notion of philosophy or philosophizing, but a multitude of diverging perspectives. Therefore, every stance on teaching philosophy reverts to a philosophical (or metaphilosophical) problem in its own right, by critically reflecting and explicitly upholding a particular way of understanding the practice and content of philosophy itself (Couló 2008).

requisite philosophical knowledge to find those authors or texts better suited to deepening and refining the students' spontaneous proposals. Intuitive ideas and responses can be analyzed and critically reflected on by drawing on the significant aspects of the philosophical canon.

A further asset of problem-based classes comes from the fact that it makes it easier to foster students' interest. Teachers should bear in mind that this approach does not imply their presenting a problem (and then giving a pre-established answer for it), but in generating the conditions for students to feel challenged by a problematic situation, and uncomfortable with their previous (seemingly obvious) ideas.

Therefore, philosophers who teach philosophy to nonphilosophers should adopt a particular point of view, a shift of emphasis. They would not seek to *simplify* philosophy (by taking out "difficult" texts or issues), but to *restructure* both teaching content and methodology. While carrying out academic philosophy or teaching philosophy to philosophers, we bring educational, scientific, or other questions to be analyzed, deconstructed or criticized with regard to philosophical texts, traditions or points of view. When teaching philosophy to nonphilosophers, we should bring philosophical tools and traditions to scientific, artistic or educational undertakings. In teaching POS to science teachers, therefore, topics should be selected concerning the particular needs and interests of science teachers, with an aim of helping them become excellent science teachers, and not mediocre philosophers of science. It would be important to pay attention to the fact that their needs and interests differ from the needs and interests of philosophers. A serious consideration of the aims and ends of philosophy teaching within a science teaching educational program implies that the choice of philosophical problems and the selection of material from the history of philosophy and philosophical texts should be relevant to those ends. Besides being relevant, these topics should not require a wide previous philosophical background, or assume proficiency in philosophical methods, tools or skills. POS courses for science teachers may have to resign some important issues (from a philosophical point of view), because addressing them with a reasonable depth would be beyond its scope and competency.

As stated before, controversy seems to be the hallmark of philosophical practice. Philosophy may be distinguished from other disciplines and professions because

> […] it is rampant with antagonistic thesis and approaches. [In the philosophical world], there is no possibility for pluralistic dialogue in a wide scale; there are no shared criteria to identify problems, methods, solutions or viable ends; it is impossible to reach a satisfyingly wide and stable common consensus. (Rabossi 2008 *my translation from original Spanish*)

On the contrary, science instruction tends to highlight current knowledge and models. When taught ahistorically, a strong sense of progress and timelessness tends to attach to scientific content. Even in a historically sensitive curriculum, scientists and science educators can place even very specific questions in a larger picture that can be perceived as relevant by nonspecialist audiences. Likewise, history can provide examples of difficult or seemingly esoteric questions being resolved and contributing directly or indirectly to wider social concerns (Kitcher 2011).

Philosophy teaching presents an opposite picture: for every problem, there are usually one or more different or even contradictory potential solutions. For every argument, one or several counterarguments can be found. While science teaching emphasizes convergence, philosophical practice works well with divergence. At this point, it is important to remember that there is a significant difference between philosophy teaching and science teaching with regard to the expected outcome. Usually, science teachers would aim to change their students' outlook from a naïve, common sense concept toward the more complex, sophisticated and sometimes completely incompatible ideas upheld within the pertinent scientific theories or models. However, philosophy classrooms do not aim at the same kind of belief change. Rather, they seek to help their students to question and doubt those beliefs, enriching the context in which they hold them, making more explicit the tacit assumptions that underlie them and the consequences that may follow. Correspondingly, they aim to help students to hone the arguments that justify their beliefs, and at heightening not just tolerance but appreciation of disagreement and controversy over those beliefs. There is no intrinsic failure in a philosophy student maintaining a belief that is roughly the same before and after taking that course. However, it would be a failure if he did not acquire a subtler, more aware, more autonomous, more critical attitude toward it or did not have better reasons for changing or confirming his belief. Triviality and paucity of ideas or questions should be abandoned, and especially inconsistency should be acknowledged as an insurmountable obstacle to keeping beliefs that had not resisted the test of eristic dialogue.

There is no distinctive subject matter in philosophy that could be deemed more important than another irrespective of the particular philosophical stance from which it were chosen. There is a myriad of philosophers and philosophical texts that are worthy of consideration. Thus, we face a dilemma: which questions should be addressed, and which should be left aside? Which are the philosophers, the texts, the controversies that should be preferred over other equally worthy (or even worthier, from a strictly philosophical point of view) philosophers, texts or controversies?

A possible way of dealing with this problem is to take into consideration those questions that underlie issues that are included in most NOS curricula (for a comprehensive discussion, see Matthews 2012, 2014; and especially Hodson 2014 and Lederman et al. 2014). However, given the plurality of stances sketched above, how can we organize the different philosophical perspectives? By way of an example, let us refer to the question of non-epistemic values in science. This issue is frequently included as part of nature of science issues that should be addressed within the curriculum (Couló 2014; Jones et al. 2010). Ethics of inquiry, ethical dilemmas grounded in the very nature of the research that is attempted, the ethical, social or political relevance of research, can be addressed in science education. However, the issue of whether there is such a thing as freedom of will underlies any question about responsibility and accountability. Therefore, we need to select and organize different philosophical stances and arguments. The notion of "*philosophical apory*" coined by Nicholas Rescher (1985, 2006)[10] may be of help in this instance. An

[10] In the original Greek "*aporía*" means "impassable", from the negative particle "*a*" and "*póros*", "passage". It usually refers to a contradiction or paradox, or, subjectively, a puzzle conducive to a state of grave uncertainty or doubt. In *elenchus*, one of the forms of the Socratic Method as depicted

apory is a "collective inconsistency among individually plausible contentions" (Rescher 2006, p. 20) that structure philosophical positions by showing their mutual interrelationships, even when they are not easily recognizable at first sight. In this way, a space of deliberation and argument may be defined around a specific issue. Apories bring inconsistencies to light, which means that they force us to choose between possible stances or arguments, although they do *not* force us to choose *one* specific stance or argument. For instance, Rescher presents the following apory:

1. All human acts are causally determined
2. Men [*sic*] can and do make free acts of choice
3. A genuinely free act cannot be causally determined (for if it is so determined then the act is not free by virtue of this very fact).

These three theses represent an inconsistent triad in which consistency can be restored by any of three distinct approaches:

Deny (1):	"Voluntarism" – the exemption of free acts of the will from causal determination (Descartes)
Deny (2):	"Determinism" of the will by causal constraints (Spinoza)
Deny (3):	"Compatibilism" of free action and causal determination, for example, via a theory that distinguishes between inner and outer causal determination and sees the former sort of determination as compatible with freedom (Leibniz). (Rescher 1985, p. 27)

Note that for every possible way out of the inconsistency there is at least one philosophical stance to be taken into consideration. In this case, Rescher cites only modern philosophers, but it would not be hard to find other philosophers (of a different historical period, of a different gender, of a different philosophical persuasion, etc.) who address these questions.

Now, if this issue were to be addressed from a critical/problem-based perspective, then the first step would be to *elicit* the kind of problem we want to address. This means designing a learning situation where student teachers themselves can find and pose the problem as a first step for further discussion and conceptualization, rather than the teacher explaining the problem, and then enouncing a predetermined response. A highly evocative stimulus should be presented, in a reassuring environment, so that everyone can feel comfortable asking seemingly obvious questions or advancing tentative ideas. Teachers can help reflection along by asking about assumptions and about the reasons behind the proposed ideas. A thought experiment, a story, a picture, a poem, a film clip, a historical or contemporary case or situation, can trigger debate and analysis. Depending on the audience and the context, the problem of free will can be initiated by the idea of psychohistory (in Asimov's *Foundation*), or by the "Ludovico treatment" in *The clockwork orange* (either the book or the film), or by the precogs in *Minority report* (either the film or the short story by Philip K. Dick). In a course for science teachers, one could debate

by Plato, *aporía* refers to the moment when someone's position is refuted, and no apparent way out is envisioned. Rescher coins a stipulative definition for this term.

Newcomb's Paradox (Nozick 1969), or analyze the last instalment of Canadian TV-series *Re-Genesis*, where not only free will but also scientific and social responsibility constitute the core of the episode.[11] By debating any of these productions, the teacher could help students to put the question into words, refine its expression, and explore the subjacent assumptions. Then, a subsequent philosophical frame for conceptualization and argumentation through selected readings, writing assignments and class interaction can be set. Students can learn to develop and refine their intuitive answers. They can read philosophical texts, not just for their intrinsic value, but also as a means of critically analyzing alternative stances, comparing them, and justifying their personal choices. This kind of learning experience fosters critical, argumentative, and communication skills. Moreover, this approach opens up opportunities for interdisciplinary work between courses on different subjects to plan a core problem or project (Perrenoud 1999) to be addressed from their relevant perspectives.

24.5 Conclusion

This chapter could only raise a few problems that arise from how to teach a POS course for science teacher students. The discussion is certainly far from addressing all the relevant issues. Nevertheless, I hope it is a useful contribution for a growing interaction between science educators and philosophy educators. We share the interest in providing the best possible education for science and philosophy teachers. I believe that expanding the space for dialogue between science educators and teachers, and philosophy educators and teachers would be an enriching factor for both practices. We need to learn each other's languages of practice, so that we can communicate and understand one another. We can work together in our shared interests (science and education), recognizing and gaining from our different contributions, without losing our own identities or blurring the lines between our different fields.

References

Adúriz-Bravo, A. (2004). Methodology and politics: A proposal to teach the structuring ideas of the philosophy of science through the pendulum. *Science & Education, 13*(7), 717–731.

Adúriz-Bravo, A. (2011). Epistemología para el profesorado de física: Operaciones transpositivas y creación de una "actividad metacientífica escolar". *Revista de Enseñanza de la Física, 24*(1), 7–20.

Adúriz-Bravo, A. (2014). Revisiting school scientific argumentation from the perspective of the history and philosophy of science. In M. R. Matthews (Ed.), *International handbook of research in history, philosophy and science teaching* (pp. 1443–1472). Dordrecht: Springer.

[11] Good ideas for this type of resource can be found in Baggini (2006).

Adúriz-Bravo, A. (2017). Desafíos de la enseñanza de la epistemología al profesorado de ciencias. In Z. Monroy Nasr, R. León Sánchez, G. Alvarez Díaz de León (Eds.), *Obstáculos epistemológicos en la enseñanza y el aprendizaje de la filosofía y de la ciencia*. México: Facultad de Psicología y DGAPA, UNAM.

Baggini, J. (2006). *The pig that wants to be eaten (and 99 other thought experiments)*. London: Granta.

Biesta, G. (2012). The future of teacher education: Evidence, competence or wisdom? *RoSE – Research on Steiner Education, 3*(1), 8–21.

Burbules, N. C. (1993). *Dialogue in teaching: theory and practice*. New York: Teachers College Press.

Clough, M. (2004). Teaching the nature of science to secondary and post-secondary students: Questions rather than tenets. The Pantaneto Forum, Issue 25, January . http://www.pantaneto.co.uk/issue25/front25.htm. Republished (2008) in *California Journal of Science Education, 8* (2): 31–40.

Couló, A. C. (2008). Enseñar filosofía: Supuestos y decisiones metafilosóficos. In *IX Coloquio Internacional Bariloche de Filosofía: Resúmenes*, 26. San Carlos de Bariloche: Fundación Bariloche.

Couló, A. C. (2014). Philosophical dimensions of social and ethical issues in school science education: Values in science and in science classrooms. In M. Matthews (Ed.), *International handbook of research in history, philosophy and science teaching* (pp. 1087–1117). Dordrecht: Springer.

Couló, A. C. (2015). Enseñar filosofía a docentes de ciencias naturales en formación. In A. Cerletti & A. Couló (Eds.), *Didácticas de la filosofía: Entre enseñar y aprender a filosofar* (pp. 47–61). Buenos Aires: Noveduc.

Couló, A. C. (2016). Values in science and in science classrooms. In M. A. Peters (Ed.), *Encyclopedia of educational philosophy and theory* (pp. 1316–1323). Singapore: Springer.

Darling-Hammond, L. (2008). Knowledge for teaching: What do we know? In M. Cochran-Smith, S. H. Feiman-Nemser, J. McIntyre, & K. Demers (Eds.), *Handbook of research on teacher education* (pp. 1316–1323). New York: Routledge.

Davson-Galle, P. (2008). Why compulsory science education should not include philosophy of science. *Science & Education, 17*, 677–716.

Erduran, S., & María del Pilar Jiménez-Aleixandre, M. (Eds.), (2008). *Argumentation in science education: Perspectives from classroom-based research*. Dordrecht: Springer.

Forge, J. (2008). *The responsible scientist: A philosophical inquiry*. Pittsburgh: University of Pittsburgh Press.

Grüne-Yanoff, T. (2014). Teaching philosophy of science to scientists: Why, what and how. *European Journal for Philosophy of Science, 4*, 115–134.

Hayden, M. J. (2012). What do philosophers of education do? An empirical study of philosophy of education journals. *Studies in Philosophy and Education, 31*, 1–27.

Hempel, C. G. (1966). *Philosophy of natural science*. Princeton: Prentice-Hall.

Hodson, D. (2014). Nature of Science in the Science curriculum: origin, development, implications and shifting emphases. In M. R. Matthews (Ed.), *International handbook of research in history, philosophy and science teaching* (pp. 911–970). Dordrecht: Springer.

Jones, A., McKim, A., & Reiss, M. (2010). *Ethics in the science and technology classroom*. Rotterdam: Sense Publishers.

Kant, I. ([1800] 1992). The Jäsche logic. In Tr. and ed. M. Young, *Lectures on logic*: *The Cambridge edition of the works of Immanuel Kant*. Cambridge: Cambridge University Press.

Kitcher, P. (2001). *Science, truth and democracy*. Oxford: Oxford University Press.

Kitcher, P. (2011). Philosophy inside out. *Metaphilosophy, 42*(3), 248–260.

Lacey, H. (1999). *Is science value free? Values and scientific understanding*. London: Routledge.

Lécourt, D. (2000). *Rapport au Ministre de l'Éducation Nationale sur l'enseignement de la philosophie des sciences*. http://www.education.gouv.fr/cid1946/l-enseignement-de-la-philosophie-des-sciences.html. Accessed March 2014.

Lederman, N. G., Bartos, S. A., & Lederman, J. S. (2014). The development, use, and interpretation of Nature of Science assessments. In M. R. Matthews (Ed.), *International handbook of research in history, philosophy and science teaching* (pp. 971–996). Dordrecht: Springer.

Lee, M.-H., Wu, Y.-T., & Tsai, C.-C. (2009). Research trends in science education from 2003 to 2007. A content analysis of publications in selected journals. *International Journal of Science Education, 31*(15), 1999–2020.

Longino, H. (1990). *Science as social knowledge: Values and objectivity in scientific inquiry*. Princeton: Princeton University Press.

Matthews, M. R. (2012). Changing the focus: From nature of science (NOS) to features of science (FOS). In M. S. Khine (Ed.), *Advances in nature of science research: Concepts and methodologies* (pp. 3–26). Dordrecht: Springer.

Matthews, M. R. (2015). *Science teaching: The contribution of history and philosophy of science* (2nd ed.). New York: Routledge.

Matthews, M. R. (Ed.). (2009). A special issue: Science, worldviews and education. *Science & Education, 18*(6–7), 641–666.

Matthews, M. R. (2014). *International handbook of research in history, philosophy and science teaching*. Dordrecht: Springer.

McKim, A. (2010). Bioethics education. In A. Jones, A. McKim, & M. Reiss (Eds.), *Ethics in the science and technology classroom* (pp. 19–36). Rotterdam: Sense Publishers.

McLeod, M. (2015). Modernizing philosophy of science for the philosopher and student alike. *Metascience, 24*(3), 507–510.

Obiols, G. (2008). *Una introducción a la enseñanza de la Filosofía*. Buenos Aires: Libros del Zorzal.

Nietzsche, F. (1910). On the future of our educational institutions. In O. Levy (Ed.), *The complete works* (Vol. 5, pp. 3–142). Edinburgh: Foulis.

Nola, R., & Irzik, G. (2005). *Philosophy, science, education and culture*. Dordrecht: Springer.

Nozick, R. (1969). Newcomb's problem and two principles of choice. In N. Rescher (Ed.), *Essays in honour of Carl G. Hempel* (pp. 114–146). Dordrecht: Springer.

Perrenoud, P. (1999). Apprendre à l'école à travers des projets: Pourquoi? Comment? http://www.unige.ch/fapse/SSE/teachers/perrenoud/php_main/php_1999/1999_17.html. Accesed Jan 2016.

Rabossi, E. (1993). Enseñar filosofía y aprender a filosofar: Nuevas reflexiones. In G. Obiols & E. Rabossi (Eds.), *La Filosofía y el filosofar: Problemas en su enseñanza* (pp. 13–23). Buenos Aires: CEAL.

Rabossi, E. (2000). Sobre planes de estudio, enfoques de la filosofía y perfiles profesionales. In G. Obiols & E. Rabossi (Eds.), *La enseñanza de la filosofía en debate* (pp. 98–108). Buenos Aires: Novedades Educativas.

Rabossi, E. (2008). *En el comienzo Dios creó el Canon: Biblia berolinensis*. Buenos Aires: Gedisa.

Ratcliffe, M., & Grace, M. (2003). *Science education for citizenship: Teaching socio-scientific issues*. Maidenhead: Open University Press.

Rescher, N. (1985). *The strife of systems: An essay on the grounds and implications of philosophical diversity*. Pittsburgh: University of Pittsburgh Press. Spanish Edition: Rescher, N. (1995). *La lucha de los sistemas* (A. G. Sienra, Trans.). México: UNAM.

Rescher, N. (2006). *Philosophical dialectics. An essay on metaphilosophy*. Albany: State University of New York Press.

Schulz, R. M. (2014). Philosophy of education and science education: A vital but underdeveloped relationship. In M. Matthews (Ed.), *International handbook of research in history, philosophy and science teaching* (pp. 1259–1316). Dordrecht: Springer.

Siegel, H. (1989). The rationality of science, critical thinking, and science education. *Synthese, 80*(1), 9–41.

Stengel, B. S. (2002). Cause for worry or agenda for action? *Educational Theory, 53*(3), 281–290.

Tiedemann, M. (2012). The principle of a problem-based approach and its consequences for teaching philosophy and "Ethik". Analytic Teaching and Philosophical Praxis, 33(1), http://journal.viterbo.edu/index.php/atpp/article/view/908/672. Accessed Mar 2016.

UNESCO. (2007). *La philosophie, une école de la liberté. Enseignement de la philosophie et apprentissage du philosopher : État des lieux et regards pour l'avenir*. Trans. in English as *Philosophy, a school of freedom. Teaching philosophy and learning to philosophize: Status and prospects*. http://unesdoc.unesco.org/images/0015/001536/153601f.pdf (Fr) or http://unesdoc.unesco.org/images/0015/001541/154173e.pdf (Eng). Accessed Apr 2012.
UNESCO. (2009). *Teaching philosophy in Latin America and the Caribbean*. Paris: UNESCO.
UNESCO. (2011). *Teaching philosophy in Europe and North America*. Paris: UNESCO.

Chapter 25
Epistemological Debate Underlying Computer Simulations Used in Science Teaching: The Designers' Perspective

M. Eugenia Seoane, Irene Arriassecq, and Ileana M. Greca

25.1 Introduction

The advent of the computer has brought about profound changes in the way we see the world. Virtually all fields of study reflected such changes; almost all disciplines use, in one way or another, computers and specialized software. Durán (2015, p. 88) points out that the change introduced into the scientific activity by the use of computers was originally related to the simplification of complex calculations. But, he highlights, the complexity achieved by the development of computer simulations and the wide variety of techniques associated with them clearly show that it would be simplistic to regard computer simulations only as calculations that aim to obtain results of the equations that shape a phenomenon. In fact, their use has encouraged research and the production of scientific knowledge, and we can state that much of the emergent knowledge we now have about the world – the application of theories to the real world, of special interest to us citizens – is related to them. Within this more complex view on simulations in the scientific sphere, many epistemologists claim that their use involves a new way of scientific production (Galison 1996; Winsberg 1999; Durán 2015).

If we assume that it is vital for teachers to receive training in aspects regarding how science evolves (NRC 2012; Acevedo Díaz 2008) and that simulations play a

M. E. Seoane (✉) · I. Arriassecq
ECienTec, Faculty of Exact Sciences, National University of the Center of the Province of Buenos Aires, Buenos Aires, Argentina
e-mail: mseoane@exa.unicen.edu.ar; irenearr@exa.unicen.edu.ar

I. M. Greca
University of Burgos, Burgos, Spain
e-mail: imgreca@ubu.es

M. E. B. Prestes, C. C. Silva (eds.), *Teaching Science with Context*,
Science: Philosophy, History and Education,
https://doi.org/10.1007/978-3-319-74036-2_25

key role in scientific activity, which crucially influences the world around us, the education of scientifically literate citizens should involve learning about simulations, their potentialities and limitations. Although over the last 30 years, simulations have been considered one of the best facilitators of computer-based teaching (Bayraktar 2002), the literature in science teaching has given little attention to the epistemological aspects involved (Koponen and Tala 2014). This research study belongs to a larger project aimed precisely at identifying the epistemological aspects necessary to be taught to secondary students and how these should be introduced, based on the opinions of the many relevant actors (epistemologists, simulation designers, science teaching researchers, and teachers). We consider the reflections offered by three renowned simulation designers and then analyze the definition of computer simulations they give, the underlying notion of reality, the role assigned to explanation, to prediction and to empirical validation, and finally the use of simulations in science teaching.

25.2 Computer Simulations in Scientific Research

In scientific research, computer simulations were originally used to study complex systems that were impossible study in an accurate analytical form because of their characteristics. However, nowadays, they are also used when no theories are available to help explain a phenomenon.

Because of the large variety of uses they offer, it is difficult to give an unambiguous definition. A study performed by Ören (2011a, b) found more than a hundred definitions of computer simulations. Nevertheless, from an operational point of view, we can define them as the representation of the dynamic behavior of a system according to an approximate (mathematical) model used to implement it on the computer (Ören 2011a). Therefore, simulations may be broadly defined as transformations of mathematical models (which may or may not be derived from theories or established laws) into discrete algorithms, which may be implemented on a computer, and which imitate the behavior of a system. To transform equations into computationally manageable algorithms, there are many methods available, such as MonteCarlo or the finite difference method. It should be noted that such transformation is neither a simple nor an obvious process (Winsberg 2010, pp. 10–17): the transformation into discrete equations is sometimes computationally expensive and the model needs to be simplified by either excluding factors or reducing degrees of freedom. In such cases, to capture elements essential for the comprehension of a phenomenon, it is necessary to include simplified mathematical relationships, not directly derived from the physical model and which sometimes do not result from theoretical considerations but from "intuition" about the phenomenon.

25.2.1 Epistemological Issues Associated with Computer Simulations

The fact of regarding simulations as a new way of scientific production has generated many epistemological discussions. We provide below a brief outline of these discussions based on the work performed by Greca et al. (2014).

The main issue relates to the status of simulations as models and their relationship with theory and experimentation. The "semantic view" on models – models as a representation that gives meaning to mathematical formalism and relegates the role of theory – has been reviewed over the last few years (Cartwright 1999; Sismondo 1999). Models are therefore considered to be something different from theory and, being partially independent of the theory and of the world, have an autonomous component that allows their use as exploratory instruments in both domains (Morgan and Morrison 1999). As a result of this mediating role, models would enable us to deal with questions that theory by itself would fail to answer. From this point of view, we can analyze the role of simulations as models as they allow the perfecting of those with which to describe phenomena by helping to evaluate parameters to determine the most relevant ones. However, some epistemologists deny this role. For example, Simpson argues for the need to distinguish between simulations and scientific models, given that the first employ computer models that are functional and ontologically independent of scientific models (Durán 2015). Another issue concerns their classification: Can they be considered theory? Are they a different way of performing experiments? Working on a model involves exploring the relationships between input and output data to make verifiable predictions or better models (Lenhard 2010). Researchers experiment with simulations by changing, adding, and adapting parameters, and by running simulations repeatedly (Winsberg 2003; Lenhard 2010) to compare and adjust the behavior of these models with the phenomenon they should simulate. It follows then that simulations have a relationship with both theory and experimentation (Winsberg 2003).

Simulations are executed using the computer program that allows the parameters to be modified. Therefore, as the computational experiment is based on symbols and digits and so is not directly compared with the real world, simulations, in contrast to "traditional" experiments, fail the acid test for theories (Guillemot 2010). In fact, in a computational model, the hypothesis to be tested (for instance, by modifying a parameter) and the numerical experiments are arranged on a continuum: the hypothesis should first be turned into algorithms and inserted into the model to then be simulated by running the model. The experiment is still a virtual one.

Another important epistemological issue relates to the idea that the massive use of simulations entails a change in the meaning and the objective of the explanations (Johnson and Lenhard 2011). Advocates of this idea argue that computational models reproduce a predictive answer and not a mimetic model of the mechanisms that generate the real phenomenon. Regarding this, as Johnson and Lenhard (op. cit.) argue, another difference has to do with transparency: whereas mathematical traditional models are aimed at providing transparent causal explanations, computational models are opaque. What happens inside the computational model is unclear and, in some cases, the reason why the model is consistent with reality is only partly explicable. In the case of simulations, many different models may match the experimental data because consistency may be achieved in multiple ways.

On the other hand, the use of simulations focuses on predictions about the behavior of systems, only some of which may be experimentally confirmed. In addition, some predictive models are validated in the absence of verification, sometimes because they use simulation methods employed by other verified models (Kuppers and Lenhard 2005). This follows the current validation practices in physics based on the results of experimentation. Another aspect related to validation is the availability of public simulation codes, on the rise in recent years, in addition to the already available commercial codes. This turns many of such codes into black boxes. Users can obtain a good imitation of the phenomena, but the underlying mathematical model of the system is unknown.

Finally, we should distinguish between simulations – understood as processes – and other processes, such as emulation, that are also implemented on a computer (Durán 2015, p. 95). An emulator is a program that imitates the behavior of another program, ensuring an advantage as regards hardware or software and implementing a behavior identical to that of the inner architecture of machines. However, a simulation may be defined as the representation of the dynamic behavior of a system according to a (mathematical) approximate model that is implemented on a computer. In computational science, emulation and simulation are tools that provide different work platforms, but, in current philosophical debates, emulation is the process of hardware or software imitation, and simulation is the representation of the behavior of a system based on a model with certain characteristics. We see the same distinction in science teaching: virtual laboratories "mimic" real ones, and computer simulations represent the behavior, as determined by theory, of a phenomenon.

25.2.2 Simulations in Science Teaching

In science teaching, simulations are regarded as learning facilitators because they seem more efficient than other traditional practices of fostering the learning of concepts, conceptual change, and the development of procedural skills (Smetana and Bell 2012).

However, we should distinguish between scientific simulations, earlier discussed, and simulations in science teaching. In contrast to the first ones, educational simulations may be defined as "interactive learning environments in which a model simulates the characteristics of a system following the actions performed by the students" (Kirschner and Huisman 1998). The main difference is that whereas the aim of scientific simulations is to gain insight into complex phenomena and processes through models based on known theoretical assumptions or other information sources, educational simulations are intended for students to understand the underlying model and, based on that model, to understand the theoretical principles. For this to happen, students need certain guidelines (de Jong and van Joolingen 1998).

There are two ways of working with educational simulations (Doerr 1997). On the one hand is exploration, through which students explore the consequences of their actions within the boundaries of the simulation that the teacher or an expert has created for some specific knowledge representation through an ideal model. On the other hand, is modeling, which resembles the scientific way of working. For this, students use programs (such as *Modellus*, or *Stella*) that allow them to create simulations. To build these simulations, students should identify the relevant variables, quantify the relationships among them, and assess the validity of the model. This kind of use allows students to express their own concepts and learn from the process of representing them. Although, in addition to being crucial for scientific activity, educational simulations have proved to be efficient educational tools, their use is very limited, at least in secondary education. In a comprehensive study of the international literature concerning simulations in physics teaching at primary and secondary levels in 2003–2013, López et al. (2016) found only 21 articles that address this topic. The main use of simulations in those studies referred to was exploration; only two articles dealt with modeling, which is more related to scientific activity. It should be noted that none of them presented a discussion of epistemological aspects regarding the use of simulations.

During 2010, the Conectar Igualdad[1] program was created in Argentina to effectively deal with the use and understanding of ICT. It was aimed at providing computers to all students and teachers from state secondary schools, special schools, and teacher training colleges throughout the country. The objective was also to train teachers in the use of this tool and to develop educational proposals that encourage their incorporation in the teaching–learning process. Each computer is equipped with an extensive list of resources – of which *Modellus* is one – and possible ways to exploit them. However, teachers seem to lack training in the use of these tools, at a conceptual and at an epistemological level. An epistemological discussion is also missing from the work in science teaching carried out by novel researchers (Seoane et al. 2015), as suggested by the literature in this field (Greca et al. 2014; López et al. 2016). It follows from this that, in the case of simulations used in physics, it would be relevant to learn from designers themselves about their epistemological view on simulations, the objectives they consider for their design, their use in the classroom, and their valuation of such at secondary and university levels. This study should help to design more effective didactical strategies for teaching using physics simulations.

[1] http://www.conectarigualdad.gob.ar/Official website of Conectar Igualdad program.

25.3 Method

25.3.1 Case Study

In this work, we present a study case with the aim of "reconstructing" reality as observed by the actors of a previously defined system (Hernández Sampieri et al. 2010).

The criteria used for the selection of the participants was based on the need to analyze the perspective of the different actors who research, design, use, and implement computer simulations, to interpret their conceptualizations (Seoane et al. 2014). To analyze the answers provided during the interviews, a phenomenographic study was conducted taking into account the definition of simulations given, the notion of reality, the role assigned to the explanation, to the prediction and to the empirical validation, and the view on the use of simulations in science teaching. For this, we selected three researchers who design computer simulations for science teaching at secondary and university levels. They use the Java platform to design these simulations, which can be downloaded and used on free or private operating systems.

For the selection, we considered their educational background, their contributions to science teaching, and the impact of the material produced on the sphere of science education.

Francisco Esquembre[2] is Doctor in Mathematics and works for Universidad de Murcia. His academic experience involves the field of the differential equation, dynamic systems, and numerical analysis. He has participated in many projects in mathematics and the use of the computer as a tool to improve science teaching and learning. His research work includes computer-mediated teaching and learning, in addition to modeling and simulation of scientific processes for teaching purposes. He has been a member of the Conceptual Learning of Science and Visual Thinking and Learning Group (CoLoS) since 1989 and actively takes part of the Open Source Physics project. Esquembre develops computer models on Easy Java Simulations (EJS). EJS is an authoring tool designed to help teachers and students to create interactive simulations in Java in a fairly simple way. It can be used as a modeling tool (at university level) or as an exploration tool, drawing on simulations previously created by teachers.

Ángel Franco García is Doctor in Physics Science and teacher at Universidad del País Vasco. He currently runs the Interactive Internet Course on Physics.[3] This is a general physics course that deals with concepts ranging from simple ones such as rectilinear movement to more complex ones such as the energy bands of solids. The interactivity is achieved through 545 applets (in Java programming language) embedded in his web pages that are simulations of physics systems, laboratory practices, etc. This course has been awarded an Honorable Mention at the Ninth Annual

[2] http://www.um.es/fem/PersonalWiki/Official website of Francisco Esquembre.

[3] http://www.sc.ehu.es/sbweb/fisica/Physics on the computer. Interactive Internet Course on Physics.

Course on Software (1998) organized by *Computers in physics* magazine. The applets designed in Java allow working on physics problems for secondary and university levels. Each content item presented in the course has applets, the mathematical formulas represented as programming codes and interactive videos. These applets allow only exploration on the part of the students.

Vítor Duarte Teodoro is Licentiate in Physics and Doctor in Educational Sciences. He is the author of dozens of educational software titles for physics and mathematics teaching published in Portugal and other countries. Modellus[4] is one of such designs, which is used in the physics curriculum in several countries (especially in Great Britain, in advanced physics) and is available in seven languages. Duarte has coauthored more than 20 school textbooks. He is currently taking part in various European research and development projects within the framework of the Information Society Technologies program. He is also a member of several international networks, including the CoLoS Group. Modellus is a free application that enables both secondary or university teachers and students to model physics phenomena in an interactive way and in a standard mathematical notation. This allows the creation of animations with interactive objects and the analysis of experimental data based on images, animations, graphs, and tables. In addition, it can be used as a tool for the exploration of simulations created by teachers.

To analyze the interviews, we transcribed the designers' answers, data collection, and the creation of categories.

25.3.2 Analysis Technique

Phenomenography is an approach within qualitative research aimed at identifying qualitative different ways in which people experience, understand, and perceive phenomena. To draw on experience, interviews are usually carried out as a means of obtaining information (González Ugalde 2014). The basis of phenomenographic research is not formed by the phenomenon explored or the people who experience it, but instead by the relationship between both of them; that is, by how the phenomenon is experienced (comprehended or perceived) (Marton 1986).

For this analysis, we wrote the script for a semistructured interview for the three computer simulation designers mentioned above. The interview questions were based, in broad terms, on the epistemological discussions about the role of simulations in scientific activity.

The interviews were transcribed, and the aspects considered relevant were selected. These aspects were then grouped into categories assessed by two of the authors of this work independently of each other. In cases of disagreement, agreement was reached by consensus. Within the phenomenographic tradition, the

---

[4] http://modellus.co/index.php?lang=es Modellus' official site, which offers a description of the software, its applications, and its theoretical framework.

categories were identified at a semantic level (Braun and Clarke 2006), without further inferences and only based on what the interviewees explicitly stated.

25.4 Results

A twofold objective guided the analysis of each category: identifying conceptual aspects common to the three interviewees' discourse, and recording conceptual diversity. To identify who gave each of the answers reported below, a code has been assigned to each designer: Francisco Esquembre (D1), Ángel Franco García (D2), and Vítor Duarte Teodoro (D3). To represent the way in which the phenomena were understood by the designers, the following categories were created:

Category 1: *Computer simulations are models*

The three respondents agreed on the fact that computer simulations are mathematical models the researcher designs to study a particular phenomenon, although they identified two phases on a continuum: a modeling phase and the simulation itself that is its implementation on the computer (D1: "A computer simulation is the implementation of a basically mathematical model on a computer program"). They stressed the difference between simulations that belong to the scientific field and those that belong to the educational one. As it was pointed out in Sect. 2.2, educational simulations have different objectives such as the exploration of actions within the model, the formulation of new questions, and the modification of parameters (D2: "The simulation in education is different from the other; the ones I have designed are visual and interactive").

Category 2: *Simulations necessarily derive from theory*

According to the three designers, simulations are created upon theoretical knowledge we possess (D2: "Every simulation always derives from theory"). For them, this is true for both scientific simulations and educational ones.

Category 3: *Computer simulations allow experimentation*

The designers agreed on the fact that simulations allow experimentation when parameters are added and adapted to explore the phenomenon under study and the computer model is run to be compared with the phenomenon it is supposed to simulate (D2: "…computer models are tested and then you try with a real model or real experiments; first you try with a model that is basically computational." D1: "With simulations, we can simulate experiments we have never done before and get results that can predict or represent new situations…").

Category 4: *Simulations require empirical validation*

The interviewees expressed the need for empirical validation as a way of authenticating results (D3: "*…simulations must match reality*"). This is connected with what Winsberg (2003) suggests: the fundamental need for the "calibration" of the simulation by comparing it with what we know about the phenomenon or with the results obtained after contrasting it with reality.

Category 5: *Simulations are not a substitute for experimentation*

As regards simulations in education, both D2 and D3 unequivocally stated that they could never replace experimental activity in the classroom (D2: "...although the gap between real and virtual experience is smaller and smaller, simulations cannot replace real experience; reality is richer than the virtuality").

Category 6: *Computer simulations allow the explanation and prediction of phenomena*

The interviewees pointed out that simulations should be understood as models that predict and adequately solve new problems in scientific activity (D1: "... a simulation should be expected to predict or solve new situations and help us understand the phenomenon better"). When designing a simulation based on the mathematical model derived from theory and testing it against unknown situations, the process itself helps us to explain the phenomena better. This idea is shared by the three designers, who considered this aspect fundamental to both scientific and educational simulations.

Category 7: *Computer models may prove opaque*

The computer models these designers develop are executed on Java platform. D2, in particular, builds computer models to explore the actions within the boundaries of the simulation he has written or to modify variables in the mathematical model by incorporating new equations. These simulations become in this way "black boxes" for the users – a good imitation of the phenomenon is obtained, but the underlying mathematical model is unknown to the user (the student). This designer was aware of this problem, but considered that to be able to design computer models, the student should possess in-depth knowledge in areas such as physics, mathematics, and programming, which does not seem to be the case (D2: "...I work with ready-made closed simulations. I use them all the time in my classes...").

On the other hand, researcher D1 teaches his students how to model, but in his case, they are advanced university students. This way of working is more similar to scientific activity (D1: "Rather than use simulations other people have created for them, I make them construct simulations").

Category 8: *Simulations cause a change in the way in which scientific knowledge is produced*

The designers considered that simulations have changed the way of working in science and that they are more than a potent tool to solve equations (D2: "...simulations have evolved; in addition to solving non-linear equations, simulations are seen as the only way of operating big programs such as those of fluids. We nowadays try with computer models first before moving on to real models or real experiments". D3: "...they are regarded as the third scientific method").

Category 9: *Simulations are an important tool in science teaching*

The three designers were of the opinion that simulations facilitate science learning because they allow the student to work on a phenomenon from three perspectives: the physics model, the mathematical model, and the virtual model in an interrelated way, which helps to understand the relationship between the mathematical model (or equation) and the physics phenomenon they explain (or describe) (D1: "…with simulations students recognize the modeling process: for example, creating a mathematical model of a physics problem". D2: "…it is rewarding to work with a mathematical model, a physics model, and a virtual model". D3: "it helps students to understand; it makes them visualize abstract concepts"). In this way, equations are no longer isolated symbols and become a form of representation of the physics phenomenon.

Category 10: *The use of simulations in education requires important epistemological aspects to be considered*

The designers were aware of the fact that problems associated with epistemological views may arise in the didactical use of simulations, D3 pointed out that students should comprehend the meaning of a scientific explanation, of a mathematical model, and the representation rules. This relates to another highlighted aspect that the interviewees observed in students: the difficulty in distinguishing between virtuality and reality (D1: "Students believe the computer is always right", "…if the computer says so…" "Being themselves the ones who design simulations should help them understand what happens". D2: "…the use of traditional laboratories [together with simulations] should avoid that problem…").

25.5 Analysis of the Results

We have been able to identify in the answers some common conceptualizations:

1. Simulations derive from theories and precede experimentation, within the simulation itself and, in many cases, also within real experiments. In addition, simulations, just like traditional scientific models, should help to both explain and predict new problems, and a continuum can be identified ranging from the design of the mathematical model of the phenomenon, based on a theory (classic modeling), to the implementation on the computer (with the decisions regarding the computer model and the necessary simplifications involved). It follows that for these designers, computer simulations are one more type of model that the researcher designs to study a particular phenomenon. This idea of simulations as models, with the limitations implied, is absent from science education practice.
2. According to the designers, even if we adopt a relatively classic view of simulations as models, simulations are considered to have changed scientific activity and to have become more than a mere efficient calculus tool.

3. The designers highlighted the differences that exist between computer simulations in science and in science education. In the case of scientific simulations, scientists design complex models that translate into computational programs, based on theoretical and empirical assumptions (and sometimes, intuitions) about the systems, phenomena or processes they want to better understand, and they experiment with those simulations to gain new knowledge. On the other hand, typical educational simulations, such as applets, are learning environments, designed by teachers or professional programmers, based on known physics laws that simulated systems, phenomena or processes (already known). Through their manipulation, students may learn or practice some new (for them) concepts or laws. This difference, so clear to the designers but apparently not to teachers, is a concept barely considered in class and this could therefore lead students to develop a distorted view of scientific activity.
4. The designers also stress that when students use some programs specifically designed to help them to create simulations (such as Modellus or EJZ), greater conceptualization is fostered when studying natural phenomena in class, because this process allows students to "see" the interrelation among the physics, the mathematical, and the virtual model. The fact that these simulations articulate the conceptual and the mathematical description of the phenomenon during the construction and visualization phases is crucial for understanding physics theories. However, this aspect (and this type of simulation) is generally disregarded in secondary school.
5. Another relevant teaching aspect stressed by the designers is that the empirical validation of the results of the simulations must be essential both in science and in science education.
6. In science education, it is also necessary to avoid the confusion between reality and virtuality. The designers considered it useless to use simulations for exploration in replacement of experimentation, something that often occurs in science teaching. Two of them also stressed the importance of going beyond the exploratory use of simulations and actively promoting the development of their teaching potential.
7. The interviewees also demonstrated a deep epistemological reflection on simulations and their use in science teaching, something missing from research in the area and the everyday use of educational simulations.

25.6 Final Comments

The designers of simulations or of programs for the design of simulations in science education convey epistemological views on simulations and their use in science that are more complex than those found in studies on science teaching carried out by novel researchers, or in articles about science teaching at secondary school. It should be noted that their views, both epistemological and didactic, seem relevant to the development of new teaching proposals that would help to avoid a distorted view of scientific activity and also to assume a critical attitude when implementing computer simulations in science teaching and learning processes.

References

Acevedo Díaz, J. A. (2008). El estado actual de la naturaleza de la ciencia en la didáctica de las ciencias. *Revista Eureka Enseñanza de Divulgación Científica, 5*(2), 134–169.

Bayraktar, S. (2002). A meta-analysis of the effectiveness of computer-assisted instruction in science education. *Journal of Research on Technology in Education, 34*(2), 173–188.

Braun, V., & Clarke, V. (2006). Using thematic analysis in psychology. *Qualitative Research in Psychology, 3*(2), 77–101.

Cartwright, N. (1999). Models and the limits of theory: Quantum Hamiltonians and the BCS model of superconductivity. In M. S. Morgan & M. Morrison (Eds.), *Models as mediators: Perspective on natural and social science* (pp. 241–281). Cambridge: Cambridge University Press.

de Jong, T., & van Joolingen, W. R. (1998). Scientific discovery learning with computer simulations of conceptual domains. *Review of Educational Research, 68*(2), 179–201.

Doerr, H. (1997). Experiment, simulation and analysis: An integrated instructional approach to the concept of force. *International Journal of Science Education, 19*(3), 265–282.

Durán, J. (2015). Nociones de simulación computacional: simulaciones y modelos científicos. *Argumentos de Razón Técnica, 18*, 87–110.

Galison, P. (1996). Computer simulation and the trading zone. In P. Galison & D. J. Stump (Eds.), *The disunity of science: Boundaries, contexts, and power* (pp. 118–157). Stanford: Stanford University Press.

González Ugalde, C. (2014). Investigación fenomenográfica. *Revista Internacional de Investigación en Educación, 7*(14), 141–158.

Greca, I. M., Seoane, M. E., & Arriassecq, I. (2014). Epistemological issues concerning computer simulations in science and their implications for science education. *Science & Education, 23*(1), 897–921.

Guillemot, H. (2010). Connections between simulations and observation in climate computer modeling. Scientist's practices and "bottom-up epistemology" lessons. *Studies in History and Philosophy of Modern Physics, 41*, 242–252.

Hernández Sampieri, R., Collado, C. F., & Lucio, P. B. (2010). *Metodología de la Investigación* (5th ed.). Santa Fe: McGraw-Hill.

Johnson, A., & Lenhard, J. (2011). Toward a new culture of prediction: Computational modeling in the era of desktop computing. In A. Nordmann et al. (Eds.), *Science transformed? Debating claims of an epochal break* (pp. 189–200). Pittsburgh: University of Pittsburgh Press.

Kirschner, P., & Huisman, W. (1998). Dry laboratories in science education: Computer-based practical work. *International Journal of Science Education, 20*, 665–682.

Koponen, I. T., & Tala, S. (2014). Generative modelling in physics and in physics education: From aspects of research practices to suggestions for education. In M. R. Matthews (Ed.), *International handbook of research in history, philosophy and science teaching* (pp. 1143–1169). Dordrecht: Springer.

Kuppers, G., & Lenhard, J. (2005). Validation of simulation: Patterns in the social and natural sciences. *Journal of Artificial Societies and Social Simulation, 8*(4), 1–13. http://jasss.soc.surrey.ac.uk/8/4/3.html. Accessed 20 Jul 2016.

Lenhard, J. (2010). Computation and simulation. In R. Frodeman, J. T. Klein, & C. Mitcham (Eds.), *The Oxford handbook on interdisciplinarity* (pp. 246–258). Oxford: Oxford University Press.

López, S., Veit, E. A., & Araujo, I. S. (2016). Una revisión de la literatura sobre el uso de la modelación y simulación computacional para la enseñanza de la física en la educación básica y media. *Revista Brasileira de Ensino de Física, 38*(2), 1–16.

Marton, F. (1986). Phenomenography: A research approach to investigating different understandings of 25 reality. *Journal of Thought, 21*(3), 28–49.

Morgan, M. S., & Morrison, M. (Eds.). (1999). *Models as mediators: Perspective on natural and social science*. Cambridge: Cambridge University Press.

National Research Council. (2012). *A framework for K-12 science education: Practices, crosscutting concepts, and core ideas*. Washington, DC: The National Academies Press.

Ören, T. I. (2011a). The many facets of simulation through a collection of about 100 definitions. *SCS M&S Magazine, 2*(2), 82–92.

Ören, T. I. (2011b). A critical review of definitions and about 400 types of modeling and simulation. *SCS M&S Magazine, 2*(3), 142–151.

Seoane, M. E., Arriassecq, I., & Greca, I. (2014). Implicancias epistemológicas en la utilización de simulaciones computacionales en diversas disciplinas científicas. *VIII Reunião Latino-Americana sobre Ensino de Física*, 3–7 November 2014, Salvador, Bahía, Brasil.

Seoane, M. E., Arriassecq, I., & Greca, I. (2015). Simulaciones computacionales: un análisis fenomenográfico. *Revista Enseñanza de la Física, 27*, 289–296.

Sismondo, S. (1999). Models, simulations and their objects. *Science in Context, 12*(2), 247–260.

Smetana, L. K., & Bell, R. L. (2012). Computer simulations to support science instruction and learning: A critical review of the literature. *International Journal of Science Education, 34*(9), 1337–1370.

Winsberg, E. (1999). Sanctioning models: The epistemology of simulation. *Science in Context, 12*(2), 275–292.

Winsberg, E. (2003). Simulated experiments: Methodology for a virtual world. *Philosophy of Science, 70*, 105–125.

Winsberg, E. (2010). *Science in the age of computer simulation*. Chicago: The University of Chicago Press.

Part V
Sociology of Science

Chapter 26
The Nature of Science in Secondary School Geology: Studying Recontextualizing Processes

Ana M. Morais, Sílvia Castro, Sílvia Ferreira, and Isabel P. Neves

26.1 Introduction

The introduction of the nature of science (NOS) into science education has been defended by many authors.[1] There is today a broad definition of NOS that contains aspects such as the characteristics of scientific research, the process of construction of scientific theories, the social and intellectual circumstances where scientific knowledge has been developed, the way in which scientists work as a social group and the way in which science influences and is influenced by the social context (Hodson 2014). Mostly in its dimensions of history of science, philosophy of science, and sociology of science, NOS has been part of scientific literacy and consequently an important objective of the teaching–learning processes of the science curricula of many countries (Millar and Osborne 1998; Osborne and Dillon 2008; Hodson 2014). The Program for International Student Assessment (PISA) 2015 has also followed this direction by giving further weight to a component associated with NOS. When defining scientific literacy, "the notion of 'knowledge about science' has been specified more clearly and split into two components – procedural knowledge and epistemic knowledge" (OCDE 2016, p. 22).

John M. Ziman's conceptualization of NOS (1984, 2000) is, in epistemological terms, an important theoretical support of the study described in this chapter. Ziman

[1] E.g., Lederman (2007), Clough and Olson (2008), Matthews (2009), McComas (2014), and Taber (2017).

A. M. Morais (✉) · S. Castro · I. P. Neves
UIDEF, Instituto de Educação, Universidade de Lisboa, Lisbon, Portugal
e-mail: ammorais@ie.ulisboa.pt; imneves@ie.ulisboa.pt

S. Ferreira
UIDEF, Instituto de Educação, Universidade de Lisboa, Lisbon, Portugal

Agrupamento de Escolas de São Gonçalo, Torres Vedras, Portugal

M. E. B. Prestes, C. C. Silva (eds.), *Teaching Science with Context*,
Science: Philosophy, History and Education,
https://doi.org/10.1007/978-3-319-74036-2_26

characterized science as a complex system, which should be analyzed in terms of four metascientific dimensions: philosophical, psychological, historical, and sociological. This multidimensional theorization of science has allowed an interrelated analysis of the various metascientific disciplines.

The philosophical dimension of science refers to the methods used by scientists to make science, as is the case of observation, formulation of hypotheses, experimentation, and theorization. Science is characterized as a dynamic process of knowledge construction, which has diverse methodologies. The psychological dimension respects the psychological characteristics of scientists that influence their work such as curiosity and perseverance, but others to, not so noble but also proper to the human condition, such as intellectual dishonesty. The historical dimension of science respects the accumulation of scientific knowledge, organized in coherent theoretical schemes and divulged in publications (science archive). The sociological dimension refers to the relations among the members of the scientific community (internal sociology) and to the inter-relations between science and the society at large (external sociology). The science/technology/society (STS) relationship is, according to this conceptualization, part of the external dimension of sociology.

Basil Bernstein's model of pedagogical discourse (1990, 2000) is, in sociological terms, the main theoretical framework of this study. Through this model, Bernstein seeks to understand how the pedagogical discourse, determined by a complex set of relationships, which presuppose the intervention of various fields and contexts, is produced and reproduced.

Even though the pedagogical discourse reflects the dominant principles of society, which constitute the general regulative discourse (GRD), that discourse is not the immediate result of those principles, as recontextualizing processes may occur at the various levels of the pedagogical device. As a result of the official recontextualizing of the GRD, the official pedagogic discourse (OPD) is produced, which is part for example of curricula and syllabuses. The OPD may also be the object of a second recontextualizing process in the pedagogical recontextualizing field (e.g., departments of education, teachers' education schools, and institutions for the production of pedagogical materials). This process leads to the construction of the pedagogical discourse of reproduction (PDR), which is present for example in textbooks. The official and pedagogical recontextualizing fields are influenced by the fields of economy and symbolic control and on the whole, they define the *what* and the *how* of pedagogical discourse. The *what* refers to the knowledge and skills to be the object of the teaching–learning process and the *how* refers to the way in which the knowledge and skills are transmitted in the teaching–learning context.

The present study is part of broader research (Castro 2017) that took place in Portugal and that focused on the analysis of NOS in the curriculum and textbooks of biology and geology[2] of the tenth grade (age 15–16), i.e., of the first year of secondary school. The study follows former research developed by the ESSA

[2] Biology and Geology, although epistemologically distinct, have traditionally been part of the same discipline in Portugal (often but not always called Natural Sciences).

Group[3] (e.g., Calado and Neves 2012; Ferreira and Morais 2013). This chapter is centered on the geology section of those texts and addresses the following general research problem: What is the extent to which the message contained in the official pedagogical discourse of the syllabus of secondary school geology with regard to NOS is recontextualized in the pedagogical discourse of textbooks? From this problem, the following research questions were derived:

1. What is the message about NOS that is transmitted by the geology syllabus?
2. What is the message about NOS that is transmitted by the geology textbooks?
3. What are the recontextualizing processes that may have occurred between the pedagogical discourse of the syllabus and the pedagogical discourse of textbooks with regard to NOS?

The analysis of the message about NOS that is transmitted by the geology syllabus and textbooks, and of the recontextualizing processes that might have occurred between the message of the syllabus and the message of the textbooks, provides the basis for a reflection on the consequences of those processes for the teaching–learning of NOS in science education.

26.2 Methodology

The biology and geology syllabus for the tenth grade (DES 2001) is divided into two main sections that correspond to the scientific areas of biology and geology. Each one of these sections contains two parts: the "syllabus presentation," with the general guidelines and the "development of the syllabus," with specific guidelines to operationalize the general principles. The analysis of the syllabus was centered on both parts for the area of geology, so that a comparison between them could be made.

The analysis of textbooks was centered on the area of geology of the two textbooks for tenth grade biology and geology (textbooks A and B), which had been more widely selected across the whole country by the teachers/schools, in the academic year 2013/2014. That analysis included both the part that was directed to the students (corpus of the textbook) and the teachers' support materials. This chapter gives global results of these two parts of each one of the textbooks.

The NOS was therefore analyzed in the area of geology of the discipline of biology and geology of the tenth grade, both at the level of the official pedagogical discourse present in the syllabus and at the level of the pedagogical discourse of reproduction transmitted in textbooks. Figure 26.1 shows the various dimensions of analysis that were considered in the study described in the chapter. The analysis was focused on dimensions related to the *what* and the *how* of pedagogical discourse, where the *what* refers to the metascientific knowledge to be transmitted-acquired

[3] The ESSA Group – Sociological Studies of the Classroom – is a research group of the Institute of Education, University of Lisbon. http://essa.ie.ulisboa.pt

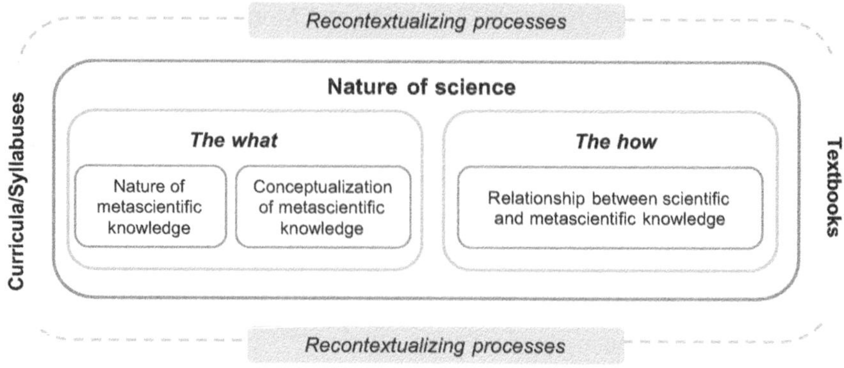

Fig. 26.1 Dimensions of analysis of the nature of science in curricula/syllabuses and textbooks. (Adapted from Castro 2017)

and the *how* refers to the way in which the transmission–acquisition of that knowledge should take place. The analysis of metascientific knowledge took into account the nature of that knowledge, which is the dimension of science construction to which it refers, and also its conceptualization level. The analysis of the *how* took into account the degree of relation between scientific knowledge and metascientific knowledge when representing a relationship between knowledge within a same discipline (intradisciplinary relationship).

To characterize the message contained in the syllabus and in the textbooks these texts were divided into units of analysis.[4] A unit of analysis was considered as an excerpt of the text with one or more sentences which together have a given semantic meaning (Gall et al. 2007). Segmentation of text into units of analysis differed according to the nature of that text – syllabuses and textbooks. In the case of textbooks broader units of analysis were considered. On the basis of the trend shown by the analysis of all units of analysis of each syllabus section and of each one of the textbooks, it was possible to infer their respective messages about the NOS in the various dimensions under study (Fig. 26.1). The study follows an approach that combines quantitative and qualitative methods of analysis (Morais and Neves 2010).

Several instruments were constructed to analyze each one of the units of analysis. One of such instruments was developed to be a referential of analysis to identify the dimensions of science construction and the level of complexity of the metascientific knowledge related to each one of those dimensions. That knowledge was separated by its level of complexity into two groups, simple knowledge and complex knowledge, according to the distinction between facts, simple concepts, complex concepts and unifying themes/theories made by several authors (Anderson

[4]The general guidelines of the syllabus were divided into 44 units of analysis and the specific guidelines into 272. In textbook A 531 units were analyzed, 276 of which were part of the corpus of the textbook. In textbook B 426 units were analyzed, 360 of which were part of the corpus of the manual.

Table 26.1 Excerpt from the referential instrument for the nature of metascientific knowledge – philosophical dimension and external sociological dimension of science

Simple knowledge	Complex knowledge
(Facts and simple concepts)	(Complex concepts and unifying themes /theories)
Philosophical dimension	
Science as a dynamic process of knowledge construction that contains various methodologies	
The construction of scientific knowledge uses methods and principles based on gathering, organization, and interpretation of data obtained by various methods	The construction of scientific knowledge makes use of models that are representations of the world through which it is sought to simplify reality, so that such reality can be analyzed
To answer the same problem distinct hypotheses may co-exist and through tests and/or analysis of data obtained from reality may be supported or refuted	All scientific knowledge is fallible, meaning that is only valid until it is not refuted by experience, and as a consequence scientific knowledge is not absolute
In science, new data lead to the reformulation of concepts and theories	The scientific knowledge produced is part of broader theoretical frameworks or unifying themes
External sociological dimension	
Inter-relation between science, technology and society	
Scientific research and also knowledge production and scientific predictions influence society and/or the environment/human species – Sc-S relationship	The inter-relation that develops between science, technology and society originates a STS cycle – Sc-T-S relationship
The development of scientific knowledge permits the development of new technologies – Sc-T relationship	Socio-scientific controversies are generated by social impacts of scientific and technological innovations which divide both science community and society in general and which involve scientists, political decision-makers and groups of citizens
The development of technology leads to further science research and consequently to the development of science –T-Sc relationship	

Source: Adapted from Castro (2017) and Ferreira and Morais (2013)

et al. 2001; Brandwein et al. 1980; Cantu and Herron 1978). Table 26.1 shows an excerpt of this instrument for the philosophical and external sociological dimensions.

The conceptualization of metascientific knowledge was analyzed by an instrument constructed with four degrees of complexity, related to the four dimensions of science construction. The descriptors of that scale, considered in an increasing order of complexity, are focused on factual knowledge (degree 1), simple concepts (degree 2), complex concepts[5] (degree 3), and unifying themes and theories (degree 4). The

[5]The simple concepts correspond to concrete concepts proposed by Cantu and Herron (1978) and are those that have a low level of abstraction, defining attributes and examples that are observable. The complex concepts correspond to abstract concepts and "are those that do not have perceptible instances or have relevant or defining attributes that are not perceptible" Cantu and Herron (1978, p. 135).

Table 26.2 Excerpt of the instrument for characterizing the conceptualization level of metascientific knowledge with regard to the philosophical dimension and examples of units of analysis

Degree 1	Degree 2	Degree 3	Degree 4
Factual metascientific knowledge of the philosophical dimension of science is referred to, corresponding to concrete, observable or perceptible information	Simple metascientific knowledge of the philosophical dimension of science is referred to, corresponding to concepts with a low level of abstraction and perceptible characteristics	Complex metascientific knowledge of the philosophical dimension of science is referred to, corresponding to concepts with a high level of abstraction and nonperceptible characteristics	Complex metascientific knowledge of the philosophical dimension of science is referred to, corresponding to structuring ideas and theories

Degree	*Units of analysis:*
Degree 2	[1] The geologists work directly in all places they may have access, in the whole world: from the icy peaks of high mountains and the active volcanoes to the deep oceans. Moreover, the geologists have to rely on their indirect observations, by using sensible measurement instruments and creating models. (*Syllabus, Geology section, 10th grade*)
Degree 4	[2] The explanations given are part of a catastrophic line of thought. For some scientists, the dinosaurs' extinction would have been caused by the impact of a meteorite, whose crater would have formed near the Gulf of Mexico. [...] However, some other scientists, namely the paleontologists, state that there is no need to use these catastrophic explanations to explain dinosaurs' extinction. (*Textbook, Geology section, 10th grade*)

Source: Adapted from Castro (2017)

descriptors are similar for all dimensions of science construction. Table 26.2 shows an excerpt of the instrument for the philosophical dimension of science and examples of units of analysis classified by making use of this instrument.

Excerpt [1] presents simple metascientific knowledge (see Table 26.1), at the level of simple concepts (degree 2) associated with the philosophical dimension, namely the following: the construction of scientific knowledge is made with the help of practical or field work, which implies the use of measurement instruments and/or equipment and/or specific technics. Excerpt [2] is focused on complex knowledge associated with the philosophical dimension with the highest degree of complexity (degree 4): the scientific knowledge produced is inserted in broader theoretical frameworks or in unifying themes (see Table 26.1). Knowledge associated with the internal sociological dimension, but with a lower degree of complexity is also present in this excerpt: there are sometimes different theories to answer the same problem inside the scientific community (Castro 2017).

The analysis of the relationship between scientific and metascientific knowledge was made through an instrument with a four-degree scale constructed on the basis of Bernstein's concept of classification (1990, 2000). In this particular case, classification refers to the existence of fairly strong boundaries between scientific and metascientific knowledge. The extreme value of the strongest classification (degree

Table 26.3 Excerpt of the instrument for characterizing the relationship between scientific and metascientific knowledge and examples of units of analysis

Degree 1	Degree 2	Degree 3	Degree 4
C^{++}	C^{+}	C^{-}	C^{--}
There is a focus on scientific knowledge only	There is a focus on metascientific knowledge, but the relationship between that knowledge and the scientific knowledge is not made	There is a relationship between metascientific and scientific knowledge where scientific knowledge is given more status in that relationship	There is a relationship between metascientific and scientific knowledge where both types of knowledge have the same status in that relationship
Degree	*Units of analysis:*		
C^{++}	[3] The remains of the dinosaurs' presence on Earth's surface can be found in sedimentary rocks. These rocks are characterized by their frequent presence in layers. It should be referred to that there are other rocks such as magmatic and metamorphic rocks, which together with the sedimentary rocks are part of the rocks cycle. (*Syllabus, Geology section, 10th grade*)		
C^{--}	[4] […] we know that the temperature increases with the increasing of depth, that the temperature inside the Earth is high and we believe that we also know the internal structure and composition of our planet. This knowledge is a result of the interpretation of data obtained by geophysical methods: electrical (conductivity), magnetic (magnetism), radioactive (radioactivity), gravimetric (isostasy and gravity anomalies), seismic (behavior of seismic waves) and geothermic (volcanism, Earth's internal heat). These are the indirect methods that the geologists and the geophysicists use to study the Earth's crust, mantle, and core. (*Syllabus, Geology section, 10th grade*)		

Source: Adapted from Castro (2017)

1/C^{++}) corresponds to a situation where there is no relationship between these two types of knowledge to a point that there is not even reference to metascientific knowledge. The extreme value of the weakest classification (degree 4/C^{--}) corresponds to a situation where there is a strong relationship between these two types of knowledge, which are given equal status. Table 26.3 shows an excerpt from the instrument and examples of units of analysis classified with this instrument.

Excerpt [3] is only focused on scientific knowledge, related to sedimentary rocks, and for that reason it was classified with degree 1 (very strong classification – C^{++}). Excerpt [4] calls for a relationship between scientific and metascientific knowledge, at the level of the philosophical dimension, respectively related to Earth's internal structure and indirect methods used by geologists. Both types of knowledge have the same status in the relationship and for that reason this excerpt was classified with degree 4 (very weak classification – C^{--}).

The following illustrative example shows how the same unit of analysis was classified in terms of all metascientific dimensions:

> [5] "The idea that Man's history on Earth had been preceded by another one history, previous to man's existence, began to become evident by the end of the XVIII century. The stratified sedimentary rocks contained often a thickness and richness of fossils which suggest an extremely slow deposition, which on its turn implied the acceptance of long chronologies. However not all defenders of a long scale of time accepted the sole acting of slow and gradual causes. For many that immense period of time might have been interrupted by

violent catastrophes. [...] Recalling the main ideas defended by Cuvier may raise some questions with educational interest: how is it possible that the same objects and phenomena can be interpreted by two distinct models? [...] There is nowadays a renovated interest for the catastrophist conceptions under the designation of neocatastrophism: what is the reason for that reappearance?" (*Syllabus, Geology section, 10th grade*)

On the basis of the referential instrument for metascientific knowledge (of which Table 26.1 is an example), this unit of analysis contains, with regard to the *what* of the OPD, knowledge of the philosophical and historical dimensions. For each one of these dimensions, excerpt [5] calls for simple metascientific concepts and for this reason both dimensions were classified with degree 2. With regard to the *how* of the OPD, excerpt [5] calls for a relationship between scientific and metascientific knowledge where both have the same status and for this reason that relationship was classified with degree C^{--}.

26.3 Results

The graph of Fig. 26.2 shows the results of the analysis of the metascientific knowledge in the geology syllabus section and also of textbooks with regard to the dimensions of science construction: philosophical (PhD), historical (HD), psychological (PD) and sociological, internal (SID) and external (SED).

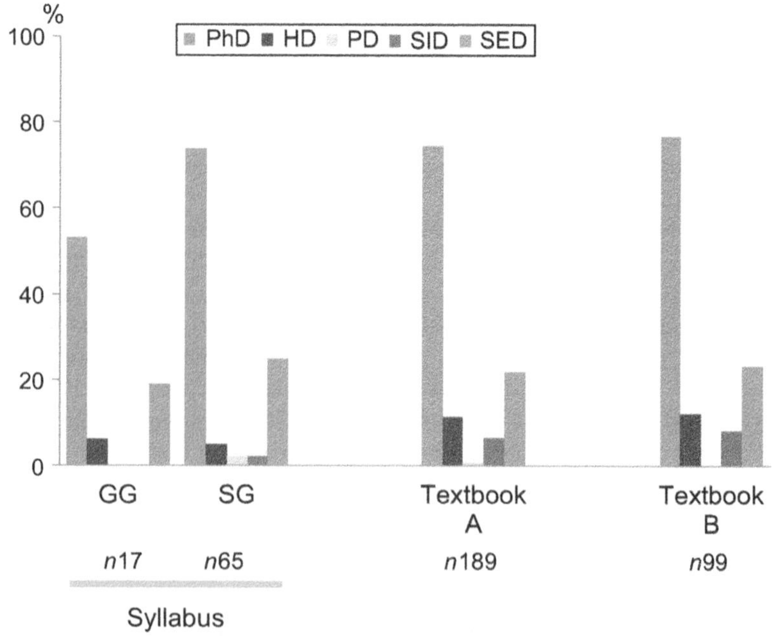

Fig. 26.2 Nature of metascientific knowledge in the geology section of the general (GG) and specific (SG) guidelines of the syllabus and of textbooks. *n* total number of units of analysis studied. (Adapted from Castro 2017)

These results show the prevalence of the philosophical dimension in both the syllabus and the two textbooks. Science methodologies are the most relevant aspects of science construction in the texts analyzed. They are followed by the science, technology and society relationships, i.e., the external sociological dimension, which comes out as the second most present dimension of science construction. The historical dimension of science comes out as the third most represented dimension. In general, the relationships inside the scientific community (internal sociological dimension) and mostly the scientists' psychological characteristics (psychological dimension) are the less valued dimensions.

The results also show that recontextualizing processes occurred within the syllabus and between the syllabus and the textbooks. When passing from general to specific guidelines of the syllabus, there is a valuing of the philosophical dimension and the presence, although small, of the psychological and internal sociological dimensions. When passing from the syllabus' specific guidelines to the textbooks, the emphasis given to the philosophical and to the external sociological dimensions is similar, but there is a valuing of the historical and internal sociological dimensions of science.

The results on the degree of conceptualization of metascientific knowledge in the syllabus and textbooks are shown in Fig. 26.3, when metascientific knowledge is considered as a whole, independently of the dimension of science construction to which those results refer.[6] The results show that in all texts analyzed most metascientific knowledge corresponds to simple concepts (degree 2). In the syllabus and in textbook B there is no reference to metascientific knowledge with a factual character (degree 1) nor to complex metascientific knowledge relative to structuring ideas and theories (degree 4).

These data show some discontinuities between the messages of the various texts, when the level of conceptualization of metascientific knowledge is considered. In the syllabus, when passing from the directions given on general guidelines to their concretization at the level of the specific guidelines, decreases the level of conceptualization of knowledge with a lower percentage of complex metascientific knowledge (degree 3). On the other hand, when the message contained in the specific guidelines of the syllabus is compared with the textbooks' message, the level of conceptualization tends in general to be lower in textbooks, particularly in textbook B.

The graph of Fig. 26.4 shows the results of the analysis of the relationship between scientific and metascientific knowledge. The results of this analysis refer only to the cases where metascientific knowledge is present and for that reason classification C^{++} is not considered (this value refers only to scientific knowledge).

In the case of the syllabus, the relationship between scientific and metascientific knowledge (weaker classifications, C^{-} and C^{--}) is mostly present in the specific

[6] When a same unit of analysis contained references to several dimensions of science construction, each one of them was considered as a separate reference. When a same unit of analysis contained several references to the same dimension of science construction with different degrees of complexity, the reference with the highest degree was considered.

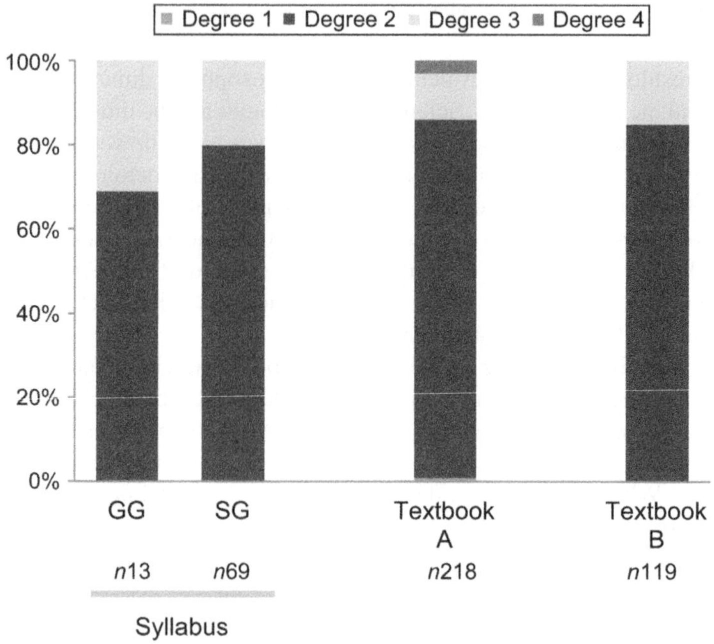

Fig. 26.3 Conceptualization of metascientific knowledge in the geology section, in general (GG) and specific (SG) guidelines of the syllabus and also in textbooks. *n* total number of references to metascientific knowledge. (Adapted from Castro 2017)

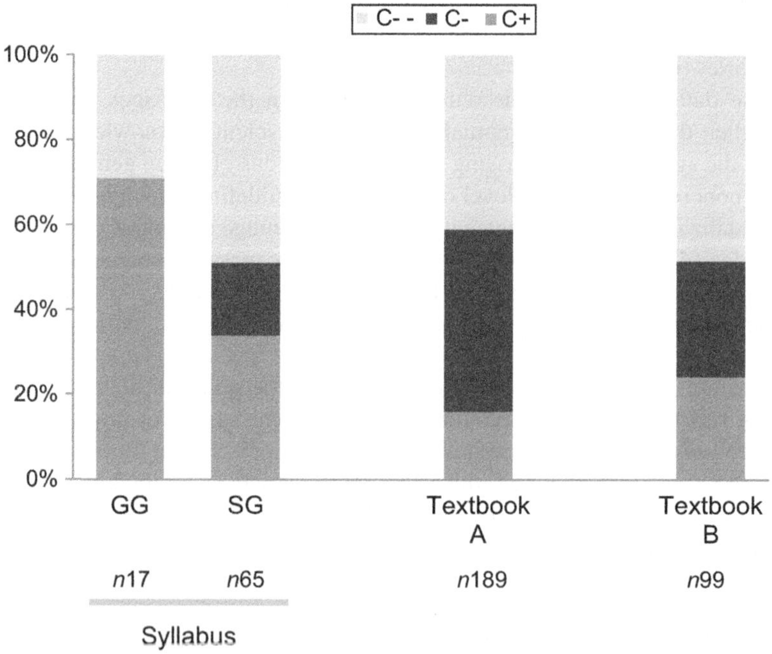

Fig. 26.4 Relationship between scientific and metascientific knowledge in the geology section, in general (GG) and specific (SG) guidelines of the syllabus and also in textbooks. *n* total number of units of analysis studied. (Adapted from Castro 2017)

guidelines having less expression in the general guidelines. In the case of textbooks, this intradisciplinary relationship is a little better represented in textbook A. Thus, when passing from the general guidelines to the specific guidelines of the geology syllabus a recontextualization of the pedagogical discourse did occur, in the direction of a strengthening of the relationship between scientific and metascientific knowledge (a smaller percentage of units classified with C^+ and a greater percentage of units classified with C^- and C^{--}). The relationship between these two types of knowledge is further strengthened when passing from the syllabus to textbooks, something that is more evident in textbook A. It is important to point out the situation where there is a relationship between the two types of knowledge, but where higher status is given to scientific knowledge (C^-). This situation is either absent (general guidelines) or undervalued (specific guidelines) in the syllabus and it is better represented in textbooks, particularly in textbook A.

26.4 Conclusions

This study was centered on NOS in the teaching/learning context of geology of the Portuguese secondary school. The message contained in the pedagogical discourse of the syllabus and textbooks was analyzed to characterize the nature and conceptualization of the metascientific knowledge and of its relationship with scientific knowledge, and to evaluate the recontextualizing processes that might have occurred between and within those texts.

The results of the study show that, with regard to geology of both the syllabus and textbooks of the discipline of biology and geology of tenth grade, the global message about NOS privileges the methodology of science (an aspect of the philosophical dimension) followed by its external sociology. Little emphasis is given to other important aspects of the NOS, particularly at the level of the psychological and internal sociological dimensions. The results also show a low level of conceptualization of the metascientific knowledge, which mostly corresponds to simple concepts with a low level of abstraction. These results are similar to the results of other national studies (Calado and Neves 2012; Castro 2006; Ferreira and Morais 2013) that have shown that, whenever the NOS is present in science education, the external sociological (STS) and the philosophical dimensions tend to be the most valued perspectives, although having, in general, a low level of conceptualization in the curriculum. Furthermore, international studies (McComas and Olson 1998; Vesterinen et al. 2009) have shown that in the science curricula of several countries the most emphasized aspects of the NOS are those related to philosophy and history of science.

Whenever the NOS is present in the syllabus and in textbooks, the results show an intradisciplinary relation between scientific and metascientific knowledge, namely in the cases of the syllabus-specific guidelines and in textbooks. These results are in accordance with the ideas of several authors (Aydin and Tortumlu 2015; Taber 2017), when they point out the integration of the NOS into the teaching–learning process of scientific knowledge as the most favorable approach to promoting the understanding of the NOS by students.

The conceptualization level of metascientific knowledge and the degree of relationship between scientific and metascientific knowledge, constitute a basis to appreciate the level of conceptual demand of scientific learning in the context of NOS (Ferreira et al. 2015; Morais and Neves 2016). The results of the present study could lead us to think that the valuing of the relationship between scientific and metascientific knowledge would have given a high contribution to raise the desirable level of conceptual demand of scientific learning. However, the low level of conceptualization of metascientific knowledge does limit that contribution. From this point of view, we can state that the curriculum texts analyzed in this study may limit the access of all students to a broad understanding of science construction, and, in this way, they do not promote a high level of scientific literacy.

The recontextualizing processes between the official pedagogical discourse (syllabus) and the pedagogical discourse of reproduction (textbooks), with regard to the dimensions of the NOS studied, vary in direction and degree. In terms of the nature of metascientific knowledge, there is a slightly valuing of some dimensions of science construction when passing from general to specific guidelines of the syllabus and from the syllabus-specific guidelines to the textbooks. In terms of the complexity of metascientific knowledge there is, in general, little difference between the two parts of the syllabus and between the syllabus and textbooks. The low level of conceptualization of metascientific knowledge is kept in all texts. However, stronger recontextualizing processes are evident at the level of the intradisciplinary relation between scientific and metascientific knowledge. This intradisciplinary relation becomes stronger when passing from general to specific guidelines of the syllabus and also when passing from the syllabus to textbooks. A recontextualization process also occurs when the reference is the status given to scientific knowledge in such relation. The case where scientific knowledge has more status in the relationship between scientific and metascientific knowledge (weak classification – C^-) is either absent or barely represented in the syllabus, but is valued in textbooks, particularly in textbook A. This situation, which was considered the most favorable to high science understanding according to the theoretical framework of the study, was therefore not valued by the authors of both the syllabus and one of the textbooks.

An important aspect that should be highlighted and which is related to recontextualization processes refers to the incoherence that may exist within the official pedagogical discourse between the two main sections of the syllabus and also between the official pedagogical discourse of the syllabus and the pedagogical discourse of textbooks. These incoherencies may be a cause of difficulty for textbook authors when interpreting the syllabus and for teachers when implementing the syllabus and textbooks. Textbooks authors' recontextualization of syllabuses can be a major problem in education whenever it decreases the scientific level in any of the dimensions of analysis of NOS (textbooks sometimes do increase that level), because, as research has been showing, teachers mainly base their practices on textbooks, rarely consulting syllabuses (Cavadas and Guimarães 2012; Valverde et al. 2002).

However, it should be stressed that a sound teachers' education focused on the integration of the NOS into science education and on the importance of conceptual demand in promoting scientific literacy may lead teachers to recontextualize the message of syllabuses and textbooks, in the direction of raising the conceptual level of these curriculum texts. In raising questions related to the construction of syllabuses and textbooks and their relationship with teachers' practices, the study points to the crucial importance of teachers' education. In this, it follows many other studies (Hodson 2014; Irzik and Nola 2011; McComas 2014).

In theoretical terms, the study raises questions related to the importance of refining the message related to NOS, which is contained in curricula, syllabuses, and textbooks, by pointing to the introduction in these texts of all aspects of NOS and also suggesting a conceptualized learning of metascience in its relation to science. It also points to the need for coherence not only inside each one of the pedagogical texts, but also between them (internal and external coherence).

In methodological terms, the study makes a contribution to the development of analyses centered on NOS. Contrary to other epistemological positions, Ziman's theorization about science contains a broad conceptualization, which allows a clear and detailed categorization of different aspects (metascientific dimensions) of NOS. The conceptual structure and broadness of Bernstein's theory of pedagogical discourse permit very rigorous and fine analyses of pedagogical texts and contexts, and their relationships, at different levels of the educational system. Despite the analysis focusing on the Portuguese educational system, the theorization and the instruments developed can also be used to appreciate NOS in international pedagogical texts. The use of the same methodological approach may allow comparisons between them.

References

Anderson, L. W., Krathwohl, D., Airasian, P., Cruikshank, K., Mayer, R., Pintrich, P., Raths, J., & Wittrock, M. (Eds.). (2001). *A taxonomy for learning, teaching and assessing: A revision of Bloom's taxonomy of educational objectives*. New York: Longman.

Aydin, S., & Tortumlu, S. (2015). The analysis of the changes in integration of nature of science into Turkish high school chemistry textbooks: Is there any development? *Chemistry Education Research and Practice, 16*, 786–796.

Bernstein, B. (1990). *Class, codes and control. Vol. IV: The structuring of pedagogic discourse*. London: Routledge.

Bernstein, B. (2000). *Pedagogy, symbolic control and identity: Theory, research, critique* (Revised ed.). New York: Rowman & Littlefield.

Brandwein, P., Cooper, E., Blackwood, P., Cottom-Winslow, M., Boeschen, J., Giddings, M., Romero, F., & Carin, A. (1980). *Concepts in science: Teacher's edition*. New York: Harcourt Brace Jovanovich.

Calado, S., & Neves, I. P. (2012). Currículo e manuais escolares em contexto de flexibilidade curricular: Estudo de processos de recontextualização. *Revista Portuguesa de Educação, 25*(1), 53–93.

Cantu, L. L., & Dudley Herron, J. (1978). Concrete and formal Piagetian stages and science concept attainment. *Journal of Research in Science Teaching, 15*(2), 135–143.

Castro, S. (2006). *A construção da ciência na educação científica do ensino secundário: Análise do programa de Biologia e Geologia do 10.° ano*. Dissertação Mestrado, Faculdade de Ciências, Universidade de Lisboa, Lisboa, Portugal.

Castro, S. (2017). *A construção da ciência na educação científica do ensino secundário: Estudo do discurso pedagógico do programa e de manuais escolares de Biologia e Geologia do 10.° ano e das concepções dos professores*. Tese Doutorado, Instituto de Educação, Universidade de Lisboa, Lisboa, Portugal.

Cavadas, B., & Guimarães, F. (2012). Práticas inovadoras nos manuais escolares de zoologia. In J. V. Brás & M. N. Gonçalves (Eds.), *O corpo: Memória e identidade* (pp. 77–87). Lisboa: Edições Universitárias Lusófonas.

Clough, M. P., & Olson, J. K. (2008). Teaching and assessing the nature of science: An introduction. *Journal of Science & Education, 17*(2–3), 143–145.

DES, Departamento do Ensino Secundário. (2001). *Programa de Biologia e Geologia: 10.° ou 11.° anos*. Lisboa: Ministério da Educação.

Ferreira, S., & Morais, A. M. (2013). The nature of science in science curricula: Methods and concepts of analysis. *International Journal of Science Education, 35*(16), 2670–2691.

Ferreira, S., Morais, A. M., Neves, I. P., Saraiva, L., & Castro, S. (2015). Conceptualização da construção da ciência em currículos e manuais escolares. In Conselho Nacional de Educação, CNE (Ed.), *Currículos de nível elevado no ensino das ciências* (pp. 180–238). Lisboa: CNE.

Gall, M. D., Gall, J. P., & Borg, W. R. (2007). *Educational research: An introduction* (8th ed.). Boston: Pearson/Allyn and Bacon.

Hodson, D. (2014). Nature of science in the science curriculum: Origin, development, implications and shifting emphases. In M. R. Mathews (Ed.), *International handbook of research in history, philosophy and science teaching* (pp. 911–970). Dordrecht: Springer.

Irzik, G., & Nola, R. (2011). A family resemblance approach to the nature of science for science education. *Science & Education, 20*, 591–607.

Lederman, N. G. (2007). Nature of science: Past, present, and future. In S. K. Abell & N. G. Lederman (Eds.), *Handbook of research on science education* (pp. 831–879). Mahwah: Lawrence Erlbaum Associates.

Matthews, M. R. (Ed.). (2009). *Science, worldviews and education*. Dordrecht: Springer.

McComas, W. F. (2014). Nature of science in the science curriculum and in teacher education programs in the United States. In M. R. Matthews (Ed.), *International handbook of research in history, philosophy and science teaching* (pp. 1993–2022). Dordrecht: Springer.

McComas, W. F., & Olson, J. K. (1998). The nature of science in international science education standards documents. In W. F. McComas (Ed.), *The nature of science in science education: Rationales and strategies* (pp. 41–52). Dordrecht: Kluwer.

Millar, R., & Osborne, J. (1998). *Beyond 2000: Science education for the future*. London: King's College London School of Education.

Morais, A. M., & Neves, I. P. (2010). Basil Bernstein as an inspiration for educational research: Specific methodological approaches. In P. Singh, A. R. Sadovnik, & S. F. Semel (Eds.), *Toolkits, translation devices and conceptual accounts: Essays on Basil Bernstein's sociology of knowledge* (pp. 11–32). New York: Peter Lang.

Morais, A. M., & Neves, I. P. (2016). Vertical discourses and science education: Analysing conceptual demands of educational texts. In P. Vitale & B. Exley (Eds.), *Pedagogic rights and democratic education: Bernsteinian explorations of curriculum, pedagogy and assessment* (pp. 174–191). London: Routledge.

OCDE. (2016). *PISA 2015 Assessment and analytical framework: Science, reading, mathematic and financial literacy*. https://doi.org/10.1787/9789264255425-en. Accessed 3 Dec 2016.

Osborne, J., & Dillon, J. (2008). *Science education in Europe: Critical reflections*. London: The Nuffield Foundation.

Taber, K. S. (2017). Reflecting the nature of science in science education. In K. S. Taber & B. Akpan (Eds.), *Science education: An international course companion* (pp. 23–37). Rotterdam: Sense Publishers.

Valverde, G. A., Bianchi, L. J., Wolfe, R. G., Schmidt, W. H., & Houang, R. T. (2002). *According to the book: Using TIMSS to investigate the translation of policy into practice through the world of textbooks*. Dordrecht: Kluwer Academic Publishers.

Vesterinen, V.-M., Aksela, M., & Sundberg, M. R. (2009). Nature of chemistry in the national frame curricula for upper secondary education in Finland, Norway and Sweden. *NorDiNa, 5*(2), 200–212.

Ziman, J. M. (1984). *An introduction to science studies: The philosophical and social aspects of science and technology*. Cambridge: Cambridge University Press.

Ziman, J. M. (2000). *Real science: What it is, and what it means*. Cambridge: Cambridge University Press.

Index

M. E. B. Prestes, C. C. Silva (eds.), *Teaching Science with Context*,
Science: Philosophy, History and Education,
https://doi.org/10.1007/978-3-319-74036-2

The manufacturer's authorised representative in the EU is Springer Nature Customer Service Centre GmbH, Europaplatz 3, 69115 Heidelberg, Germany. If you have any concerns regarding our products, please contact ProductSafety@springernature.com

Printed and bound by CPI Group (UK) Ltd, Croydon, CR0 4YY
30/09/2026
02225659-0013